Charles Wye Williams, D. Kinnear Clark

Fuel

Its Combustion and Economy

Charles Wye Williams, D. Kinnear Clark

Fuel

Its Combustion and Economy

ISBN/EAN: 9783744643481

Printed in Europe, USA, Canada, Australia, Japan

Cover: Foto ©berggeist007 / pixelio.de

More available books at **www.hansebooks.com**

FUEL
ITS COMBUSTION AND ECONOMY

CONSISTING OF ABRIDGMENTS OF

"TREATISE ON THE COMBUSTION OF COAL AND THE
PREVENTION OF SMOKE," BY C. W. WILLIAMS, A.I.C.E.,
AND "THE ECONOMY OF FUEL," BY
T. SYMES PRIDEAUX

WITH EXTENSIVE ADDITIONS ON

RECENT PRACTICE IN THE COMBUSTION AND ECONOMY
OF FUEL; COAL, COKE, WOOD, PEAT,
PETROLEUM, ETC.

BY THE EDITOR,

D. KINNEAR CLARK, C.E.

MEMBER OF THE INSTITUTION OF CIVIL ENGINEERS; AUTHOR OF "RAILWAY
MACHINERY;" "A MANUAL OF RULES, TABLES, AND DATA;"
"TRAMWAYS: THEIR CONSTRUCTION AND WORKING;"
EDITOR OF "ROADS AND STREETS," ETC. ETC.

Cupio Lumen

LONDON
CROSBY LOCKWOOD & CO.
7, STATIONERS' HALL COURT, LUDGATE HILL
1879

PREFACE.

In preparing a new and revised edition of the Rudimentary works of Mr. C. Wye Williams and Mr. T. Symes Prideaux on Coal and its Combustion, it has been judged advisable to condense these useful treatises, to a considerable extent, by omitting tautological and irrelevant matter, thus enhancing the value and utility of the remainder. Mr. C. W. Williams has, in his day, done useful work ; if in nothing more than his persistent enunciation of the cardinal principle that, in order to complete the combustion of coal, and prevent the formation of smoke, the elements, gas and air, must be thoroughly and promptly intermixed. The absolute necessity of such perfect intermixture is now—thanks to the assiduity of that great apostle of smoke-prevention—well understood and appreciated. In the work of Mr. T. S. Prideaux, which forms the second part of this publication, the practical value of heated air in combustion for the attainment of high temperatures is shadowed forth ; and it must be acknowledged that his previsions of the important advantages of the use of heated air in the manufacture of iron which give the keynote to his work, have been thoroughly fulfilled in the iron furnace of the present day.

In the third part of this book, the editor has endeavoured succinctly to summarise the more recent results of progress in the combustion and economical use of fuels, on the lines laid down in the treatises of Mr. Williams and Mr. Prideaux. He has prefaced the matter with a few chapters on the composition and combustion of various fuels now in use. The results of the combustion of coal on ordinary grates under ordinary boilers, under various conditions, are exemplified; with extended notices of the important series of experimental trials of steam-boilers with coals, conducted by Mr. James A. Longridge, at Newcastle-on-Tyne, and by Mr. Lavington E. Fletcher, at Wigan. The principles and employment of gas-furnaces—the furnaces of the future—are treated at considerable length, in view of their importance in the practical arts. It is hoped that the reader may trace the development of the gas-furnace in these pages with as much interest as the editor has experienced in writing on it. This admirable application of science for the generation of heat, culminated in that marvel of applied science,—the regenerative furnace of Siemens. In this furnace are happily combined the four elements of success, —fuel gasefied, spare heat economised, air heated, and intermixture of the elements completely effected. The less ambitious furnaces for heating and puddling iron have been illustrated and described with a considerable degree of detail; together with the unique and interesting results of Mr. Crampton's applications of powdered fuel.

D. K. CLARK.

8, BUCKINGHAM STREET, ADELPHI, LONDON.
December, 1878.

CONTENTS.

PART I.—ON THE COMBUSTION OF COAL, AND THE PREVENTION OF SMOKE.

By C. WYE WILLIAMS, A.I.C.E.

	PAGE
PREFACE TO THE FIRST EDITION	3
CHAPTER I.—ON THE CONSTITUENTS OF COAL, AND THE GENERATION OF COAL-GAS	6
CHAPTER II.—OF GASEOUS COMBINATIONS, AND PARTICULARLY OF THE UNION OF COAL-GAS AND AIR	10
CHAPTER III.—OF THE QUANTITY OF AIR REQUIRED FOR THE COMBUSTION OF CARBON, AFTER THE GAS HAS BEEN GENERATED	17
CHAPTER IV.—OF THE MIXING AND INCORPORATION OF AIR AND COAL-GAS	21
CHAPTER V.—OF THE PRINCIPLES ON WHICH BOILERS AND THEIR FURNACES SHOULD BE CONSTRUCTED . . .	28
CHAPTER VI.—OF THE INTRODUCTION OF THE AIR TO THE FUEL, IN A FURNACE, PRACTICALLY CONSIDERED . .	38
CHAPTER VII.—OF REGULATING THE SUPPLY OF AIR TO THE GAS BY SELF-ACTING OR OTHER MECHANICAL APPARATUS .	48
CHAPTER VIII.—OF THE PLACE MOST SUITABLE FOR INTRODUCING THE AIR TO THE GAS IN A FURNACE . . .	58
CHAPTER IX.—OF VARIOUS FURNACE ARRANGEMENTS, WITH OBSERVATIONS THEREON	63
CHAPTER X.—ON PROVIDING ADEQUATE INTERNAL SURFACE FOR TRANSMITTING THE HEAT TO THE WATER FOR EVAPORATION	94
CHAPTER XI.—OF FLAME, AND THE TEMPERATURE REQUIRED FOR ITS PRODUCTION AND CONTINUANCE, AND ITS MANAGEMENT IN THE FURNACES AND FLUES	96

PAGE

CHAPTER XII.—OF THE CIRCULATION OF WATER IN THE BOILER 99

CHAPTER XIII.—ON THE CIRCULATION OF THE WATER IN RELATION TO EVAPORATION, AND ITS INFLUENCE ON THE TRANSMISSION OF HEAT 109

CHAPTER XIV.—OF THE CIRCULATION OF THE WATER IN RELATION TO THE DURABILITY OF THE PLATES . . . 116

CHAPTER XV.—OF THE DRAUGHT 124

CHAPTER XVI.—OF THE TUBULAR SYSTEM AS APPLIED TO MARINE, LAND, AND LOCOMOTIVE BOILERS, IN REFERENCE TO THE CIRCULATION OF THE WATER AND THE PROCESS OF COMBUSTION 139

CHAPTER XVII.—ON THE USE OF HEATED AIR, AND ITS SUPPOSED VALUE IN THE FURNACES OF BOILERS . . 147

CHAPTER XVIII.—ON THE INFLUENCE OF THE WATER GENE-RATED IN FURNACES FROM THE COMBUSTION OF THE HYDROGEN OF THE GAS 151

CHAPTER XIX.—ON INCREASING THE HEAT-TRANSMITTING POWER OF THE INTERIOR PLATE-SURFACE OF BOILERS . 154

CHAPTER XX.—ON THE GENERATION AND CHARACTERISTICS OF SMOKE 166

CHAPTER XXI.—CONCLUDING REMARKS 169

APPENDIX. — EXTRACTS FROM THE SECOND REPORT OF MESSRS. LONGRIDGE, ARMSTRONG, AND RICHARDSON TO THE STEAM COAL COLLIERIES' ASSOCIATION, NEWCASTLE-ON-TYNE 174

PART II.—ON ECONOMY OF FUEL.

BY T. SYMES PRIDEAUX.

INTRODUCTION 189

CHAPTER I.—ON THE BEST MEANS OF RENDERING COMBUSTION PERFECT 194

CHAPTER II.—ON CONTRIVANCES FOR THE EMPLOYMENT OF INFERIOR KINDS OF FUEL 199

CHAPTER III.—ON THE USE OF COMPRESSED AIR IN REVER-BERATORY FURNACES 202

CHAPTER IV.—ON THE ECONOMY TO BE ATTAINED BY INCREAS-ING THE TEMPERATURE OF FURNACES 204

CHAPTER V.—ON FEEDING FURNACES WITH HOT AIR . . 209

CHAPTER VI.—ON THE MANUFACTURE OF IRON . . . 213

PART III.—FUELS: THEIR COMBUSTION AND ECONOMICAL USE.

By D. K. CLARK, M. INST. C.E.

PAGE

CHAPTER I.—CHEMICAL COMPOSITION OF FUELS, AND FORMULAS FOR COMBUSTION 229

CHAPTER II.—COAL 233

CHAPTER III.—COMBUSTION OF COAL 236

CHAPTER IV.—EVAPORATIVE PERFORMANCE OF COAL IN A MARINE BOILER AT NEWCASTLE-ON-TYNE 245

CHAPTER V.—EVAPORATIVE PERFORMANCE OF COAL IN LANCASHIRE AND GALLOWAY BOILERS AT WIGAN 250

CHAPTER VI.—COAL-BURNING IN LOCOMOTIVES 256

CHAPTER VII.—COKE 260

CHAPTER VIII.—LIGNITE, ASPHALTE, AND WOOD . . . 265

CHAPTER IX.—PEAT 268

CHAPTER X.—TAN, STRAW, AND COTTON-STALKS . . . 272

CHAPTER XI.—LIQUID FUEL—PETROLEUM 274

CHAPTER XII.—TOTAL HEAT OF COMBUSTION OF FUELS . . 276

CHAPTER XIII.—GAS-FURNACE:—FUNCTION AND OPERATION OF GAS-FURNACES 277

CHAPTER XIV.—APPLICATION OF GAS-FURNACES FOR THE MANUFACTURE OF GAS 282

CHAPTER XV.—GAS-FURNACES FOR STEAM-BOILERS . . . 286

CHAPTER XVI.—DECOMPOSITION OF FUEL IN GAZOGENES . . 292

CHAPTER XVII.—IRON-FURNACES.—ORDINARY FURNACES . . 295

CHAPTER XVIII.—UTILISING THE WASTE HEAT OF ORDINARY IRON-FURNACES BY GENERATING STEAM 301

CHAPTER XIX.—IRON-FURNACES IN WHICH WASTE HEAT IS UTILISED BY HEATING THE AIR 304

CHAPTER XX.—BLAST-FURNACES 327

CHAPTER XXI.—THE SIEMENS REGENERATIVE GAS-FURNACE . 330

CHAPTER XXII.—THE PONSARD GAS-FURNACE, WITH RECUPERATOR 357

CHAPTER XXIII.—GORMAN'S HEAT-RESTORING GAS-FURNACE . 372

CHAPTER XXIV.—WATER-GAS GENERATORS FOR HEATING PURPOSES 374

CHAPTER XXV.—POWDERED FUEL 379

INDEX 388

ILLUSTRATIONS.

FIG.   PAGE

1. Carburetted Hydrogen 12
2. Bi-carburetted Hydrogen 12
3. Atmospheric Air 13
4. Steam 14
5. Carbonic Acid 14
6. Carbonic Acid 17
7. Carbonic Oxide 18
8. Carbonic Acid 18
9. Carbonic Oxide 19
10. Flame of a Candle 24
11, 12, 13. Argand Lamp 25
14. Flame of a Candle 26
15. Diffusion-Jet 26
16. Diagram of Temperature in Boiler-Flue 31
17. Pyrometer 33
18. Section of Furnace, showing the Flame-bed 36
19. Air admitted through one orifice at the Bridge . . . 42
20. Air admitted through numerous orifices at the Bridge . 42
21. Air admitted through numerous orifices at the Door . . 43
22, 23, 24. Supply of Air to burn Gas 44
25. Gas Furnace at Treveray 45
26. Ditto ditto 46
27. Equalising the Supplies of Gas and Air 57
28. Air through orifices in the Fireplace 60
29. Ditto ditto 61
30. Argand Furnace 64
31. Ditto 65
32. Ordinary Marine Furnace 66
33. Parkes's Split Bridge 66
34. Split Bridge Modified 67
35. Split Bridge and Air at the Grate 68
36. Air at the Bridge 68

Fig.		Page
37.	Air at the Bridge, and small Grate	69
38.	Argand Furnace	69
39.	Ditto	70
40.	Ditto	70
41.	Ditto	71
42.	Ditto	71
43.	Air-box at Bridge	72
44.	Argand Furnace	72
45.	Common Furnace	73
46.	Air-box at Bridge	73
47.	Ditto	74
48.	Ditto	75
49.	Common Furnace	76
50.	Perforated Plate at Bridge	77
51.	Furnace with Supplementary Grate	78
52.	Hot-air Expedient and Split Bridge	79
53.	Ditto ditto	79
54.	Hot Air at the Bridge	80
55.	Chanter's System	80
56.	Hot Air at the Bridge	81
57.	Hot Air from the Flues	82
58.	Blast of Smoke and Air	83
59.	Ditto ditto	84
60.	Common Furnace	87
61.	Argand Furnace	87
62.	Ditto	88
63.	Proper Firing	89
63a.	Improper Firing	89
64.	Argand Furnace in a Locomotive Boiler	92
65.	Circulation of Water	99
66.	Ditto ditto	100
67.	Ditto ditto	101
68.	Ebullition	102
69.	Ditto	103
70.	Ditto	104
71.	Ditto	105
72.	Ditto	106
73.	Ditto	106
74.	Steam Rising	109
75.	Ditto	109
76.	Circulation of Steam and Water	110
77.	Ditto ditto	111
78.	Over-heated Furnace-Crown	111
79.	Circulation of Steam and Water	113

FIG.   PAGE
80. Circulation of Steam and Water 114
81. Boiler of the *Liverpool*. 118
82.   Ditto   ditto   Section 119
83.   Ditto   ditto   ditto 120
84.   Ditto   ditto   ditto 121
85, 86, 87, 88. Draught in Flues 125
89.   Ditto 126
90. Draught in a Chimney 126
91.   Ditto   ditto 127
92. First Boilers of the *Great Britain* 129
93. Boiler of the *Great Liverpool* 130
94.   Ditto   ditto 131
95. Split Draught 133
96.   Ditto 133
97. Mechanical Draught 135
98. Mechanical Draught *versus* Ordinary Draught . . . 137
99. Locomotive Boiler 140
100. Marine Boiler 141
101.   Ditto 143
102. Boiler of the *Leeds* 145
103. Volume of Air 148
104.   Ditto 149
105. Condenser 152
106, 107. Conductor Pins 155
108. Tin Boiler 156
109, 110. Ditto 157
111. Conductor Pins 161
112. Boiler of the *Royal William* 162
113.   Ditto   ditto 163
114. Lamb and Summers' Boiler 164
115. Gas, Flame, Smoke 167
116. Combustion of Carburetted-Hydrogen 170
117. Step-Grate 242
118. Liquid Fuel for Steam Boiler 274
119, 120, 121. Gas-furnaces for the Manufacture of Gas . . 284
122. Gas-furnace for Steam Boiler 288
123. Gas-furnace for an Internally-fired Boiler . . . 290
124. Old Puddling Furnace 295
125. Boetius Heating Furnace 308
126.   Ditto   ditto 309
127. Bicheroux Puddling Furnace 310
128. Casson-Dormoy Puddling Furnace 314
129. Price's Retort Furnace 318
130.   Ditto   ditto 319

FIG.   PAGE

131. Caddick and Mabery's Furnace 325
132. Siemens Regenerative Furnace. Section 336
133.   Ditto   ditto   ditto 337
134.   Ditto   ditto   ditto 338
135. Ponsard Gas-furnace 358
137.   Ditto   ditto   Recuperator 360
138.   Ditto   ditto   ditto 362
139. Gorman's Heat-restoring Gas-furnace 373
140. Crampton's Powdered-Fuel Furnace 380
141.   Ditto   ditto   Revolving Puddling

    Furnace 381
142.   Ditto   ditto   ditto   382
143. Stevenson's Coal-dust Furnace 386
144.   Ditto   ditto 387

PART I.

ON THE COMBUSTION OF COAL

AND THE

PREVENTION OF SMOKE.

By C. WYE WILLIAMS, A.I.C.E.

COMBUSTION OF COAL AND PREVENTION OF SMOKE.

———◆———

PREFACE TO THE FIRST EDITION.

Being much interested in the improvement of steam-vessels, from my connection with several steam navigation companies, and having had a longer and more extended experience in the details of their building and equipping than, perhaps, any individual director of a steam company in the kingdom, my attention has been uninterruptedly given to the subject since the year 1823, when I first established a steam company, and undertook to have the first steam-vessel constructed capable of maintaining a commercial intercourse across the Irish Channel, during the *winter* months, and which, till then, had been considered impracticable.

The result of this long experience is the finding, that, notwithstanding the improved state to which the construction and appointments of the hull and general machinery of steam-vessels have arrived, great uncertainty and risk of failure still prevail in the *use of fuel* and the *generation of steam*.

It is true, the engineer, who undertakes the construction of the engines, also undertakes that the boilers shall provide a sufficiency of steam to work them; but what that *sufficiency* means has not been decided; and, in too many instances, the absence of some fixed data on the subject leaves the evils of a deficiency of steam or a great expenditure of fuel unabated.

So long as the operations of steam-vessels were confined to coasting or short voyages, the consequences of these defects in boilers, as regards the quantity of fuel, were a mere question of pounds, shillings, and pence. When, however, those operations came to be extended to long sea voyages, these consequences took a more comprehensive range, and involved the more important question, whether such voyages *were practicable* or profitable.

From being so deeply interested in the improvement of this department of steam navigation, I have watched, with no small anxiety, the efforts of the engineers to arrive at some degree of certainty in what was admitted, on all hands, to be the most serious drawback to the successful application of steam-vessels to long sea voyages. I perceived the absence of any well-founded principle in the construction of the boiler—that the part on which most depended appeared least understood and least attended to, namely, the *furnace;* and that this was too often left to the skill (or want of it) of working boiler-makers. I saw that, although the great operations of combustion carried on in the furnace, with all that belongs to the introduction and employment of atmospheric air, were among the most difficult processes within the range of chemistry, the absence of sound scientific principles still continued to prevail; yet on these must depend the extent or perfection of the combustion in our furnaces.

Years were still passing away, and while every other department was fast approaching to perfection, all that belonged to the combustion of fuel—the production of smoke—and the wear and tear of the furnace part of the boiler, remained in the same *status quo* of uncertainty and insufficiency; and even that boilers and their furnaces, constructed within the last few years, exhibit still greater violations of chemical truths, and a greater departure from the principles on which Nature proceeds.

In the proper place I will show that, of late years, as much uncertainty as to the success of a new boiler has prevailed as when I first began operations, thirty years ago; and that few boilers, for land or marine engines, exhibit more in the way of effecting perfect combustion or economy of fuel than those of any former period since the days of Watt.

I do not affect to give any new view of the nature of combustion. What I take credit for is, the practical application, *on the large scale of the furnace,* of those chemical truths which are so well known in every *laboratory*. I also take credit for bringing together the scattered facts and illustrations of such authorities as bear on the subject before us, and so applying them as to enable practical men to understand that part which chemistry has to act in the construction, arrangements, and working of our boilers and furnaces.

C. W. WILLIAMS.

CHAPTER I.

OF THE CONSTITUENTS OF COAL, AND THE GENERATION OF COAL GAS.

In the following treatise I do not undertake to show how *the smoke* from coals can be *burned;* but I do undertake to show how *coals may be burned without smoke;* and this distinction involves the main question of economy of fuel.

When smoke is once produced in a furnace or flue, it is as impossible to burn it or convert it to heating purposes, as it would be to convert the smoke issuing from the flame of a candle to the purposes of heat or light.

When we see smoke issuing from the flame of an ill-adjusted common lamp, we also find the flame itself dull and murky, and the heat and light diminished in quantity. Do we then attempt to *burn that smoke?* No; it would be impossible. Again, when we see a well-adjusted Argand lamp burn *without producing any smoke*, we also see the flame white and clear, and the quantity of heat and light increased. In this case, do we say the lamp *burns its smoke?* No; we say the lamp *burns without smoke.* This is the fact, and it remains to be shown why the same language may not be applied to the combustion of the same coal and the same gas, in the *furnace*, as in the *lamp*.

In a treatise purporting to describe the means of obtaining the largest quantity of heat from coal, the first step is an inquiry into the varieties of that combustible and its respective constituents.

The classification of the various kinds of coal, the details of an elaborate analysis, made by Mr. Thomas Richardson, with the aid of Professor Liebig, are as follows :—

Species of Coal.	Locality.	Carbon.	Hydrogen.	Azote and Oxygen.	Ashes.
Splint . .	Wylam . .	74·823	6·180	5 085	13·912
,, .	Glasgow. .	82·924	5·491	10·457	1·128
Cannel . .	Lancashire .	83·753	5·660	8·039	2·548
,, . .	Edinburgh .	67·597	5·405	12·432	14·566
Cherry . .	Newcastle .	84·846	5·048	8·430	1·676
,, . .	Glasgow. .	81·204	5·452	11·923	1·421
Caking . .	Newcastle .	87·952	5·239	5·416	1·393
,, . .	Durham . .	83·274	5·171	9·036	2·519

The most important feature in reference to this analysis is the large proportion of hydrogen which all bituminous coal contains, and which may be estimated at 5¼ per cent.— hydrogen being the main element in the evolved gas, and by the combustion of which flame is produced.

We know, *scientifically*, that carburetted hydrogen and the other compounds of carbon require given quantities of atmospheric air to effect their combustion ; yet we adopt no means, *practically*, of ascertaining what *quantities* are supplied, and treat them as though no such proportions were necessary. We know, *scientifically*, the relative proportions in which the constituents of atmospheric air are combined ; yet, *practically*, we appear wholly indifferent to the distinct nature of these constituents, or their effects in combustion. We know, *scientifically*, that the inflammable gases are combustible only in proportion to the *degree of mixture* and union which is effected between them and the oxygen of the air; yet, *practically*, we never trouble our heads as to whether we have effected such mixture or not. These and many similar illustrations exhibit a reprehensible degree of carelessness which can only be corrected by a sounder and more scientific knowledge of the subject; and this can only be attained *through the aid of chemistry.*

The main constituents of all coal, as we see in the preceding table, are *carbon* and *hydrogen.*

In the natural state of coal, the hydrogen and carbon are united and solid. Their respective characters and modes of entering into combustion are, however, essentially different; and to our neglect of this primary distinction is referable much of the difficulty and complication which attend the use of coal on the large scale of our furnaces.

The first leading distinction is, that the bituminous portion is convertible to the purposes of heat in the *gaseous state alone;* while the carbonaceous portion, on the contrary, is combustible *only in the solid state;* and, what is essential to be borne in mind, *neither can be consumed while they remain united.*

When heat is first applied to bituminous coal, the question naturally arises, What becomes of it? or, What is its effect?

A charge of fresh coal thrown on a furnace in an active state, so far from augmenting the general temperature, becomes at once an *absorbent* of it, and the source of the *volatilisation* of the bituminous portion of the coal; in a word, of the generation of the gas. Now, volatilisation is the most cooling process of nature, by reason of the quantity of heat which is directly converted from the *sensible* to the *latent* state. So long as any of the bituminous constituents remain to be evolved from any atom or division of the coal, ts solid or carbonaceous part remains black, at a comparatively low temperature, and utterly inoperative as a heating body. In other words, the carbonaceous part has *to wait its turn* for that heat which is essential to its own combustion, and in its own peculiar way.

If this bituminous part be not consumed and turned to account, it would have been better had it not existed in the coal ; as such heat would, in that case, have been saved and become available for the business of the furnace. To this circumstance may be attributed the alleged comparatively greater heating properties of coke, or anthracite, over bituminous coal.

The point next under consideration will be the processes incident to the combustion of the *gaseous portion* of the coal, as distinct from the *carbonaceous* or *solid* portion.

CHAPTER II.

OF GASEOUS COMBINATIONS, AND PARTICULARLY OF THE UNION OF COAL GAS AND AIR.

ON the application of heat to bituminous coal, the first result is its absorption by the coal, and the disengagement of gas, from which flame is exclusively derivable.

The constituents of this gas are *hydrogen* and *carbon;* and the unions which alone concern us here are *carburetted hydrogen* and *bi-carburetted hydrogen*, commonly called olefiant gas.

Combustibility is not a quality of the combustible, *taken by itself*. It is, in the case now before us, the union of the *combustible* with *oxygen*, and which, for this reason, is called the *"supporter;"* neither of which, however, *when taken alone*, can be consumed.

To effect combustion, then, we must have a *combustible* and a *supporter* of combustion. Strictly speaking, combustion means *union ;* but it means *chemical* union.

Let us bear in mind that coal gas, whether generated in a retort or a furnace, is essentially the same. Again, that, strictly speaking, it is not inflammable ; as, *by itself*, it can neither produce flame nor permit the continuance of flame in other bodies. A lighted taper introduced into a jar of car-buretted hydrogen (coal gas), so far from inflaming the gas, is itself instantly extinguished. Effective combustion, for practical purposes, is, in truth, a question more as regards *the air* than the *gas*. Besides, we have no control over the gas, as to quantity, after having thrown the coal on the furnace, though we *can* exercise a control over that of the air, in all the essentials to perfect combustion. It is this which has done so much for the perfection of the *lamp*, and may be made equally available for the *furnace*.

The first step towards effecting the combustion of any gas, is the ascertaining of the quantity of *oxygen* with which it will chemically combine, and the quantity of *air* required for supplying such quantity of *oxygen*.

Much of the apparent complexity which exists on this head arises from the disproportion between the relative *volumes*, or *bulk*, of the constituent atoms of the several gases, as compared with their respective *weights*. For instance, an atom of *hydrogen* is *double* the bulk of an atom of *carbon vapour;* yet the latter is *six times the weight* of the former.

Again, an atom of hydrogen is double the bulk of an atom of oxygen; yet the latter is *eight* times the weight of the former.

So of the constituents of atmospheric air—nitrogen and oxygen. An atom of the former is double the bulk of an atom of the latter; yet, in weight, it is as fourteen to eight.

I have stated that there are two descriptions of hydrocarbon gases in the combustion of which we are concerned: both being generated in the furnace, and even at the same time, namely, the *carburetted* and *bi-carburetted* hydrogen gases, the proportion of the latter in coal gas being estimated at about ten per cent. For the sake of simplifying the explanation, I will confine myself to the first.

On analyzing this gas, we find it to consist of two volumes of hydrogen and one of carbon vapour; the gross bulk of these three being *condensed into the bulk of a single atom of hydrogen*, that is, into two-fifths of their previous bulk, as shown in the annexed figures. Let figure 1 represent an atom of coal gas—carburetted hydrogen—with its constituents, carbon and hydrogen; the space enclosed by the lines representing the relative size or volume of each; and the numbers representing their respective weights—hydrogen being taken *as unity* both for volume and weight.

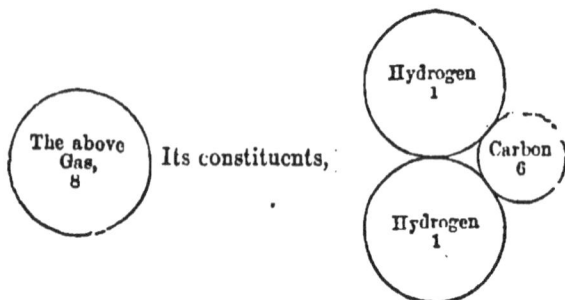

Fig 1.—Carburetted Hydrogen.

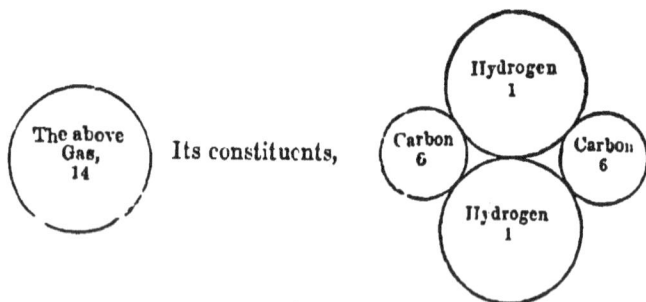

Fig. 2.—Bi-Carburetted Hydrogen.

Let us now, in the same analytical manner, examine an atom of atmospheric air, the other ingredient in combustion.

Atmospheric air is composed of two atoms of nitrogen and one atom of oxygen; each of the former being *double* the volume of an atom of the latter, while their relative weights are as fourteen to eight: the gross *volume* of the nitrogen, in air, being thus four times that of the oxygen; and in *weight*, as twenty-eight to eight, as shown in the annexed figure 3.

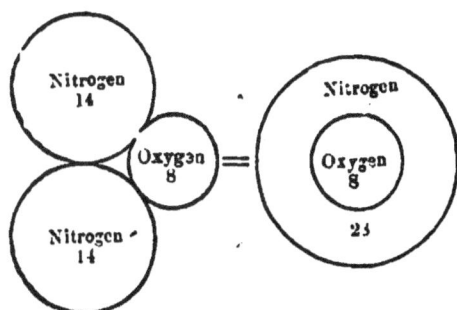

Fig. 3.—Atmospheric Air.

In the coal gas we found the constituents condensed into *two-fifths* of their gross bulk: this is not the case with *air ;* an atom of which is the same, *both as to bulk and weight*, as the sum of its constituents, as here shown. Thus, we find, the oxygen bears a proportion in volume to that of the nitrogen, as one to five; there being but 20 per cent. of oxygen in atmospheric air, and 80 per cent. of nitrogen.

We now proceed to ascertain the *separate quantity of oxygen required by each of the constituents* (of the gas), so as to effect its perfect combustion.

With respect to this reciprocal saturation, the great natural law is, that *bodies combine in certain fixed proportions only*, both in *volume* and *weight*.

On the application of heat, or what may be termed the firing or lighting the gas, when duly mixed with air, the hydrogen *separates itself from its fellow-constituent, the carbon*, and forms an union with oxygen, the produce of which is water. The saturating equivalent of an atom, or any other given quantity of hydrogen, is not *double* the volume, as in the case of the carbon, but *one-half* its volume only—the product being aqueous vapour, that is, *steam ;* the relative weights of the combining volumes being 1 of hydrogen to 8 of oxygen ; and the bulk, when combined, being two-thirds of the bulk of both taken together, as shown in the annexed figure.

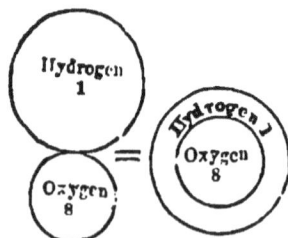

Fig. 4.—Steam.

Again, the carbon, on meeting its equivalent of oxygen, unites with it, forming carbonic acid gas, composed of *one atom* of carbon (by weight 6), and *two atoms* of oxygen (by weight 16), the latter, in volume, being double that of the former, as in the annexed figure.

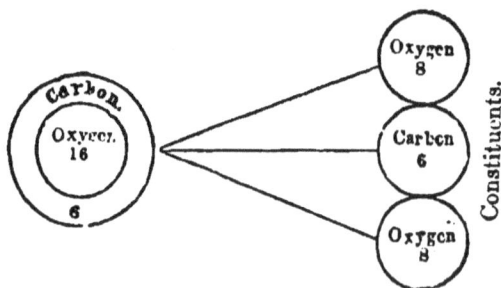

Fig. 5.—Carbonic Acid.

No facts in chemistry, therefore, can be more decidedly proved, than that one atom of hydrogen and one atom of oxygen (*the former being double the bulk of the latter*) unite in the formation of water; and, further, that one atom of carbon vapour and two atoms of oxygen (*the latter being double the bulk of the former*) unite in the formation of carbonic acid gas.

Having thus ascertained the quantity of *oxygen* required for the saturation and combustion of the two constituents of coal gas, the remaining point to be decided is, *the quantity of air that will be required to supply this quantity of oxygen.*

This is easily ascertained, seeing that we know precisely the proportion which oxygen bears, in volume, to that of the air. For, as the oxygen is but *one-fifth* of the bulk of the air, *five* volumes of the latter will necessarily be required to produce *one* of the former; and, as we want *two* volumes of oxygen for each volume of the coal gas, it follows that, *to obtain those two volumes, we must provide ten volumes of air.*

As the proportion of air required for the combustion of the *bi-carburetted hydrogen* (olefiant gas) is necessarily larger than for the *carburetted* hydrogen, a diagram of each is annexed, showing the volume of air required for combustion.

Carburetted Hydrogen.

BEFORE COMBUSTION.		ELEMENTARY MIXTURE.		PRODUCTS OF COMBUSTION.
Weight.	Atoms.	Weight.		Weight.
8 Carburetted Hydrogen	1 Carbon .	6		22 Carbonic Acid.
	1 Hydrogen	1		9 Steam.
	1 Hydrogen	1		9 Steam.
144 Atmospheric Air.	1 Oxygen .	8		
	1 Oxygen .	8		
	1 Oxygen .	8		
	1 Oxygen .	8		
	8 Nitrogen	112		112 Uncombined Nitrogen.
152		152		152

Bi-Carburetted Hydrogen.

BEFORE COMBUSTION.		ELEMENTARY MIXTURE.	PRODUCTS OF COMBUSTION.
Weight.	Atoms.	Weight.	Weight.

14 Bi-carburetted Hydrogen.
{
1 Carbon . 6 ——————— 22 Carbonic Acid.
1 Carbon . 6 ——————— 22 Carbonic Acid.
1 Hydrogen 1 ——————— 9 Steam.
1 Hydrogen 1 ——————— 9 Steam.
}

216 Atmospheric Air
{
1 Oxygen 8
1 Oxygen 8
1 Oxygen 8
1 Oxygen 8
1 Oxygen 8
1 Oxygen 8
12 Nitrogen 168 ——————— 168 Uncombined Nitrogen.
}

230   230   230

CHAPTER III.

HAVING disposed of the question of quantity, as regards the
supply of air required for the saturation and combustion of
the *gaseous* portion of coal, we have now to answer a corresponding question, with reference to the *carbonaceous* part
resting in a solid form on the bars *after* the gaseous matter
has been evolved.

Carbon is susceptible of uniting with oxygen in two proportions, by which two distinct bodies are formed, possessing
distinct chemical properties.

These proportions, in which carbon unites with oxygen,
form, first, *carbonic acid;* second, *carbonic oxide.*

Carbonic *acid*, we have seen, is a compound of one atom
of carbon with two atoms of oxygen ; while carbonic *oxide*
is composed of the same quantity of carbon with but *half*
the above quantity of oxygen, as in the annexed figures.

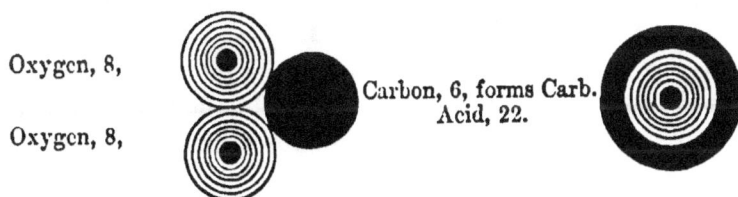

Oxygen, 8,

Oxygen, 8,

Carbon, 6, forms Carb.
Acid, 22.

Fig. 6.—Carbonic Acid.

Oxygen, 8, Carbon, 6, forms Carb. Oxide, 14.

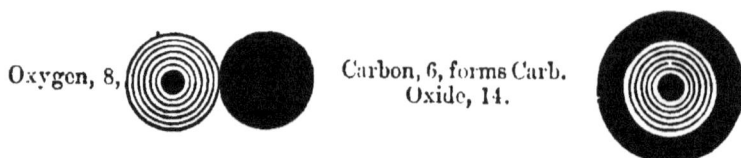

Fig. 7.—Carbonic Oxide.

Here we see that carbonic *oxide*, though containing but one-half the quantity of oxygen, is yet of the same bulk or volume as carbonic *acid*, a circumstance of considerable importance on the mere question of *draught*, and supply of air, as will be hereafter shown.

Now, the combustion of this *oxide, by its conversion into the acid*, is as distinct an operation as the combustion of the carburetted hydrogen, or any other combustible.

But the most important view of the question, and one which is little known to practitioners outside the laboratory, is as regards the *formation* of this *oxide;* and this is the part of the inquiry which most requires our attention.

The *direct* effect of the union of carbon and oxygen is the formation of carbonic *acid.* If, however, we *abstract* one of its portions of *oxygen*, the remaining proportions would then be those of carbonic *oxide.* It is equally clear, however, that if we *add* a second portion of *carbon* to carbonic *acid*, we shall arrive at the same result, namely, the having carbon and oxygen combined in equal proportions, as we see in carbonic *oxide.*

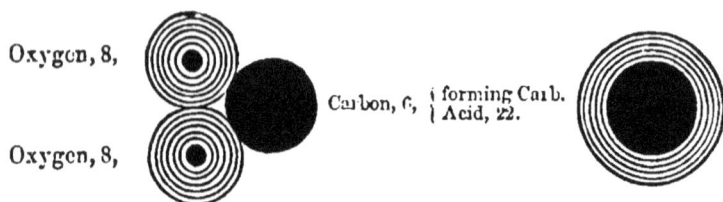

Oxygen, 8,

Carbon, 6, { forming Carb. Acid, 22.

Oxygen, 8,

Fig. 8.—Carbonic Acid.

By the addition, then, of a *second* proportion of carbon to the above, *two* volumes of carbonic *oxide* will be formed— thus :

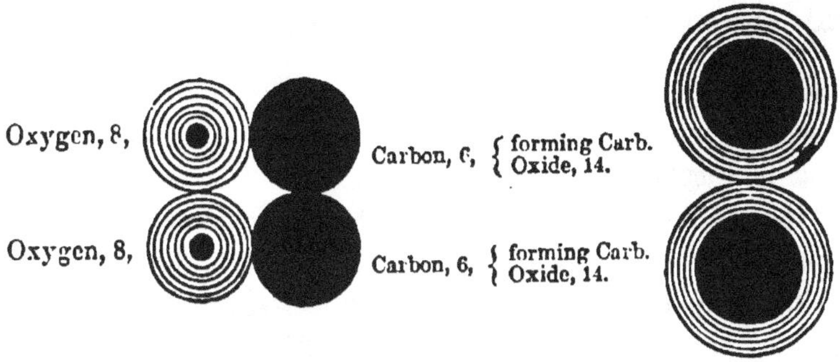

Oxygen, 8, Carbon, 6, { forming Carb. Oxide, 14.

Oxygen, 8, Carbon, 6, { forming Carb. Oxide, 14.

Fig. 9.— Carbonic Oxide.

Now, if these two volumes of carbonic oxide cannot find the oxygen required to complete their *saturating* equivalents, they pass away necessarily but *half consumed*, a circumstance which is constantly taking place in all furnaces where the air has to pass through a body of incandescent carbonaceous matter.

The most prevailing operation of the furnace, however, and by which the largest quantity of carbon is lost, in the shape of carbonic *oxide*, is thus :—The air, on entering from the ashpit, gives out its oxygen to the glowing carbon on the bars, and generates much heat in the formation of carbonic acid. This *acid*, necessarily at a very high temperature, passing upwards through the body of incandescent solid matter, takes up an additional portion of the carbon, and becomes carbonic *oxide*.

Thus, by the conversion of one volume of *acid* into two volumes of *oxide*, heat is actually absorbed, while we also lose the portion of carbon taken up during such conversion, and are deceived by imagining we have " *burned the smoke.*"

Another important peculiarity of this gas (carbonic oxide) is, that, by reason of its already possessing *one-half* its equivalent of oxygen, it inflames at a lower temperature than the ordinary *coal gas;* the consequence of which is, that the *latter*, on passing into the flues, is often cooled down below the temperature of ignition ; while the *former* is sufficiently heated, even after having reached the top of the chimney, and is there ignited on meeting the air. This is the cause of the red flame often seen at the tops of chimneys and the funnels of steam-vessels.

We may thus set it down as a certainty, that, if the carbon, either of the gas or of the solid mass on the bars, passes away in union with oxygen in any other form or proportion than that of *carbonic acid*, a commensurate loss of heating effect is the result.

CHAPTER IV.

ON THE MIXING AND INCORPORATION OF AIR AND COAL GAS.

HAVING disposed of the questions regarding the *quantity* and *quality* of the air to be admitted, our next consideration is, the effecting such a mixture as is required for effective combustion.

It seems to have been taken for granted, in practice on the large scale, that, if air, by *any means*, be introduced to " the fuel in the furnace," it will, as a matter of course, mix with the gas, or other combustible, in a proper manner, and assume the state suitable for combustion, whatever be the nature or state of such fuel.

In operating in the laboratory, when we mix a measured jar of an inflammable gas with a due complement of oxygen gas, the operation being performed leisurely, due incorporation follows, and no question as to the *want of time* arises.

In this operation the quantities are small : both bodies are gaseous : there is no disturbing influence from the presence of other matter : the relative quantities of both are in saturating proportions : and above all, are unaffected by current or draught.

But compare this deliberate laboratory operation with what takes place in the furnace. First, the quantities are large : secondly, the bodies to be consumed are partly gaseous, partly solid : thirdly, the gases evolved from the coal are part combustible and part incombustible : fourthly, they are forced into connection with a large and often overwhelming quantity of the products of combustion, chiefly carbonic acid : fifthly, the very air introduced is itself deteriorated in passing through the bars and incandescent fuel on them, and thus deprived of much of its oxygen : sixthly, and above all, instead of being allowed a suitable time, the whole are hurried away by the current or draught in large masses.

Having consulted Professor Daniell on this subject, his opinion, here given, is of importance.

OPINION.

"KING'S COLLEGE, 8th *August*, 1840.

" There can be no doubt that the affinity of hydrogen for oxygen under most circumstances is stronger than that of carbon. If a mixture of two parts of hydrogen and one of carbonic *acid* be passed through a red-hot tube, water is formed, a portion of charcoal is thrown down, and carbonic *oxide* passes over with the excess of hydrogen.

" With regard to the different forms of hydro-carbon, it is well known that the whole of the carbon is never combined with oxygen in the processes of detonation or silent combustion, *unless a large excess of oxygen be present.*

" For the complete combustion of olefiant gas, it is necessary to mix the gas with *five* times its volume of oxygen, *though three only are consumed.* If less be used, part of the carbon *escapes combination*, and is deposited as a black powder. Even subcarburetted hydrogen it is necessary to mix with more than twice its bulk of oxygen, or the same precipitation will occur.

" It is clear, therefore, that the whole of the hydrogen of any of these compounds of carbon may be combined with oxygen, while a part of their carbon may escape combustion, and *that* even when enough of oxygen is present for its saturation.

" That which takes place when the mixture is designedly made in the most perfect manner must, undoubtedly, arise in the common processes of combustion, where the mixture is fortuitous and much less intimate. Any method of ensuring the complete combustion of fuel, consisting partly of the volatile hydro-carbons, *must be founded upon the principle of producing an intimate mixture with them of atmospheric air, in excess*, in that part of the furnace to which they naturally rise. In the common construction of furnaces this is scarcely possible, as the *oxygen of the air, which passes through the fire bars*, is mostly *expended upon the solid part of the ignited fuel with which it first comes in contact.*

" J. F. DANIELL.

"To C. W. Williams, Esq., &c. &c."

Professor Daniell, in the opinion just quoted, states the true principle on which any improvement in our furnaces for insuring the complete combustion of bituminous coal must be founded, namely, the producing of an intimate previous mixture between the gaseous portion and atmospheric air.

On this head we find many convincing illustrations of what nature requires, and what a judicious mode of bringing air to the gas can effect, in the common candle, and in the Argand lamp, that I propose examining these two exemplifications of gaseous combinations and combustion, in the manner adopted by the best British and continental chemists.

Mr. Brande observes, "In a common candle, the tallow is drawn into the wick by capillary attraction, and there converted into vapour, which ascends in the form of a conical column, and has its temperature sufficiently elevated to cause it to combine with the oxygen of the surrounding atmosphere, with a temperature equivalent to a *white heat.*

But this combustion is *superficial only*, the flame being a thin film of white hot vapour, enclosing an interior portion, *which cannot burn for want of oxygen*. It is in consequence of this structure of the flame that we so materially *increase its heat*, by propelling a current of air through it by the *blow-pipe*."

Dr. Reid observes, " The flame of a candle is produced by the gas formed around the wick acting upon the oxygen of the air: *the flame is solely at the exterior* portion of the ascending gas. All *without* is merely heated air, or the products of combustion ; all *within is unconsumed gas, rising in its turn* to affect (mingle with) the oxygen of the air.

" If a glass tube be introduced within the flame of a lamp or candle (as represented in Fig. 10), part of the unconsumed gas passes through it, and may be kindled as it escapes."

All authorities agree in the main facts : *first*, that the dark part in the centre of the flame is a body of unconsumed gas *ready for combustion*, and only waiting the *preparatory step*—the *mixing*—the *getting into contact* with the oxygen of the air : *secondly*, that that portion of the gas in which the due mixing has been effected, forms but a thin film on the *outside* of such unconsumed gas : *thirdly*, that the products of combustion form the transparent envelope, which may be perceived on close inspection : *fourthly*, that the collection of gas in the *interior* of the flame cannot burn *there* for want of oxygen.

Fig. 10.—Flame of a Candle.

If, then, the unrestricted access of air to this small flame is not able, by the laws of diffusion, to form a due mixture in time for ignition, *à fortiori*, it cannot do so when the supply of air is *restricted* and that of the gas *increased*.

Dr. Reid, speaking of the Argand lamp, Fig. 11, observes,

that the intensity of the heat is augmented by causing the air to enter in the middle of a circular wick, or *series of gas-jets*, so that more gas is consumed *within a given space* than in the ordinary manner.

But why is more gas consumed *within this given space?* Solely because more capability for mixture is afforded, and a greater number of accessible points of *contact* obtained, arising out of this series of jets. This may be seen in Fig. 12, where the inner surfaces, *a a*, are shown in addition to the outer ones *b b*.

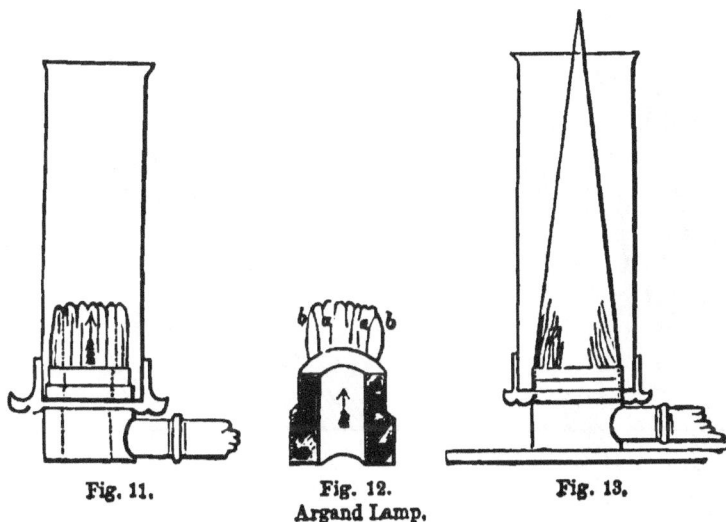

Fig. 11.   Fig. 12.   Fig. 13.
                  Argand Lamp.

" If the aperture," he observes, " by which air is admitted into the interior of the flame be closed, the flame immediately assumes the form shown in Fig. 13; part of the supply of air being thus cut off, it extends farther into the air before it meets with the oxygen necessary for its combustion."

Here we trace the *length* of the flame to the diminished rate of mixing and combustion, occasioned by the want of adequate access, within any given time, between the gas and the air, *until too late*—until the ascending current has

c

carried them beyond the temperature required for chemical action ; the carbonaceous constituent then losing its gaseous character, assuming its former colour and state of a black pulverulent body, and becoming true smoke.

In looking for a remedy for the evils arising out of the hurried state of things which the interior of a furnace

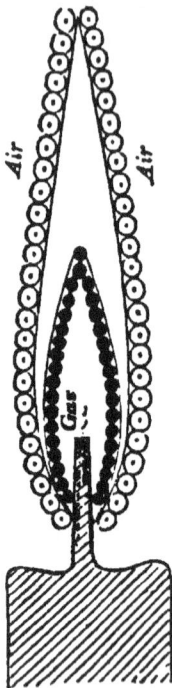

Fig. 14.—Flame of a Candle.

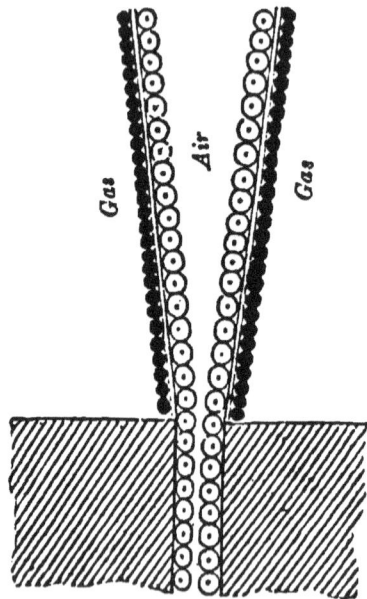

Fig. 15.—Diffusion-jet.

naturally presents, and observing the means by which the gas is effectually consumed in the Argand lamp, it seemed manifest that, if the gas in the furnace could be presented, by means of *jets*, to an adequate quantity of air, as it is in the lamp, the result would be the same. The difficulty of effecting a similar distribution of the gas in the furnace, by means of jets, however, seemed insurmountable : one alternative alone remained, namely, that, since the gas could not be introduced by jets into the body of air, *the air might be introduced by jets into the body of gas.*

This, then, is the means which I adopt, and by which I effect a complete combustion of the gases in the furnace, as we do in the lamp.

This process meets the entire difficulties of the case as to time, current, temperature, and quantity. By this means the process of diffusion is hastened without the injurious effect of cooling : and which always takes place when the air is introduced by large orifices.

The difference, then, between the application of air by means of *the jet*, and that of the ordinary action of the atmosphere, consists in the increased surface it presents for mutual contact in any given unit of time. Let Fig. 14 represent the section of a candle and Fig. 15 that of a diffusion-jet. In the former, the gas in the centre meets the air on the exterior. In the latter, the air in the centre, issuing into the atmosphere of gas, enlarges its own area for contact mechanically, and consequently, its increased measure of combustion.

Thus we see, that the value of the *jet* arises from the circumstance of its creating, *for itself*, a larger surface for contact, by which a greater number of elementary atoms of the combustible and the supporter gain access to each other in any given time.

CHAPTER V.

OF THE PRINCIPLES ON WHICH BOILERS AND THEIR FURNACES SHOULD BE CONSTRUCTED.

THE inquiry before us cannot be confined to a mere comparison of the several descriptions of boilers, mechanically considered. The merits on which, respectively, they rest their claims, must be examined with reference to other data, viz., their relation to the perfect combustion of the fuel employed—the generating the largest measure of heat —and so applying it as to produce the largest volume of steam. Apart from these considerations, indeed, there is little scope for inquiry. All boilers have their furnaces and grate-bars, on which the fuel is placed; their flues, or tubes through which the flame or gaseous products have to pass; and the chimney by which those products are to be carried away, and the necessary draught obtained.

Hitherto, those who have made boiler-making a separate branch of manufacture, have given too much attention to mere *relative proportions*. One class place reliance on enlarged grate surface; another on large absorbing surfaces; while a third demand, as the grand panacea, "*boiler-room enough*," without, however, explaining what that means. Among modern treatises on Boilers, this principle of *room enough* seems to have absorbed all other considerations, and the requisites, in general terms, are thus summed up:

1st. Sufficient amount of internal heating surface;

2nd. Sufficiently roomy furnace;

3rd. Sufficient air-space between the bars;

4th. Sufficient area in the tubes or flues; and

5th. Sufficiently large fire-bar surface.

In simpler terms, these amount to the truism—give sufficient size to all the parts, and thus avoid being deficient in any.

So gravely is this question of relative proportions insisted on, that we find many treatises on the use of Coal, and the construction of Boilers, laying down rules with mathematical precision, giving precise formulæ for their calculations; and even affecting to determine the working power of a steam-engine, by a mere reference to the size of the fire-grate, and the internal areas and surfaces of the boiler. Yet, during this apparent search after certainty, omitting all inquiry respecting the processes or operations to be carried on within them.

On a charge of coal being thrown into a furnace, the heat by which the distillatory, or gas-generating process is effected, is derived from *the remaining portion of the previous charge*, then in an incandescent state on the bars. This process corresponds with what takes place in the gas works, where the coal *inside* the retorts is acted on by the incandescent fuel *outside* of them. This demand for heat in the furnace is, however, confined to the commencement of the operation with each charge. The heat required for *continued gasification* is, or ought to be, obtained chiefly from *the flame itself;* as in the case of a candle, where the gasification of the tallow in the wick is derived from *the heat of its own flame.* This operation shows the importance of sustaining a sufficient body of incandescent fuel on the bars: in particular, when a fresh charge is about to be thrown in.

With reference to the proportions of the several parts of a furnace, we have two points requiring attention: first, *the superficial area of the grate*, for retaining the solid fuel or coke; and, second, *the sectional area of the chamber above the fuel*, for receiving the gaseous portion of the coal.

As to the *area of the grate-bars*, seeing that it is a *solid body* that is to be laid on them, requiring no more space than it actually covers at a given depth, it is alone important that it be *not too large*. On the other hand, as to the *area of the chamber* above the coal, seeing that it is to be occupied by a *gaseous body*, requiring room for its rapidly enlarging volume, it is important that it be *not too small*.

As to the best proportion for the grate, this will be the easiest of adjustment, as a little observation will soon enable the engineer to determine the extent to which he may *increase, or diminish, the length of the furnace*. In this respect, the great desideratum consists in confining that length within such limits that it shall, at *all times, be well and uniformly covered. This is the absolute condition, and sine quâ non of economy and efficiency;* yet it is the very condition which, *in practice*, is the most neglected. Indeed, the failure and uncertainty which has attended many anxiously conducted experiments has most frequently arisen from the neglect of this one condition.

If the grate-bars be not equally and well covered, the air will enter in irregular and rapid streams or masses, through the uncovered parts, and at the very time when it should be *there* most restricted. Such a state of things at once bids defiance to all regulation or control. *Now, on the control of the supply of air depends all that human skill can do in effecting perfect combustion and economy;* and, until the supply of fuel and the quantity on the bars be regulated, it will be impossible to control the admission of the air.

Of the great waste of heat and the consequent reduction of temperature in the flues, arising from the single circumstance of allowing the incandescent fuel, towards the end of the charge, to *run too low*, or be irregularly distributed, the experiment of Mr. Houldsworth, as shown in the annexed diagram, Fig. 16, is highly instructive, and merits the most attentive consideration. This experiment was made expressly

Scale of Temperature in the Flues.
in Degrees of Fahrenheit.

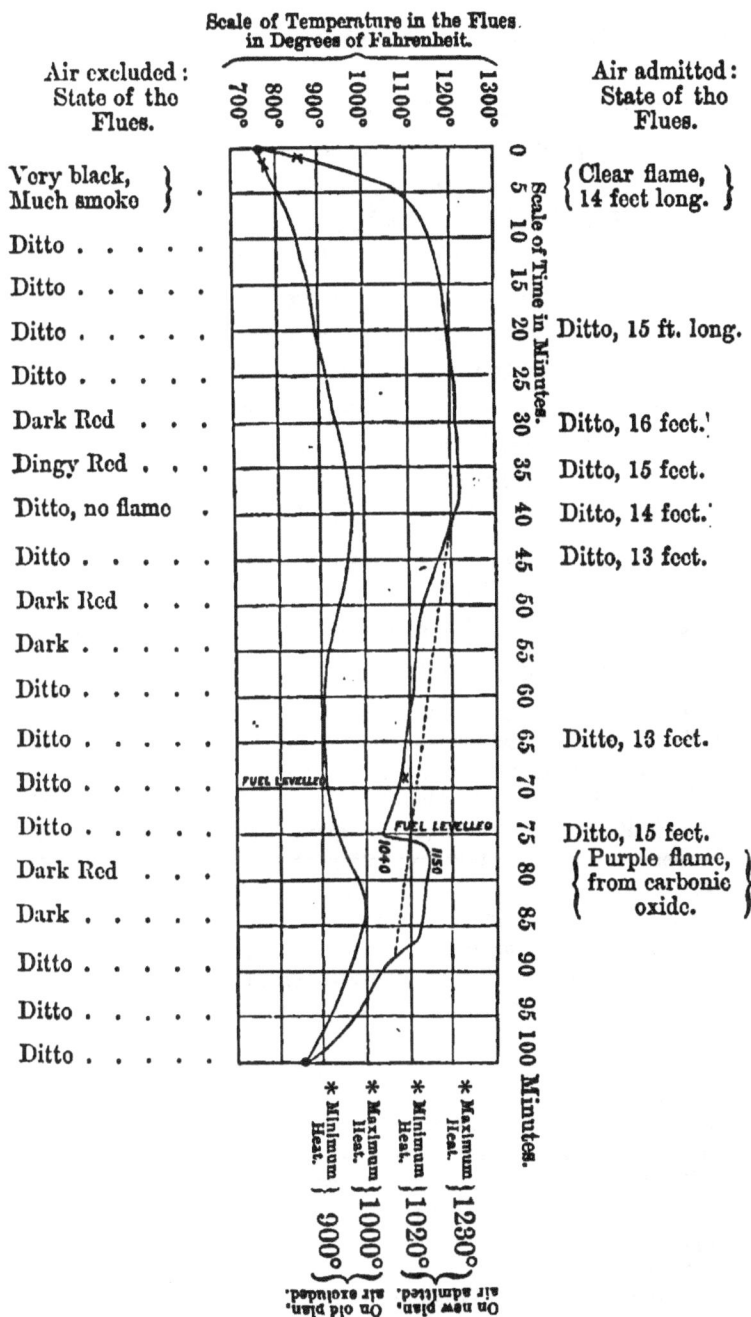

Air excluded: State of the Flues.	Air admitted: State of the Flues.
Very black, Much smoke }	{ Clear flame, 14 feet long. }
Ditto	
Ditto	
Ditto	Ditto, 15 ft. long.
Ditto	
Dark Red . . .	Ditto, 16 feet.
Dingy Red . . .	Ditto, 15 feet.
Ditto, no flame .	Ditto, 14 feet.
Ditto	Ditto, 13 feet.
Dark Red . . .	
Dark	
Ditto	
Ditto	Ditto, 13 feet.
Ditto	
Ditto	Ditto, 15 feet.
Dark Red . . .	{ Purple flame, from carbonic oxide. }
Dark	
Ditto	
Ditto	
Ditto	

Fig. 16.—Diagrams of temperature in boiler-flue.

for the British Association assembled at Manchester, in 1842.

By this diagram, it will be seen that on a charge of 3 cwt. of coal being thrown on the furnace, the temperature in the flue (as indicated by the pyrometer) rose, in 25 minutes, from 750° to 1220°, when it began to fall, and descended to 1040°, *the fuel not having been disturbed during 75 minutes.* At this stage, however, a remarkable change took place. Perceiving the temperature in the flue to have become so low, Mr. Houldsworth had *"the fuel levelled,"* that is, had it more equally distributed, and the *vacant spaces covered.* The effect was (as shown in the diagram) the sudden rise in the temperature from 1040° to 1150°, at which it continued during ten minutes, when it gradually fell to 850°.

The upper line of the diagram represents range of temperature, air being admitted.

The lower line of the same represents range of temperature, air being excluded, common plan.*

* As the use of the pyrometer is of the highest importance, not merely for experimental purposes, but for all boilers, and for general use, whenever it can be introduced, the simple but valuable instrument which is used by Mr. Houldsworth, and by which he obtained the above results, is here given from an interesting paper on "The Consumption of Fuel and the Prevention of Smoke," read before the British Association by William Fairbairn, Esq., C.E., F.R.S.

"For these experiments we are indebted to Mr. Henry Houldsworth, of Manchester; and, having been present at several of the experiments, I can vouch for the accuracy with which they were conducted, and for the very satisfactory and important results deduced therefrom.

"In giving an account of Mr. Houldsworth's experiments, it will be necessary to describe the instrument by which they were made, and also to show the methods adopted for indicating the temperature, and the changes which take place in the surrounding flues.

"The apparatus consists of a simple pyrometer, with a small bar of copper or iron (*a* in Fig. 17) fixed at the extreme end of the boiler, and projecting through the brick-work in front, where it is joined to the arm of an index lever *b*, to which it gives motion when it expands or contracts by the heat of the flue.

It is here to be observed that when a charge is nearly exhausted, or begins to *burn in holes*, the evil increases

" The instrument being thus prepared, and the bar supported by iron pegs driven into the side walls of the flue, the lever (which is kept tight upon the bar at the point *e* by means of a small weight over the pulley at *d*) is attached, and motion ensues. The long arm of the lever at *d* gives motion to the sliding rod and pencil *f*, and by thus pressing on the periphery of a slowly revolving cylinder, a line is

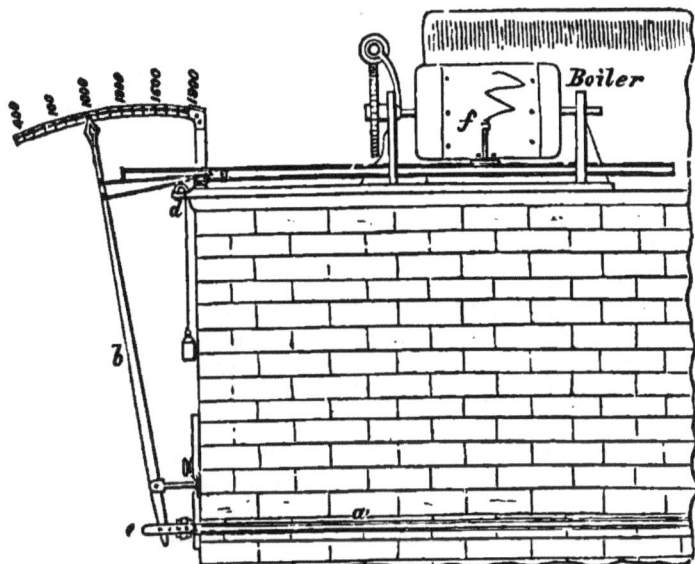

Fig. 17.—Pyrometer.

inscribed corresponding with the measurements of the long arm of the lever, and indicating the variable degrees of temperature by the expansion and contraction of the bar. Upon the cylinder is fixed a sheet of paper, on which a daily record of the temperature becomes inscribed and on which are exhibited the change as well as the intensity of heat in the flues at every moment of time. In using this instrument it has been usual to fix it at the medium temperature of 1000°, which, it will be observed, is an assumed degree of the intensity of heat, but a sufficiently near approximation to the actual temperature for the purpose of *ascertaining the variations which take place in all the different stages of combustion consequent upon the acts of charging, stirring, and raking the fires.*"

Mr. Fairbairn then gives two interesting diagrams exemplifying the

itself by the accelerated rapidity with which the air enlarges the orifices it has thus made for its own admission, causing a still more rapid combustion of the fuel around the uncovered parts, and at the very time when these orifices should have been closed.

Had it been possible, in Mr. Houldsworth's experiment,

result of experiments made by the aid of the pyrometer, and continues :—

"On a careful examination of the diagrams, it will be found that the first was traced without any admixture of air except that taken through the grate-bars; the other was inscribed with an opening for the admission of air through a diffusing plate behind the bridge, as recommended by Mr. C. W. Williams. The latter, No. II., presents very different figures : the maximum and minimum points of temperature being much wider apart in the one than the other, as also the fluctuations which indicate a much higher temperature, reaching as high as 1400°, and seldom descending lower than 1000°, giving the mean of 1160°.

"Now, on comparing No. II. with No. I., where no air is admitted, it will be found that the whole of the tracings exhibit a descending temperature, seldom rising above 1100° and often descending below 900°, the mean of which is 975°. This depression indicates a defective state in the process, and although a greater quantity of coal was consumed (2000 lbs. in 396 minutes in the No. II. experiment, and 1840 lbs. in 406 minutes in No. I.), yet the disparity is too great when the difference of temperature and loss of heat are taken into consideration. As a further proof of the imperfections of No. I. diagram, it is only necessary to compare the quantities of water evaporated in each, in order to ascertain the difference, where in No. I. experiment 5·05 lbs. of water are evaporated to the pound of coal, and in No. II. one-half more, or 7·7 lbs. is the result.

"Mr. Houldsworth estimates the advantages gained by the admission of air (when properly regulated) at 35 per cent., and when passed through a fixed aperture of 43 square inches, at 34 per cent. This is a near approximation to the mean of five experiments, which, according to the preceding table, gives 33½ per cent., which probably approaches as near the maximum as can be expected under all the changes and vicissitudes which take place in general practice."

Here are practical results from unexceptionable quarters, and although they have been so many years before the public, nevertheless, smoke-burning observations and hot-air fallacies continue to be listened to, and dearly paid for.

to have preserved the fuel continuously, and *uniformly* spread, throughout the charge of 100 minutes, the diagram would have indicated a more uniform line of temperature, as *marked by the dotted line*, and, consequently, have produced a higher average range of heat in the flue.

It is true, by mechanical contrivances, by which the fuel is thinly and continuously spread over a large surface, there would be less tendency to the formation of dense smoke, because the quantity of air introduced over that extended surface being so much greater than is chemically required, the volume of flame is considerably reduced, and consequently the volume of smoke. We must not, however, deceive ourselves in this matter. The avoidance of dense smoke by these means must be attended with the production of less available flame and heat, relatively with the area on which the fuel is spread, from the extended and attenuated temperature in the furnace chamber.

Having spoken of the *grate-bar surface*, and what is placed on it, we have next to consider the *chamber part* of the furnace, and what is formed therein. In marine and cylindrical land boilers, this chamber is invariably made *too shallow and too restricted.*

The proportions allowed are indeed so limited as to give it rather the character of *a large tube*, whose only function should be, the allowing the combustible gases to *pass through it*, rather than that of *a chamber*, in which a series of consecutive chemical processes were to be conducted. Such furnaces, by their diminished areas, have also this injurious tendency,—that they increase the already too great rapidity of the current through them.

The constructing the furnace chamber so shallow, and with such inadequate capacity, appears to have arisen from the idea, that the nearer the body to be heated was brought to the source of heat, the greater would be the quantity received. This is no doubt true when we present a body to

36 COMBUSTION OF COAL AND SMOKE PREVENTION.

be heated in front of a fire. When, however, the approach of the colder body will have the direct effect of interfering with the processes of nature (as in gaseous combustion), it

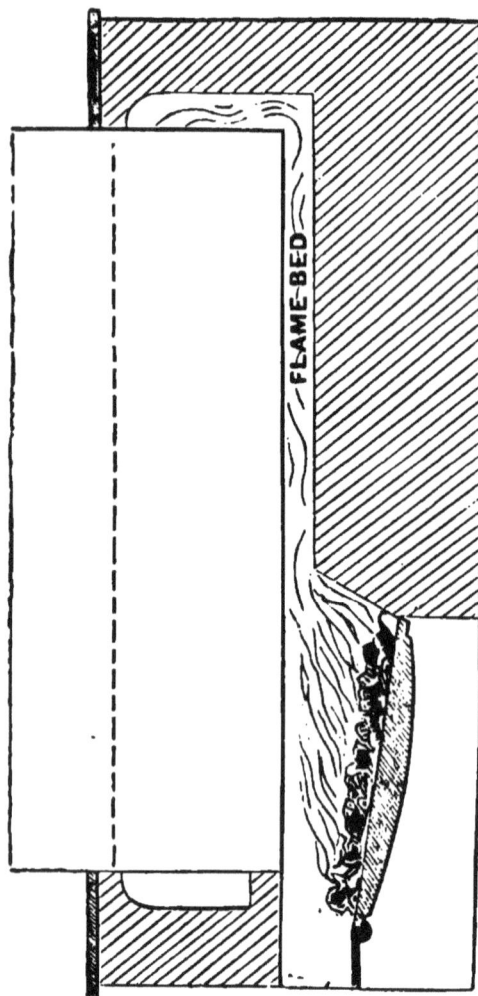

Fig. 18.—Section of Furnace, showing the Flame-bed.

must manifestly be injurious. *Absolute contact* with flame should be avoided where the object is *to obtain all the heat* which could be produced by the combustion of the entire of the constituents of the fuel.

So much, however, has the supposed value of near approach, and even impact, prevailed, that we find the space behind the bridge, frequently made but a few inches deep, and bearing the orthodox title of the *flame-bed*, as in Fig. 18. Sounder views, however, have shown that it should have been made capacious, and the impact of the flame avoided.

As a general rule, deduced from practice, it may be stated, that the depth between the bars and the crown of the furnace should not be less than *two feet six inches* where the grate is but four feet long; increasing in the same ratio where the length is greater: and, secondly, that the depth below the bars should not be less, although depth is not there so essential either practically or chemically.

CHAPTER VI.

OF THE INTRODUCTION OF THE AIR TO THE FUEL IN A FURNACE, PRACTICALLY CONSIDERED.

The Coke.—Were there nothing else requiring attention, in the use of coal, than the combustion of its fixed *carbon* (as in the fire-box of a locomotive) nothing further would be necessary than supplying the air through the grate-bars to the fuel on them. In the use of *coal*, however, as there is *the gas* also to be generated and consumed, any excess of air, or its injudicious introduction, though it might not affect the combustion of the carbon, must necessarily interfere with the quantity introduced for the use of that gas.

As to the *quantity* of air chemically required for the *coke*, or fixed portion of the coal, after the gas has been expelled, it has already been shown that every 6 lbs. of carbon requires 16 lbs. of oxygen. Now, the volume of atmospheric air which contains 16 lbs. of oxygen is estimated at about 900 cubic feet, at ordinary temperature. Taking, then, bituminous coal as containing 80 per cent. of carbon, we have 1600 lbs. of coke (the produce of 20 cwt. of coals) requiring its equivalent of oxygen, and which will be equal to 240,000 cubic feet of air; since as 6: 900: : 16: 240,000. This great quantity of air required for the exclusive use of the *coke on the bars*, must, therefore, be passed upwards, from the ash-pit, the product being transparent carbonic acid gas, of a high temperature.

In supplying the air to the coke, and to avoid the admission of a larger quantity than is legitimately required for its own combustion, the principal point requiring attention is the *preserving of a uniform and sufficient body of fuel on the bars*, as noticed in the last chapter.

The Gaseous Portion.—It has been shown that each cubic foot of gas requires, absolutely, the oxygen of ten cubic feet of atmospheric air. By the proceeds of the Gas Companies, we learn, that 10,000 cubic feet of gas are produced from each ton of bituminous coal: this necessarily requires no less than 100,000 cubic feet of air. Adding this to the 240,000 cubic feet required for the coke, we have a gross volume of 340,000 cubic feet as the minimum quantity absolutely required for the combustion of *each ton of coal*, independently of that excess which will always be found to pass beyond what is chemically required.*

Before a fresh charge of coal is thrown in, there will be, or should be, as already observed, a sufficient body of clear and highly heated coke remaining on the bars. *After* the charge has been made, a large volume of gas will be generated; and, consequently, an equivalent quantity of pure air will be required for its combustion. Now, at this stage of the process, and by reason of the mass of fresh fuel thrown in, the passage of the air through it must then, necessarily, be the most restricted. Thus the smallest quantity of air would be enabled to gain admission, simultaneously, with the greatest demand for it; and the largest generation of gas, simultaneously, with the most restricted means of enabling the air to obtain access. Were there no other considerations, these alone would be sufficient to show the absolute necessity of *providing some other channel* for the introduction of the air for the gas, and the impossibility of introducing the requisite quantity in that direction.

* The volume 340,000 cubic feet per ton is equivalent to 152 cubic feet per pound of coal.—(ED.)

The introducing of the required quantity of air will necessarily depend, first, on the *area of the orifice* through which it enters ; and secondly, *the velocity* at which it passes through that area. It has been stated that the aperture for the admission of the required quantity should average from *one-half* to *one square inch for each square foot of grate-bar surface.* So entirely disproportioned, however, is the area here stated, that it would not supply one-fourth the quantity *absolutely required ;* much less that additional quantity which we have seen must of necessity pass with it.

There seems, then, to have been some serious oversight in making these calculations. Practice and experiment prove that instead of an area of *one square inch*, no less than *from four to six square inches for each square foot* of furnace will be required, according to the gas-generative quality of the coal, and the extent of the draught in each particular case.

In examining the tables of results supplied by experimenters, the cause of their error may be traced to a mistake in the estimated velocity of the heated gaseous matter passing through furnaces to the chimney shafts. As this has, in many instances, been adopted on the supposed authority of Dr. Ure, it is right to state, that the error appears to have originated in taking what that accurate chemist and experimenter had given,—not as *practical*, but as *theoretic* results.

Now, suppose a furnace measuring $4 \times 2.5 = 10$ square feet of surface, and with moderate draught, this will be adequate to the combustion of 2 cwt. of coal per hour ;—the gas from which will require 10,000 cubic feet of air. To supply this quantity, within the hour, will require the following relative areas of admission. and velocity of current, viz. :—

Velocity of current per second of air entering the furnace.			Area of Aperture, in square inches, per foot of furnace.	
If at 6·66 feet per second, will require			6	square inches.
„ 10	„	„	4	„
„ 20	„	„	2	„
„ 40	„	„	1	„

From this we see the absolute necessity of ascertaining the practical rate of current of the air *when entering*, before we can decide on the necessary area for its admission. Hitherto no estimate has been made respecting these proportions on which reliance can be placed.

With reference to the *mode* of introducing the air, it is not a little remarkable that many overlook, or even dispute the difference in effect, when it is introduced through *one*, or *numerous* orifices. In illustration, then, of the effect of introducing the air *in a divided form*, let us take the case of a boiler furnace of modern and approved form, where the air enters by a *single orifice*, and compare it with that shown where it enters through a hundred or more orifices.

In the first example (if the body of air be not too great), the effect may be favourable, to some extent, in preventing the generation of dense smoke. Inasmuch, however, as the quantity of air thus introduced is chemically inadequate to the combustion of the gas, much of the latter must escape *unconsumed*, though not in the form of smoke, but as a light coloured vapour.

The body of air, by passing through a single aperture, produces the action of a strong current, and obtains a direction and velocity antagonistic to that *lateral motion* of its particles which is the very element of diffusion. In this case, passing along the flue, the stream of air pursues its own course at the lower level, A, while the heated products fill the *upper one* at B. It is here evident, according to the laws of motion, that the two forces, *acting in the same direction*, prevent the two bodies impelled by them (the air and the gas) from amalgamating.

Now, instead of a single aperture, let the air enter through
a hundred or more apertures, as in Fig. 20.   Here the force

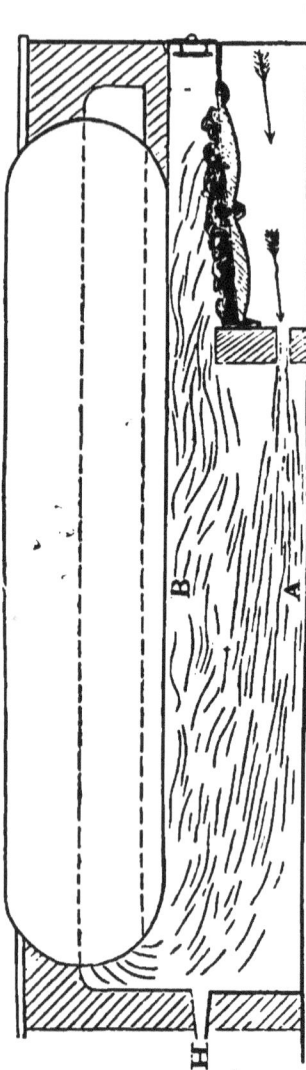

Fig. 19.—Air admitted through one orifice at the bridge.

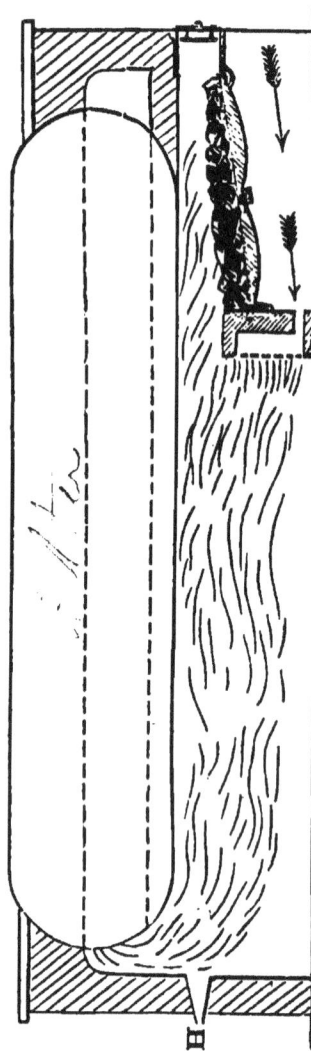

Fig. 20.—Air admitted through numerous orifices at the bridge.

and direction of *the current* will be avoided, and the required
diffusive action produced on passing the bridge.   Instead of

tho refrigeratory influence of the air, as in the first case, there will be a succession of igniting atoms, or groups, which Sir H. Davy calls "explosive mixtures," each producing combustion with its high temperature. These are distinctly perceptible from the sight holes at H.

The same results will follow, whether the single orifice or the numerous orifices are placed at the *bridge end*, or at the *door* of a furnace, as in Fig. 21. In this case, the diffusion will be more immediate and effective.

Of the advantageous effect produced by *mechanical agency*, in promoting immediate diffusion between the air and the gas, the following experiments are quite conclusive.

Let Figures 22, 23, and 24 represent each a tin apparatus, with its glass chimney, similar to the ordinary Argand burner, —the gas is admitted the same way in all three—the difference to be noted is, *in the manner in which the air is admitted.* In all these cases, the quantity of both gas and air was the same.

In Fig. 22, no air is admitted from below ; and the gas, consequently, does not meet with any until it reaches the top of the glass, where it is ignited, producing a dark smoky flame.

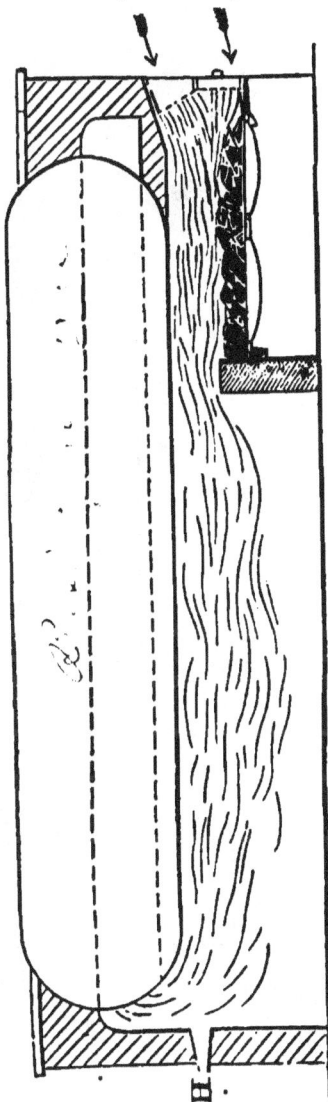

Fig. 21.—Air admitted through numerous orifices at the door.

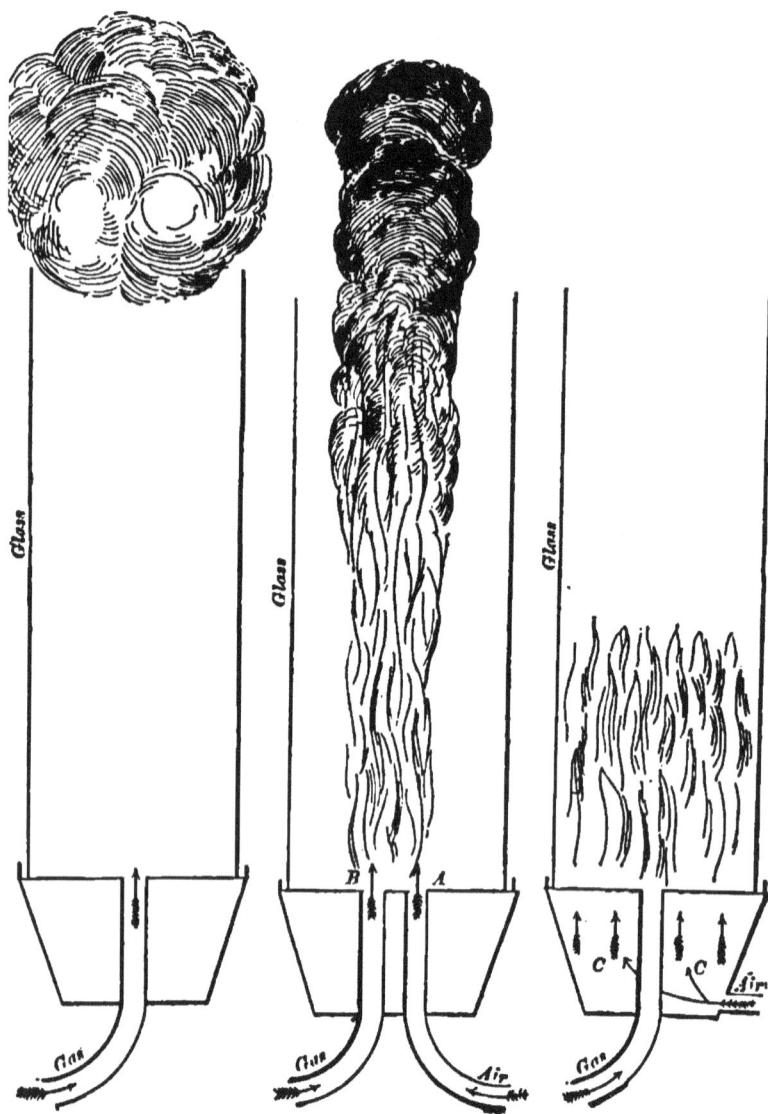

Figs. 22, 23, 24.—Supply of air to burn gas.

In Fig. 23, air is admitted from below, and rises through the orifice at A, concurrently with the gas at the orifice B. On being ignited, one long flame is produced, of a dark colour, and ending in a smoky top.

In Fig. 24, the air is introduced from below, and into the

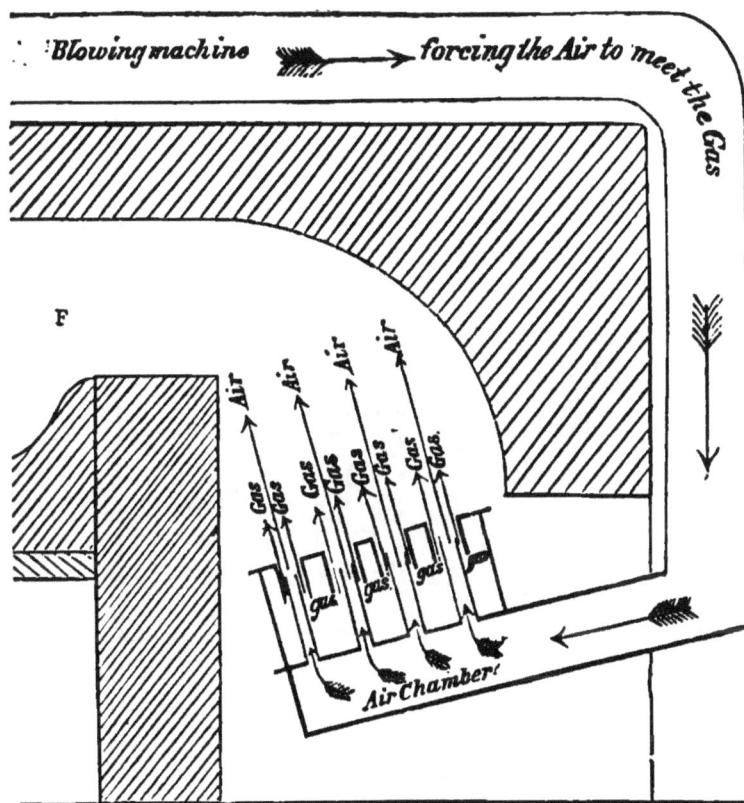

Fig. 25.—Gas furnace at Treveray.

chamber C, C, from which it issues through a perforated plate, like the rose of a watering pot; thus producing immediate mixture with the gas. On being ignited, a short, clear, and brilliant flame was produced, as in the ordinary Argand gas burner.

The *heating powers* of the flames were then tested, by

placing a vessel of cold water over each. When over Fig. 23, it required 14 minutes to raise the water to 200°, whereas, over Fig. 24, it reached 200° in 9 minutes.

Now, the difference of effect produced in those three experiments corresponds with what takes place in furnaces and

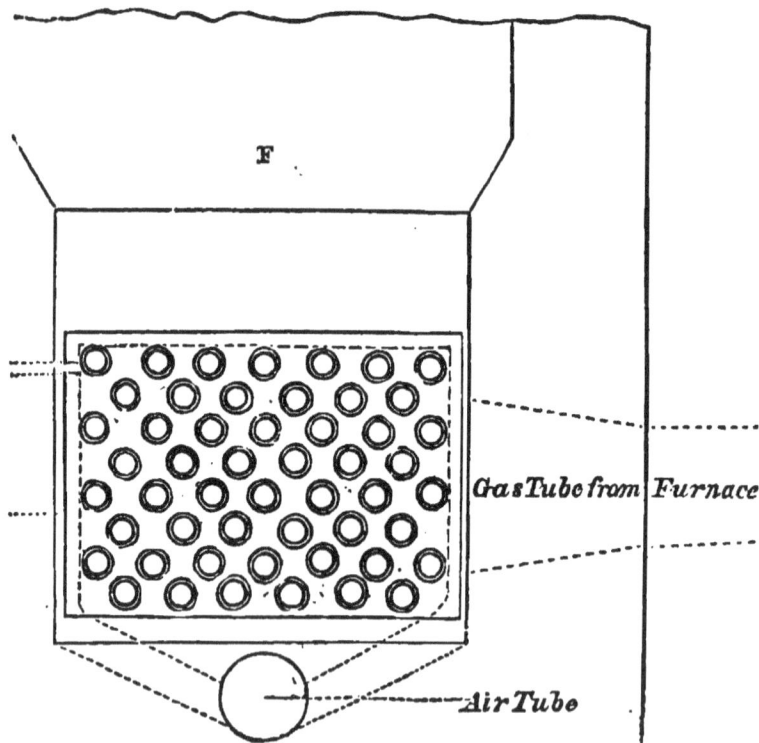

Fig. 26.—Gas furnace at Treveray.

their flues, when the air is excluded, and when it is admitted through a single or through numerous orifices.

Of the importance of *mechanical agency*, in promoting the rapid diffusion or mixture of the air and the gas, the modes adopted on the continent for rendering the coke gas, or *carbonic oxide*, available, are conclusive and instructive.

M. Péclet has given ample details of the mode of effecting

the combustion of this gas (the existence of which has, for a long time, been practically ignored in this country), in the manufacture of iron, and even in the puddling furnaces, where the most intense heat is required.

M. Péclet states that the process at Treveray, in France (see Figs. 25 and 26), is preferable to that adopted in Germany, and for the following reasons, which are quite to the point of our present inquiry.

1st. The air and the gas are better incorporated.

2nd. The relative quantities of the gas brought into contact with the air are more easily regulated.

3rd. Combustion is effected by the introduction of the smallest excess of air.

In the apparatus, as shown in the section, Fig. 25, 50 jets of air issue, each in the centre of 50 jets of the gas (carbonic oxide), led from the cupolas of the smelting furnaces. On examination of the process here exhibited, the mixing and combustion, it will be seen, takes place *on the instant*, and before the flame and heat enter the chamber of the furnace at F. By this arrangement, M. Péclet observes, "the highest temperature that the arts can require is here obtained."

CHAPTER VII.

OF REGULATING THE SUPPLY OF AIR TO THE GAS BY SELF-ACTING OR OTHER MECHANICAL APPARATUS.

Much has been urged on the necessity for regulating the supply of air entering the furnace, as a means of preventing an excess at one time, or insufficient quantity at another. The theory is plausible. Practice, however, when tested by the aid of a pyrometer, and on the large scale of the furnace, has invariably proved its unsoundness and futility.

In the report made to the Dublin Steam Company, in 1842, by Mr. Josiah Parkes (the Patentee of the Split-bridge), an engineer well qualified for such an inquiry, he observes: "During the above-named experiments, I made numerous essays of the effect produced by shutting off the admission of air to the gases, after the *visible* inflammable gases had ceased to come over, and when the fuel on the grate was clear and incandescent. *In such cases I always found the entire stoppage of air to be followed by diminished heat in the flues, and by diminished evaporation;* for at these times, *carbonic oxide* continued to be formed; a gas which, though colourless, was converted, by a due mixture of the atmospheric air, into flame, possessing, evidently, a high intensity of heat, and producing much useful effect. The calorific value of this gas is lost *when the air is excluded,* although its non-combustion is not attended with the production of visible smoke."

During these investigations it was ascertained, that *the*

SUPPLY OF AIR TO GAS.

appearance or non-appearance of visible smoke was no test, either for or against the admission of air—*as to quantity.* Mr. Parkes on this head observes : " The consequences of regulating and varying the quantity of air admitted so as to suit the varying state of the furnace, as regards the quantity of gas given off, also occupied my close attention. It is quite certain that, to effect the perfect combustion of all the combustible gases produced in a furnace, a large demand for air (distinct from the air entering the grate) *always exists:* also, that by entirely *excluding air*, smoke is produced, and the heat diminished in all states of the fire. Thus, with correctly assigned proportions once ascertained, no attention is required on the part of the fireman in regulating the admission of air. On looking through the sight holes, it was manifest, that, as a stream of either carburetted hydrogen, or carbonic oxide gas, was at all times generated and passing over ; so there was necessarily a corresponding demand for air ; and when supplied, a continuous stream of visible flame."

This is conclusive on the point of regulating the supply of air, or shutting it off at any period of a change.

In addition to this inquiry, Sir Robert Kane (one of the highest chemical authorities of the day) was also engaged, and made an elaborate investigation and report on the subject.

REPORT TO THE DIRECTORS OF THE CITY OF DUBLIN STEAM-PACKET
COMPANY.

Gentlemen,—In accordance with your request, that we should proceed to examine into the construction and performance of the Marine Boiler Furnaces erected at your works in Liverpool, upon the principle of the patent of Mr. Williams, we have to report, that we have carefully inspected the operation of these furnaces in their several parts, and also some others constructed in a similar manner, upon a large working scale, which are now in actual use in various parts of the town ; and that we have instituted several series of experiments and observations upon the temperature produced by those furnaces, and the manner in which the fuel is consumed in them.

In deducing from those experiments and observations the conclusions which will be found embodied in this report, we have taken into careful consideration the general chemical principles upon which combustion must be carried on, so as to effect the greatest economy of heat and the fuel: and we have examined how far those principles are attended to in the construction of the various kinds of furnaces that have been proposed for practical use.

The conclusion to which we have arrived, and which we believe to be established by very decisive evidence, as well of a practical as of a theoretical kind, may be briefly expressed as follows:

1st. That, in the combustion of coals, a large quantity of gaseous and inflammable material is given out, which, in furnaces of the ordinary construction, is, in great measure, lost for heating purposes, and gives rise to the great body of smoke which, in manufacturing towns, produces much inconvenience.

2nd. That the proportion which the *gaseous* and volatile portion of the fuel bears to that which is *fixed*, and capable of complete combustion on a common furnace grate, may be considered as *one-fourth*, in the case of ordinary coal.

3rd. That the air for the combustion of this gaseous combustible material cannot, with advantage, be introduced either through the interstices of the fire bars, or the door by opening it. In the former case, the air is deprived of its oxygen by passing through the solid fuel, and then only helps to carry off the combustible gases, before they can be burned; and, in the latter case, the air which would enter, by reason of its proportionate mass, would produce a cooling influence, and cannot conveniently be mixed so as properly to support the combustion of the gases.

4th. That the combustion of the gaseous materials of the fuel is best accomplished by introducing, through a number of thin or small orifices, the necessary supply of air, so that it may enter *in a divided form* and *rapidly* mix with the heated gases in such proportions as to effect their complete combustion.

5th. That, in burning *coke*, or when coal has been burned down to a *clear red fire*, although the combustion on the grate may appear to be perfect, and little or no flame may be produced, and no smoke whatever made, *there may be a great amount of useful heat lost*, owing to the formation of *carbonic oxide*, which, not finding a fresh supply of air at a proper place, necessarily passes off unburned.

6th. That under the common arrangements of boiler furnaces, where there is intense combustion on the *fire-grate*, and but little in the *flues*, the differences of temperature in and around the various parts of the boiler are greater; and, consequently, the boiler is most subject to the results of unequal temperatures. On the other hand, when the process of combustion is spread through the flues, as well as over the fire-grate, the temperature remains most uniform throughout, and the boiler and its settings must be least liable to injury.

7th. That the heat produced by the combustion of the inflammable gases and vapours from the fuel, in flues or chambers behind the bridge, must be considerable, and can be advantageously applied to boilers, the length of which may be commensurate with that of the heated flues.

In further substantiation of these conclusions, we will describe the results of our experiments made with the marine boilers fitted up with air-apertures on Mr. Williams's plan, in order to determine how far, in practice, the scientific principles of combustion may be economically carried out.

EXPERIMENTS WITH COAL.

EXPERIMENT 1.

When the fire was charged with *coal*, and air admitted only in the ordinary way (the passage to the air-distributors being closed), the entire interior of the flues was filled with a *dense black smoke*, which poured out from the orifice of the chimney in great quantity, and as observed through the sight-holes. The mean temperature of the flues in this experiment being found to be 650°.

EXPERIMENT 2.

The furnace being charged in the same manner with coal, and the supply of air by the dividing apparatus fully let on, the smoke instantly disappeared. Nothing visible passed from the chimney. The flues became filled with a clear yellow flame, which wound round at a maximum distance of thirty feet, and the mean temperature at the turn of the flue was found to be 1211°.

Hence, the quantity of heat conveyed to the water through the flues was nearly doubled by introducing the air in this divided manner; and whilst the fuel remained the same, the combustion was rendered perfect, and no smoke produced.

EXPERIMENT 3.

The furnace being charged with coal exactly as before, the passage to the air-apertures was *one-half* closed. A grey smoke issued from the chimney. The flues were occupied by a lurid flame, occasionally, of nearly forty feet in length; the mean temperature of the flues being found to be 985°.

Thus, with half the supply of air, a mean condition was obtained between the dense black smoke and imperfect combustion of the first experiment, and the vivid combustion and perfect absence of smoke of the second.

EXPERIMENTS WITH COKE.

Having thus tested the circumstances of the combustion of *coal*, under different conditions of the furnace, we next proceeded to ascertain the exact circumstances of the combustion of *coke*.

EXPERIMENT 4.

The furnace being fully charged with *coke* (from the Gas Works), and the *air-aperture closed*, so that it burned as in an ordinary furnace, the flues were dark, but a bluish-yellow flame extended under the boiler to the back, a space of ten feet. The mean temperature of the flue was then found to be 702°.

The *coal*, under the same circumstances, having given a mean temperature of 650°, a difference of 52°, heating power, was thus shown in favour of *coke*, and which agrees with results obtained by others with furnaces of the ordinary construction.

EXPERIMENT 5.

The furnace being again charged with coke, the air-aperture was opened *one-half*. The flues then became occupied with a flame of various tints,—blue, yellow, and rose-coloured—produced by the combustion of carbonic-oxide and various other gaseous products. This flame extended through twenty-five feet. The mean temperature of the flue was then found to be 1010°.

Thus, even with coke, the increase of available heating power, produced by the admission of air on Mr. Williams's plan, was found to be 300°, or three-tenths of the entire.

EXPERIMENT 6.

The furnace being again charged with coke, and the air-aperture *fully opened*, the flame in the flue shortened to about fifteen feet, and the mean temperature of the flue became 852°.

Hence it appeared, that there had been a larger quantity of air admitted in this last case than was necessary for the combustion of the gases from the *coke;* and hence a cooling effect had been produced, such as to neutralise one-half of the advantage which would have otherwise been gained.

It results from these experiments,—

1st. That the air-aperture of the furnace was sufficient for the proper combustion of *coals*, but was one-half too large for *coke*.

2nd. That by the use of the air-apertures, in the case of coals, all smoke is prevented, and the useful effect of the fuel much increased.

3rd. That, even when *coke* is used, the heating effect is also much increased by the admission of air by apertures behind or at the bridge, but it required only *one-half* of the air which is necessary for *coal*. If, however, it be supplied with the quantity best adapted for *coal*, one-half of the advantage is again lost by the cooling power of this excess of air.

4th. Since, in all ordinary cases of practice, fresh fuel is added in moderate quantities, at short intervals of time, it was not found necessary to alter the rate of admission of the air by valves or other mechanism. A uniform current, admitting a quantity of air intermediate to that necessary for coal alone, will abundantly suffice for the perfect combustion of the fuel, and need not require any extra attention on the part of the workmen.

In conclusion, we have to state as our opinion, that the arrangement of furnace and admission of distributed air on Mr. Williams's plan, fulfils the conditions of complete combustion in the highest degree, as far as is compatible with the varieties which exist in the construction of boilers, the peculiar character of the coal employed, and the nature of the draught ; the formation of smoke is prevented ; and the economy of fuel we cannot consider as being less than an average of one-fifth of the entire in the case of *coke*, and of one-third of the entire when *coal* is used.

We are, Gentlemen,
Your obedient Servants,

ROBERT KANE, M.D., M.R.I.A.,
Professor of Natural Philosophy to the ROYAL DUBLIN SOCIETY, *and Professor of Chemistry to the Apothecaries' Hall of Ireland.*

R. H. BRETT, Ph.D., F.L.S.,
Professor of Chemistry to the Liverpool Collegiate Institution.

The inference from these chemical investigations is, that *there is no interval from the beginning to the end of a charge, when there is not a large body of combustible gas generated in the furnace, and a large supply of atmospheric air required.* The advocates of self-acting valves have overlooked the chemical fact, that as soon as the coal gas (carburetted hydrogen) ceases to be evolved, the fuel on the bars would then be in an incandescent state, and precisely in the condition to furnish a copious generation of the other gas—the *coke gas,* or *carbonic oxide ;* but which had not hitherto been noticed by any writer *in connection with boiler furnaces.*

Now, as this latter gas requires (for equal volumes) one-half the quantity of air of the former, it is equally necessary that such be supplied, or the heating power of carbon would be lost. The characteristics of this gas have already been given.

With reference to the progressive rate of generation of the gas in a furnace, and the consequent demand for atmospheric air, the length of the flame (when the air is properly supplied) furnishes the best evidence. The following tabular view of the result of numerous accurate experiments, made many years back, and expressly to ascertain the *rate of evolution* of the gases, throughout a charge of 40 minutes' duration, is conclusive :—

Time.	Thermometric Temperature in Flues.	Length of Flame in Feet.
Charge made	466	10
2 minutes	462	14
4 „	490	18
6 „	508	22
8 „	518	26
10 „	524	26
12 „	528	28
14 „	534	28
16 „	540	28
18 „	540	28
20 „	540	26
22 „	536	24
24 „	524	24
26 „	508	22
28 „	494	22
30 „	486	18
32 „	476	22
34 „	468	14
36 „	464	14
38 „	460	12
40 „	460	10 *

* The bulb of the thermometer was here inserted in the flue, so far as to prevent the mercury rising above 600°—the highest range we

We here see, that so far from the quantity of gas generated being *greatest at first*, and ceasing when the charge was one-half exhausted, it is just the reverse. In fact, any one who has observed the indications of the pyrometer in the flue, and has looked into a furnace in action, must have observed that, there being much coal in the moisture to be evaporated, it required a considerable time before the full supply of gas was being generated, and the temperature in the flue had risen to the maximum. Further, that when the first half of the charge was exhausted, the greatest quantity of gas was then momentarily evolved—the longest flame existing in the flue—and the highest temperature indicated by the pyrometer; consequently, the fullest supply of air was then required.

The following experiment is also in point here : This was made with a larger charge of coal, and during 60 minutes (the bars being kept well covered), the object being to ascertain the relative quantity of *each kind of gas* evolved; and thus form a guide to the quantity of air required, at the several intervals, from the beginning to the end of a charge. [The observations were taken from two sight apertures : one at the back end of the boiler, and the other at the front, looking into the flue.] When the supply of carburetted hydrogen gas was nearly exhausted, the distinct flames, and their two distinct colours and characteristics, might clearly be distinguished. The following Table will present a view of the relative quantities of the two gases (carbonic acid and carbonic oxide, or coke gas) produced during the progress :—

---

see being 540°—when the charge was half expended. The absolute heat in the flue was, however, considerably higher, as ascertained by the melting points of a series of metallic alloys, prepared by Sir Robert Kane, expressly for the purpose. By these, inserted in the flue, it was found that the absolute heat escaping *at the foot of the funnel*, was at least 750°.

Time in minutes.	Coal Gas.	Coke Gas.	Total length of Flame in feet.
Charge of coal . .	none . . .	10	. . . 10
5 minutes . . .	10	. . . none	. . . 10
10 „	14	. . . none	. . . 14
15 „	18	. . . none	. . . 18
20 „	22	. . . none	. . . 22
25 „	22	. . . none	. . . 22
30 „	18	. . . none	. . . 18
35 „	14	. . . none	. . . 14
40 „	10	. . . 4	. . . 14
45 „	5	. . . 8	. . . 13
50 „	none	. . . 12	. . . 12
55 „	none	. . . 10	. . . 10
60 „	none	. . . 10	. . . 10

Here, column 4 may be taken as indicating the gross quantities of combustible gases evolved, and requiring a supply of air. In numerous other furnaces, in which the air was properly introduced, and the fuel properly covering the bars, the flame was seen during a large portion of an hour's charge, extending along the side flues from twenty to thirty feet. The quantity of the *coke gas* will be in proportion to the thickness or body of the fuel, and its state of incandescence.

With the view of accommodating the supply of fuel to the demand for air, the best practical mode is the *equalising of the quantity of gas requiring such supply.* This was done effectually thirty years back, by arranging the furnaces so that each pair shall be connected with one common flue. This arrangement, for alternate firing, adopted among others in the steamer *Royal William* (as hereafter shown), is every way satisfactory. A similar arrangement has been introduced in Her Majesty's steamers *Hermes*, *Spitfire*, and *Firefly*, as described in Tredgold's work; nothing, however, is there shown as to the means for introducing the air, and, consequently, the value of this flue arrangement is lost.

Fig. 27, taken from Péclet's work, shows a similar mode adopted in France, for equalising the supply and demand of

gas and air. It will be manifest that, assuming the furnaces to be charged alternately, the quantity of gas behind
the bridge will be *the mean* of that generated in both
furnaces.

Another and a very effectual mode of equalising the

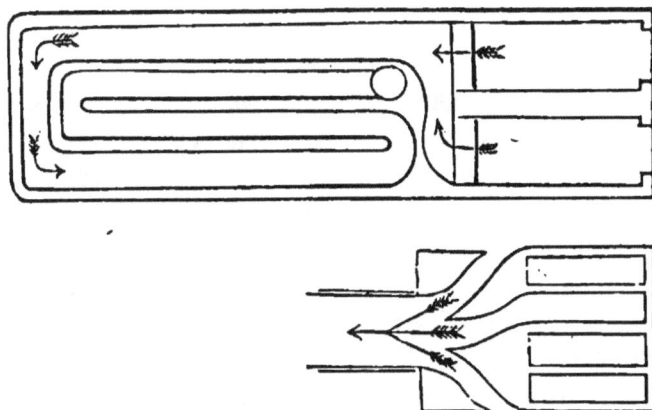

Fig. 27.—Equalising the supplies of gas and air.

supply of gas, and thus practically equalising the supply of
air, is by charging the furnace-grate alternately, *first on the
one side, and then on the other.* Where the furnace is wide
enough, this is very effective.

CHAPTER VIII.

OF THE PLACE MOST SUITABLE FOR INTRODUCING THE AIR TO THE GAS IN A FURNACE.

THE plan adopted by Mr. Parkes of introducing the air through what is called the *split bridge*, as hereafter shown, appears to have been among the first which recognised the providing *a separate supply of air* to the furnace gases, independently of that which passed through the fuel on the bars.

This plan was sufficiently effective, when combined with the system of small furnaces, with small charges of coal ; or large furnaces when charged heavily, with sufficient fuel for many hours' consumption, producing a uniform generation of gas during a long interval, and by the means of slow combustion. The issue of the air through the narrow orifice in the top of the bridge was, however, found to be unsuited to the large furnaces, with quick combustion and heavy charges incidental to the boilers used in steam-vessels. It was also liable to be occasionally obstructed by the stronger current of heated products crossing the aperture, in the same way as the ascent of smoke from a house-chimney is obstructed by a strong wind sweeping across it. Numerous modifications of this plan were adopted in steam-vessels, the most important of which will hereafter be given, with the view of explaining the several causes of their failure, and which it is often as important to know as those of success.

The arrangement subsequently adopted in several vessels of the Dublin Steam Company admitted the air through numerous apertures, and in a divided state. This mode was always effective when the draught was sufficient for the double supply of air, to the fuel in the bars, and the gas in the furnace chamber. The difference which attended its application was often considerable, and arose from the want of draught, or from the perverse adherence to the old and lazy method of charging the front half of the furnaces heavily, even to the doors, while leaving much of the bridge-end but thinly covered, as hereafter will be shown. Such a mode of charging the furnaces necessarily caused an irregular combustion of the fuel, and a consequent excessive admission of air, counteracting all efforts at appropriating separate supplies to the coke and the gas.

The introducing of the air to land boilers, in numerous films, or divided portions, was first practically adopted in 1841, at numerous furnaces in Manchester, and at the water-works in Liverpool, and at the stationary engine of the Liverpool and Manchester Railway, under the direction of the engineer, Mr. John Dewrance. That at the water-works, with a shaft of 150 feet high, had previously caused an intolerable nuisance; both, however, have since remained unnoticed and forgotten, even by the authorities in Liverpool, apparently from the mere circumstance of the nuisance having been effectually abated, and attention being no longer drawn to it.

With reference to *the place* for the admission of the air, it is here stated, advisedly and after much experience, that it is *a matter of perfect indifference as to effect, in what part of the furnace or flue it is introduced, provided this all-important condition be attended to, namely, that the mechanical mixture of the air and gas be continuously effected, before the temperature of the carbon of the gas (then in the state of flame) be reduced below that of ignition.* This temperature, according to Sir Humphry Davy, should not bo under 800° Fahr., since, below that, flame cannot be produced or sustained. In practice, the air has been introduced at all *parts of the furnace, and with equally good effect.* Its admission through a plate-distributor, at the back of the bridge and at the door end, effected all that could be desired.

The adoption during the last few years of the *tubular system* in marine boilers, is now to be noticed, inasmuch as it rendered a different arrangement absolutely necessary.

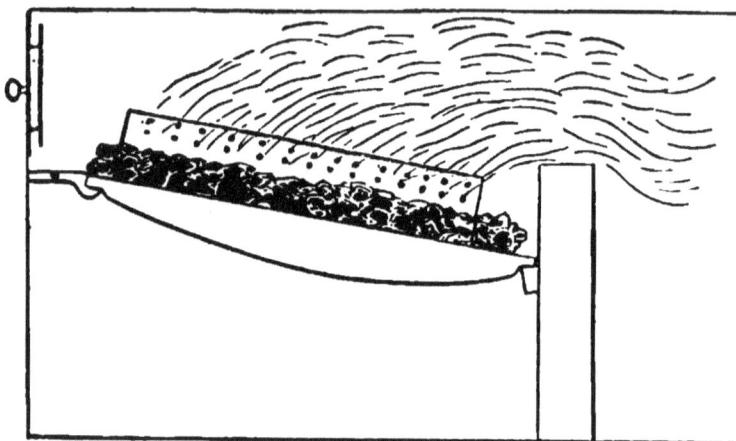

Fig. 28.—Air through orifices in the fireplace.

The chief characteristic of the tubular boiler is the *shortness of the distance, or run,* between the furnace and the tubes. The result is, the impossibility of effecting the triple duty of generating the gas, mixing it with the air, and completing the combustion within *the few feet,* and the *fraction of a second* of time, which are there available. To obtain the desired effect the air was then introduced at the *door-end* of the furnace; thus, as it were, adding the length of the furnace to the length of the run.

The main object being the introducing of the air in a

divided state to the gaseous atmosphere of the furnace chamber, the following experiment was made : The centre bar of a boiler, four feet long, was taken out, and over the vacant space an iron plate was introduced, bent in the form as shown in Fig. 28.

Here, the upper portion of the bent plate, projecting three inches above the fuel, was punched with five rows of half-inch holes, through which the air issued in 56 streams. Adequate mixture was thus instantly obtained, as in the Argand gas-burner ; the appearance, as viewed through the sight-holes at the end of the boilers, being even brilliant, and as if *streams of flame*, instead of *streams of air*, had issued from the numerous orifices. It is needless to add, that nowhere could a cooling effect be produced, notwithstanding the great volume of air so introduced.

The sectional view of the furnace, looked at from behind, as in Fig. 29, represents the character and diffusive action of the flame.

Fig. 29.—Air through orifices in the fireplace.

This led to the enlarging of the door-end of the furnaces sufficiently to admit the required number of apertures and full supply of air ; an arrangement which has been for years in successful operation, both in marine and land boilers.

On looking into the flues of land boilers, through suitably placed sight-holes, when the furnace is in full action, numerous brilliant sparks may be seen, carried through the flues with great rapidity, to the distance of ten to twenty feet before their luminous character is lost, and they become deposited in the tubes, or flues, or wherever eddies are

formed.   These sparks consist, chiefly, of particles of sand
in a state of fusion.   When these do not thus separate from
the coal, they fall on the bars, and, combining with the
ashes, form clinkers.   These particles of sand, flying off at
a high temperature, adhere to whatever they touch ; and,
with the dust, and small particles of cinders or coke, carried
onward by the current, fill up the orifices in the air-distri-
butor boxes, and, if not removed, prevent the passage of the
required quantity of air.

CHAPTER IX.

OF VARIOUS FURNACE ARRANGEMENTS, WITH OBSERVATIONS THEREON.

THE following remarks on the peculiarities of the several plans of furnaces here shown, are the results of practical observations extended over a series of years, and may here be useful, as indicating what should be avoided, as well as provided, respecting the admission of air :—

Fig. 80 represents one of the modes first adopted, under the patent for the Argand furnace of 1839 ; introducing the air in numerous jets. This was applicable to land boilers, where ample space was afforded for the perforated tubes, made of fire-clay, or cast-iron ; and was first adopted at the water-works in Liverpool. In this application the inconvenience arising from the sand and other matters in an incandescent state, adhering to and closing the orifices, was considerable. The plan, clearly described by Dr. Ure, as follows, was then substituted, and continued in active operation at those works.

"Among the fifty several inventions which have been patented for effecting this purpose, with regard to steam-boiler and other large furnaces, very few are sufficiently economical or effective. The first person who investigated this subject in a truly philosophical manner was Mr. Charles Wye Williams, managing director of the Dublin and Liverpool Steam Navigation Company, and he has also had the merit of constructing many furnaces, both for marine and land steam-engines, which thoroughly prevent the production of smoke, with increased energy of combustion, and a more

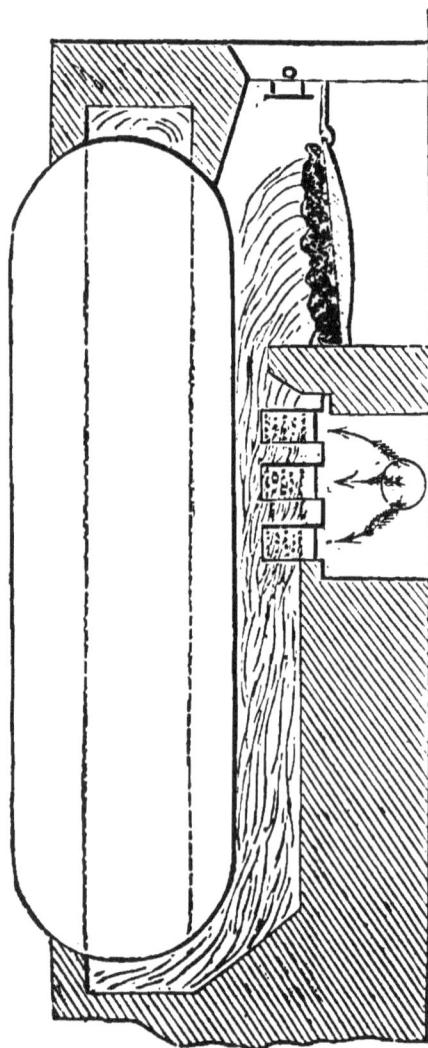

Fig. 30.—Argand Furnace.

or less considerable saving of fuel, according to the care of the stoker. The specific invention, for which he obtained a patent in 1839, consists in the introduction of a proper quantity of atmospheric air to the bridges and flame-beds of the furnaces through *a great number of small orifices*, connected with a common pipe or canal, whose area can be increased or diminished, according as the circumstances of complete combustion may require, by means of an external valve. The operation of air thus entering *in small jets* into the half-burned hydro-carburetted gases *over the fires, and in the first flue*, is their perfect oxygenation—the development

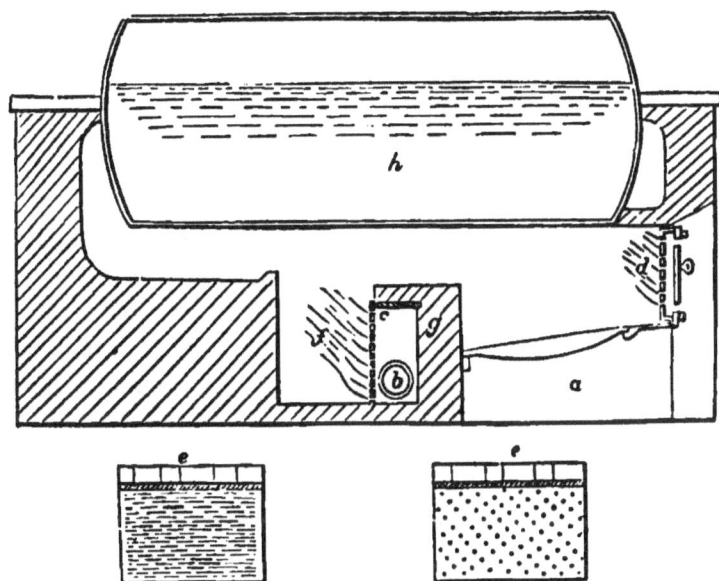

Fig. 31.—Argand Furnace.

of all the heat which that can produce, and the entire prevention of smoke. One of the many ingenious methods in which Mr. Williams has carried out the principles of what he justly calls his Argand furnace, is represented at Fig. 31, where *a* is the ash-pit of a steam-boiler furnace; *b* is the mouth of a tube which admits the external air into the chamber, or iron box of distribution *c*, placed immediately beyond the fire-bridge *g*, and before the diffusion, or mixing chamber *f*. The front box is *perforated either with round or*

oblong orifices, as shown in the two small figures *e e* beneath ;
d is the fire-door, which may have its fire-brick lining also
perforated.   In some cases the fire-door projects in front,
and it, as well as the sides and arched top of the fire-place,

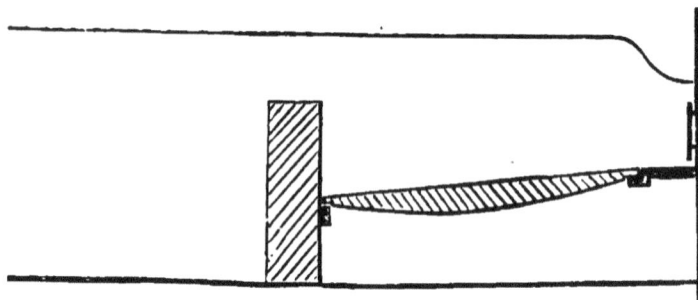

Fig. 32.—Ordinary Marine Furnace.

are constructed of perforated fire-tiles, enclosed in common
brickwork, with an intermediate space, into which the air
may be admitted in regulated quantity through a movable
valve in the door.   I have seen a fire-place of this latter
construction performing admirably, without smoke, with an

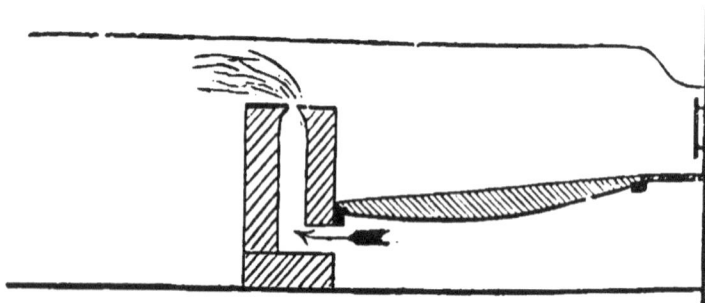

Fig. 33.—Parkes' Split Bridge.

economy of one-seventh of the coals formerly consumed in
producing a like amount of steam from an ordinary furnace."
    The following are principally connected with marine
boilers :—
    Fig. 82 represents the ordinary marine furnace.   No pro-

vision whatever is here made for the admission of air, except from the ash-pit, and through the bars, and fuel on them. It is needless to add, that, from the absence of air to the gas, a large volume of smoke must here necessarily be produced.

Fig. 33. Parkes' Split Bridge. This plan, patented in 1820, was effective when the consumption of coal and the generation of gas were small and uniform; or when the furnace was large, and heavily charged, to last for six or eight hours, with slow combustion. The generation of the gas being uniform, and the demand for air moderate, the supply through the narrow orifice in the bridge was suffi-

Fig. 34.—Split Bridge modified.

cient. This plan has formed the basis of several *re-inventions;* the patentees either not being aware of it, or not acknowledging the source of the effect for which they took credit.

Fig. 34. This adaptation of the split bridge in marine boilers was early made, by the then Engineer of the Dublin Steam Company, to avoid the collection of ashes in the lower shelf of the air orifice, by which the passage of the air was obstructed. The furnaces being charged at short intervals, and the combustion rapid, the supply of air was insufficient. The aperture at the top of the bridge was liable to be choked with ashes and small coals, occasionally thrown over.

Fig. 85.   This change was not found effective.   The
second opening for the admission of air, at the end of the
bars, was quite irregular in its action.   It was also found to
interfere with the action in the split bridge ; the air pre-
ferring, at certain states of the fuel, to enter by the open

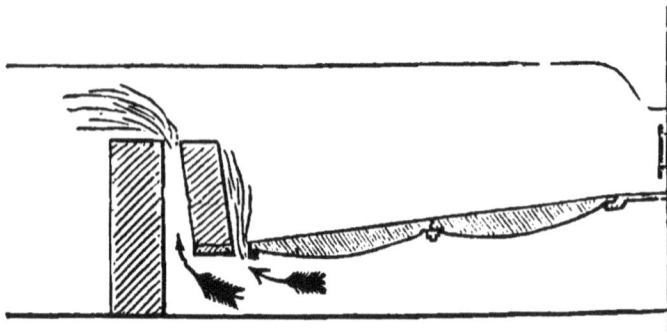

Fig. 85.—Split Bridge and Air at the Grate.

space at the end of the bars, as the nearest and hottest
course, whenever that place was uncovered.

Fig. 86.   This was adopted in a steamer of large power,
and was intended to remedy the evil as stated in the last
figure.   The aperture being made larger, the air entered

Fig. 86.—Air at the Bridge.

too much in a mass, and produced a cooling effect ; and
much fuel was also wasted by falling through into the ash-
pit.   This was subsequently altered to the plan hereafter
shown in Fig. 41 ; the bars being reduced from 7 feet
6 inches, to 6 feet, and with good effect.

Fig. 87. This arrangement remedied that of the preceding, by saving the fuel thrown to the end; and which, falling on the small supplemental grate, was there consumed. In practice, however, it was less effective as to generating steam, and irregular in its action, and was very destructive of the bars.

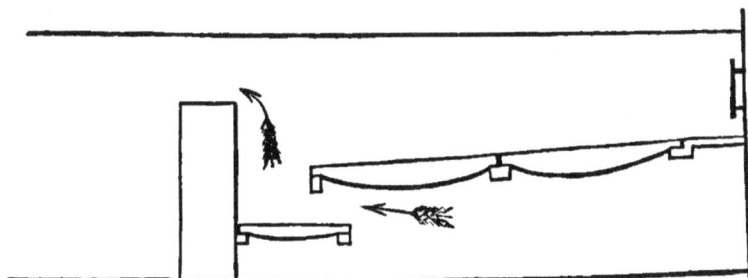

Fig. 37.—Air at the Bridge, and small Grate.

Fig. 88. This plan, adopted in 1840, was one of the first applied to marine boilers, on the principle of the Argand furnace, by which the air was made to enter in *divided streams*, through the apertures in an eight-inch tube, from behind the boiler. This plan was fully effective so long as

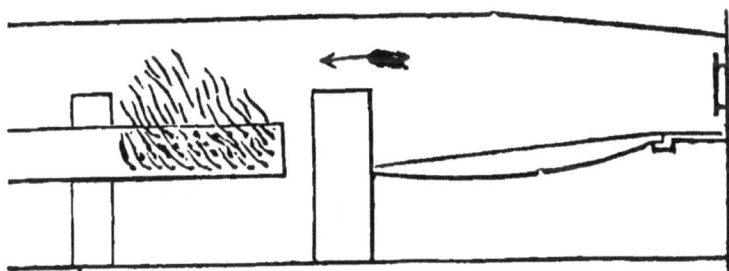

Fig. 88.—Argand Furnace.

the perforations in the tube remained open. The small orifices, each but a quarter of an inch, however, becoming covered, and closed by the sand and ashes, the supply of air was consequently diminished, and the tube became heated and destroyed.

Fig. 39. This plan, adopted in the steamer, the "*Leeds*," was very effective so long as the inclined plate and its numerous orifices remained perfect. As, however, it also became clogged, or covered with coal, thrown over during charging, it warped, and became injured.

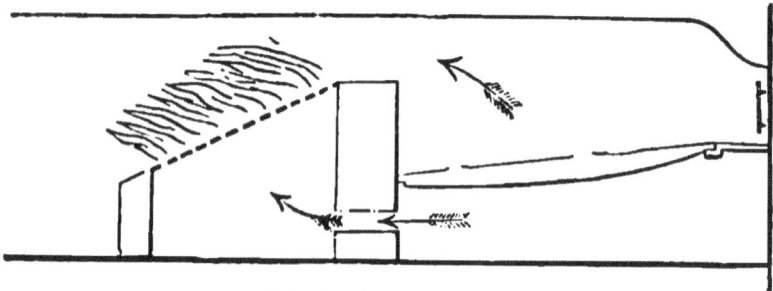

Fig. 39.—Argand Furnace.

Fig. 40. This alteration was made in the same boiler, to counteract the evil above-mentioned. The bars were shortened from 6 feet to 4 feet 6 inches. The air was here introduced through a plate pierced with half-inch holes. This was quite successful: ignition and combustion were

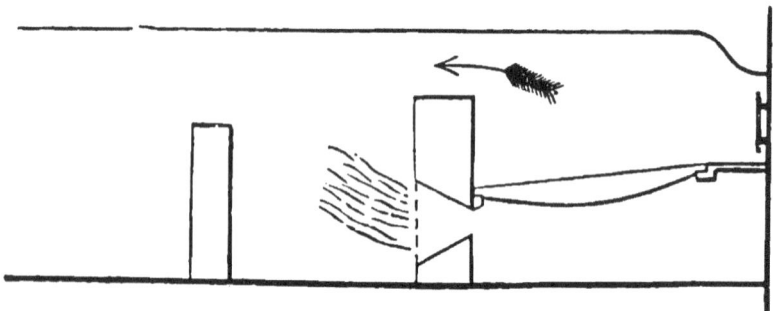

Fig. 40.—Argand Furnace.

complete; no smoke formed, and the diminished combustion of fuel was considerable. The box, however, set in the bridge, was too small, and therefore liable to become filled by the ashes carried in by the current from the ash-pit;

and the stokers neglecting to keep the air-apertures free, there was no dependence on its action.

Fig. 41. This arrangement, which remedied the above defects, was adopted in the steamer, the "*Princess*," and

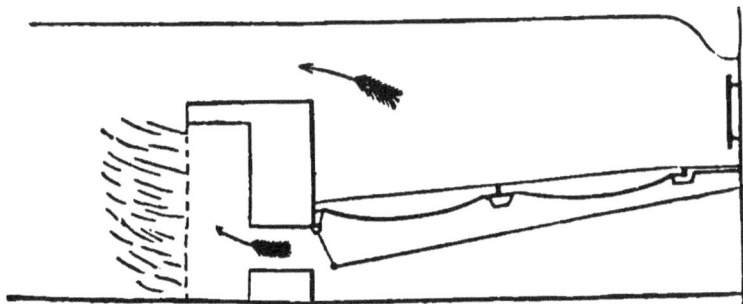

Fig. 41.—Argand Furnace.

also in the "*Oriental*" and "*Hindostan*," employed in the Mail service in the Mediterranean. Perfect combustion of the gas was effected, and, consequently, no formation of smoke. The numerous orifices are here removed from the direct action of the heat, or the liability to be choked. The

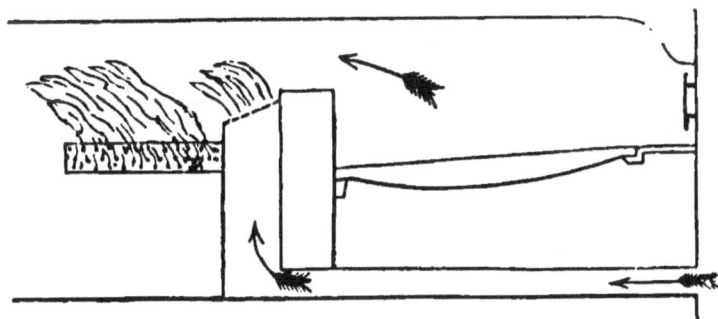

Fig. 42.—Argand Furnace.

regulating valve, originally placed on the apertures, to regulate the supply, was, after a little experience, found to be unnecessary, and was removed. This plan has become, practically, the most effective, and, during the last ten years, has been adopted in numerous marine and land boilers. The cost of the air-box was under forty shillings.

Fig. 42. In this plan, the air was introduced through a tube laid on the bottom of the ash-pit, to avoid the current

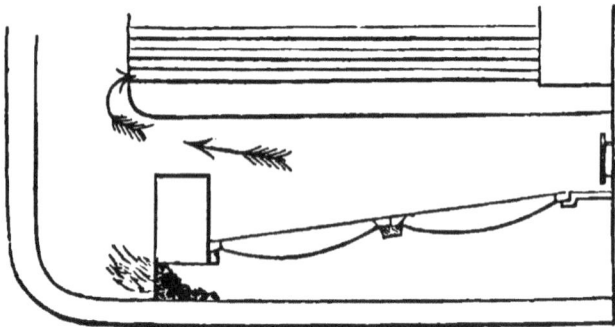

Fig. 43.—Air-box at Bridge.

of dust, and to enable the air to enter in a cooler state. This was found effective as regarded combustion, but, being still exposed to the sand, dust, and heat, as already mentioned, was subsequently altered to that of Fig. 41.

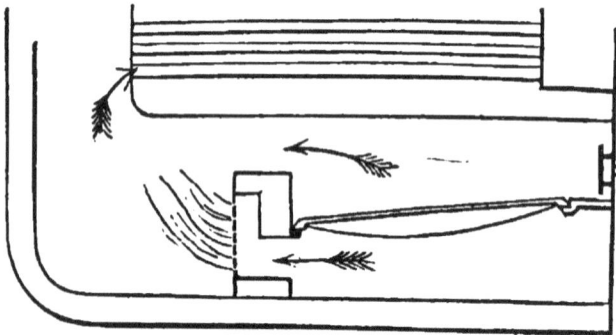

Fig. 44.—Argand Furnace.

Fig. 43. This was a tubular boiler, and is here shown as it came from the maker in 1846. It was quite ineffective, giving much smoke, the tubes also being liable to injury by the shortness of the run. The air-box in the bridge was soon filled with dust and ashes, as here shown. The grate-bar being 6 feet 10 inches long, the flame necessarily reached the tubes, doing much injury to the lower tiers. This was altered, as shown in Fig. 44.

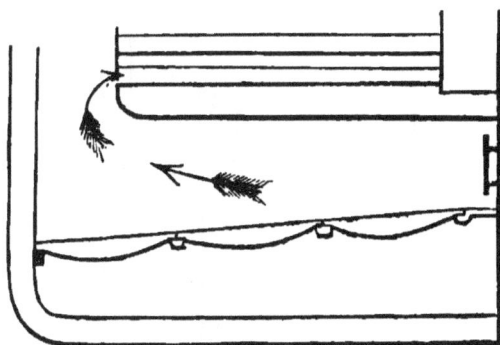

Fig. 45.—Common Furnace.

Fig. 44. This is the same boiler, the furnace alteration being attended with considerable advantage. The bars were shortened from 6 feet 10 inches, to 5 feet 8 inches. The defect of the short run, and the limited time for combustion,

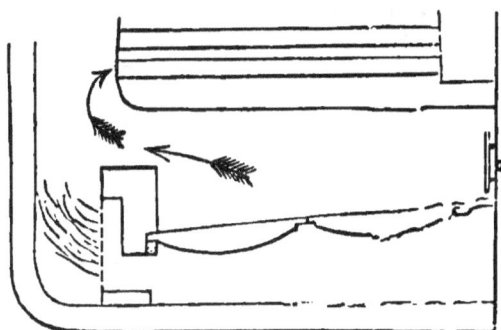

Fig. 46.—Air-box at Bridge.

ncident to tubular boilers, was, however, irremediable. The change in the length of the bars alluded to, reduced the consumption of coal considerably; smoke was, to a certain extent, avoided, and the amount of steam increased. In this boiler there were 205 tubes of 2¾-inch area. Engines 190 horse-power.

Fig. 45. This was a large steamer of 350 horse-power,

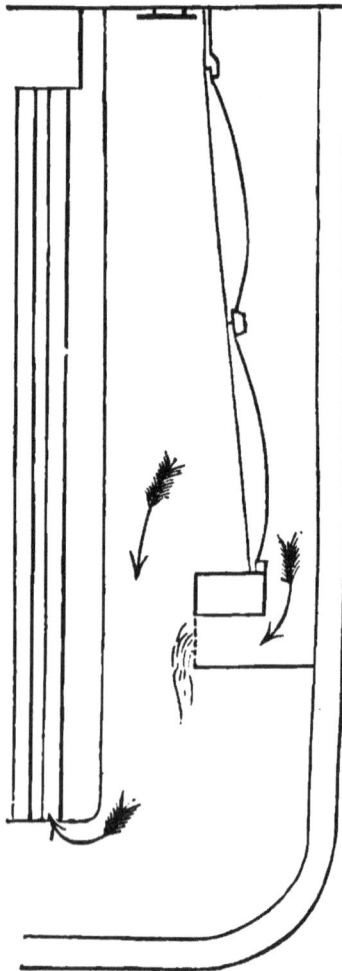

Fig. 47.—Air-box at Bridge.

with tubular boilers. The plan of furnace here shown represents it as it came from the maker. Three lengths of bars, 2 feet 8 inches each, *filled the entire space*, leaving no room for the admission of air to the gas. The consequence

was, a great consumption of fuel; a great generation of
smoke; and much inconvenience and expense, from the
destruction of the tubes and face-plate.

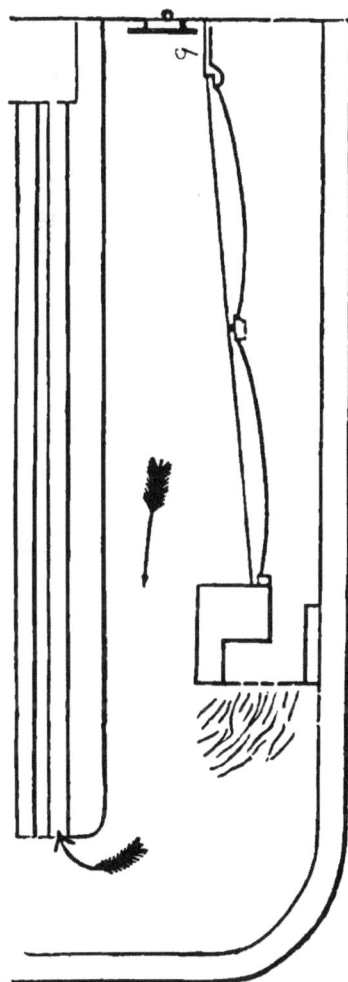

Fig. 48.—Air-box at Bridge.

Fig. 46. This is the same boiler. The bars having been
shortened, the air-box was introduced into the bridge. Notwithstanding the evils of the short run, the change here
made was satisfactory. The importance of keeping the

E 2

air-passage free from obstruction was exemplified in this case. The air-box was introduced in the *after-boiler*, leaving the *fore-boiler* as shown in Fig. 45. During the voyage, in which 90 tons 18 cwt. of coal were used in the latter, but 81 tons 15 cwt. were used in the former. The engineer reported, that "when the gases are properly consumed, the best effect is produced; good steam is obtained and less coal used."

Fig. 47. This boiler also was tubular, 17 feet 2 inches long. Engines 370 horse-power. It is here shown as it came from the maker. The grate-bars 9 feet; dead plate 9

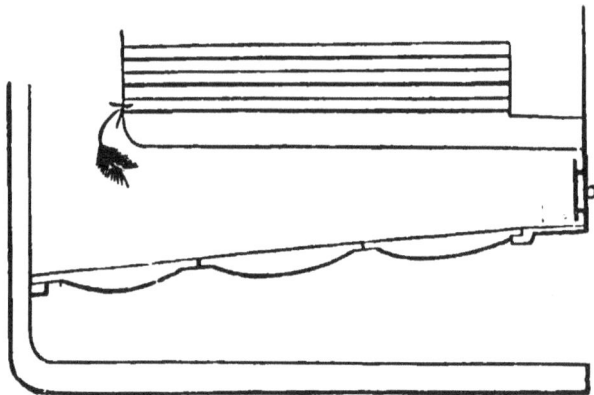

Fig. 49.—Common Furnace.

inches. The area for the admission of the air was quite inadequate to the introduction of the necessary quantity. This boiler was then altered as in Fig. 48.

Fig. 48. This is the same large steamer as in last number: the air-box being introduced into the bridge; the result was a considerable diminution in the fuel used; a better command of steam, and freedom from the nuisance of smoke.

Fig. 49. This tubular boiler is here shown as it came from the maker; grate-bars 9 feet 3 inches long, with dead-plate 12 inches. No means for admission of the air to the gas.

In this boiler the run to the end of the tubes being so short, the generation of steam depended, almost exclusively, on the large grate-surface from ten furnaces. The consumption of coal was very great, and the smoke very dense. From the shortness of the boiler there was necessarily but little room for improvement. It was altered as shown in next plan.

Fig. 50. The same boiler, altered as here described, allowing the air to enter by a *perforated* plate. The inherent defects of the short boiler, and short run, prevented the realising much advantage in this case.

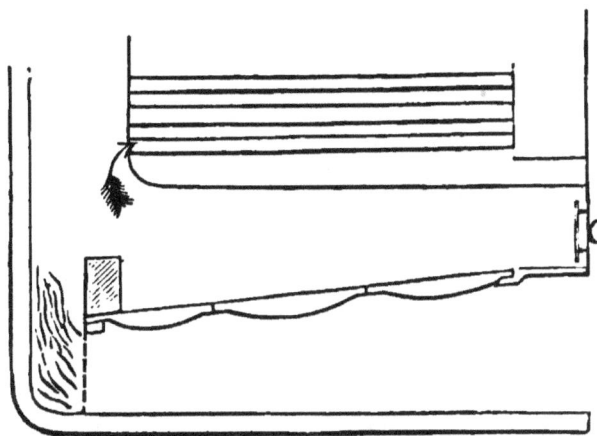

Fig. 50.—Perforated Plate at Bridge.

Fig. 51. This plan is here introduced as showing the practical error of supposing that the gases could be consumed by causing them to pass *through incandescent fuel*. The effect of this plan is to convert the gas into *carbonic oxide;* and which, from being invisible, created the impression that the "*smoke was burned.*" It is needless here to dwell on the chemical error of such an assertion. The fallacy of imagining that either gas or smoke, from a furnace, can be consumed by passing "*through, over, or among*" a body of incandescent fuel, prevailed from the days of Watt to the present. Numerous patented plans to the same effect might here be given, all having the same defect, and equally ineffective.

Fig. 52. This was one of the numerous hot-air expedients pressed upon public notice, under the illusion, that by heat-

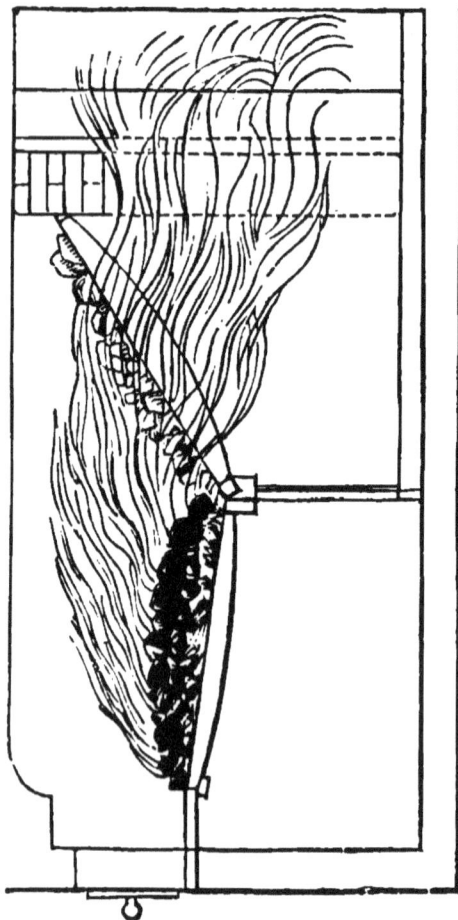

Fig. 51.—Furnace with Supplementary Grate.

ing the air, "*the smoke would be burned.*" A large hollow fire-bar, A, was placed in the centre, or sides of the furnace, with a regulating door for the admission of the air. The Admiralty having been induced to allow this plan to be

adopted in the Steam Packet, the "*Urgent*," at Woolwich,
the result was a total failure, and its consequent removal.*
The supposed heating of the air being a mere assertion,
made for the purpose of giving an appearance of novelty,

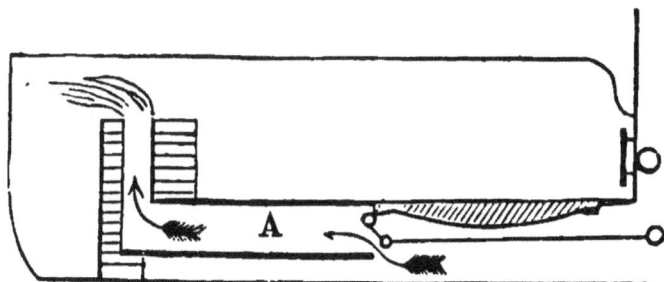

Fig. 52.—Hot-air expedient and Split Bridge.

having been wholly without effect, the result was, that it
reduced the so-called patent invention to that of Parkes'
split bridge, with all its disadvantages when applied to
marine-boilers and large furnaces.

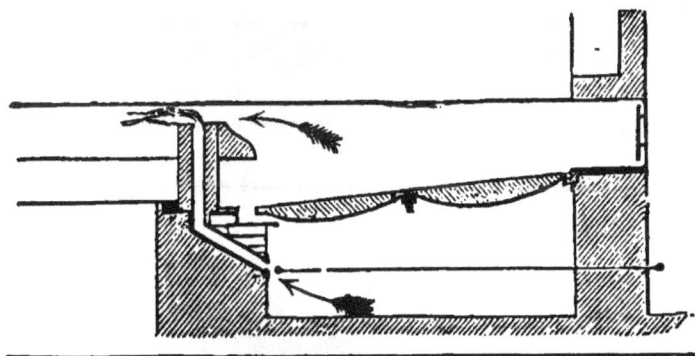

Fig. 53.—Hot-air expedient and Split Bridge.

Fig. 53. This was another modification of the split-bridge

* The *Urgent*, Captain Emerson, being then engaged in the Mail
Service at Liverpool, this steamer came under my notice. For the
purpose of testing the effects of this hollow-bar, I had an experiment
made to ascertain the extent to which the air might be heated, and
found no perceptible increase of heat could be obtained by it.

plan.   Mr. West, in his published Report, on the methods submitted to the Public Meeting at Leeds, in 1842, described

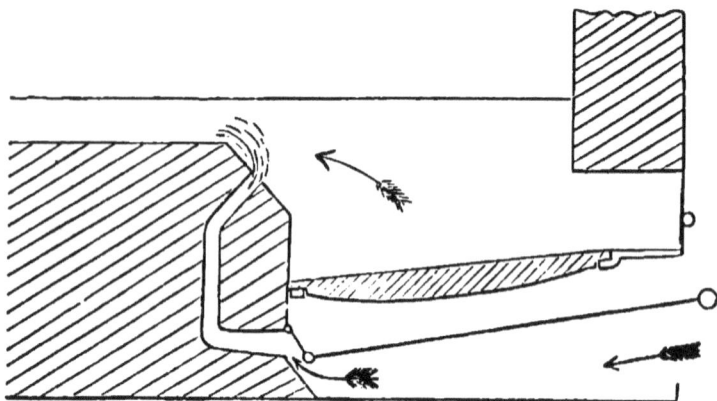

Fig. 54.—Hot Air at the Bridge.

this in the following terms : " It consists of a regulating valve, by which air is admitted into a passage through the

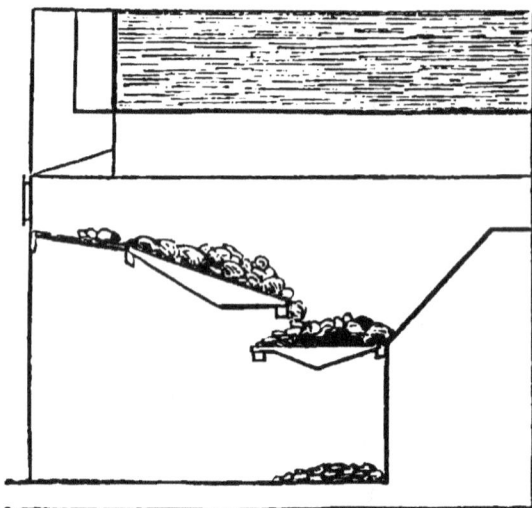

Fig. 55.—Chanter's System.

bridge (the split bridge of Parkes' expired patent), for four hours after first firing.   By this time, the coal is coked, and tho valvo shut the remainder of tho day." It is manifest there is nothing in this plan beyond the split bridge, accompanied with tho modo of firing and slow continuous combustion applicable to it.

Fig. 54. This is but another modification of tho split bridge, though announced as a plan for *heating the air,* by its passage through a body of hot brick-work. This plan, M. Péclet obscrvcs, was adopted in Franco, but abandoned.

Fig. 55. M. Péclot gives this as one of Chantor's patonts,

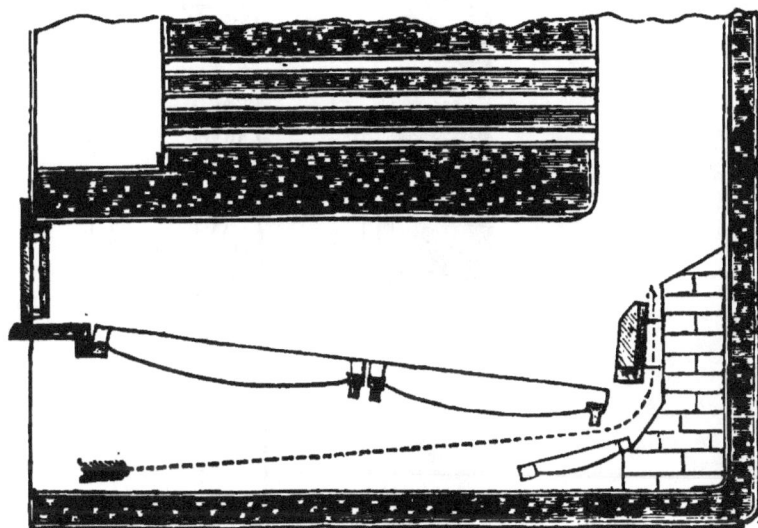

Fig. 56.—Hot Air at the Bridge.

which was also tricd and abandoned in Franco. It will bo scen that this is but a modification of the former plans.

Fig. 56. This is another of tho so-called hot-air plans, although it is nothing but the split bridge with a supplemental grate, as adopted by Chantor and othcrs. Tho patentee professes to have tho air "*intensely heated,*" by the handful of scoria, or cinders, which fall on tho supplemental grate.

Fig. 57. This is another of the hot-air plans, as given

in Mr. West's Summary. The air is here supposed to

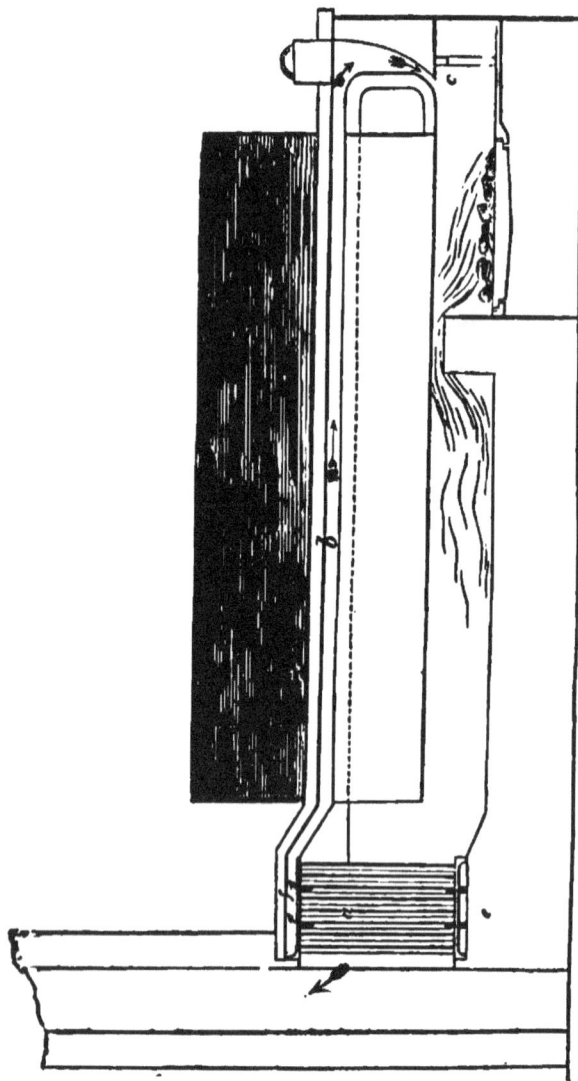

Fig. 57.—Hot Air from the Flues.

be heated by passing through the vertical tubes *a*, placed in the flue, and thence through the passage *b*, entering the furnace by a single orifice, *c*. It is only necessary to observe

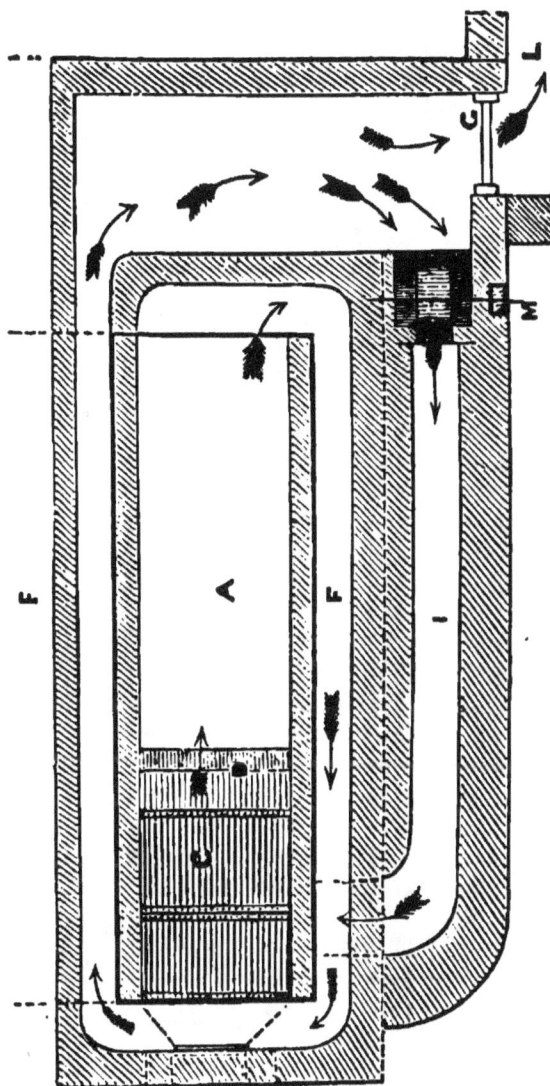

Fig. 66.—Blast of Smoke and Air.

that it would be impossible that one-fourth part of the required quantity of air could there obtain access, unless by so enlarging the orifice as to produce a cooling effect, by its then entering *en masse.*

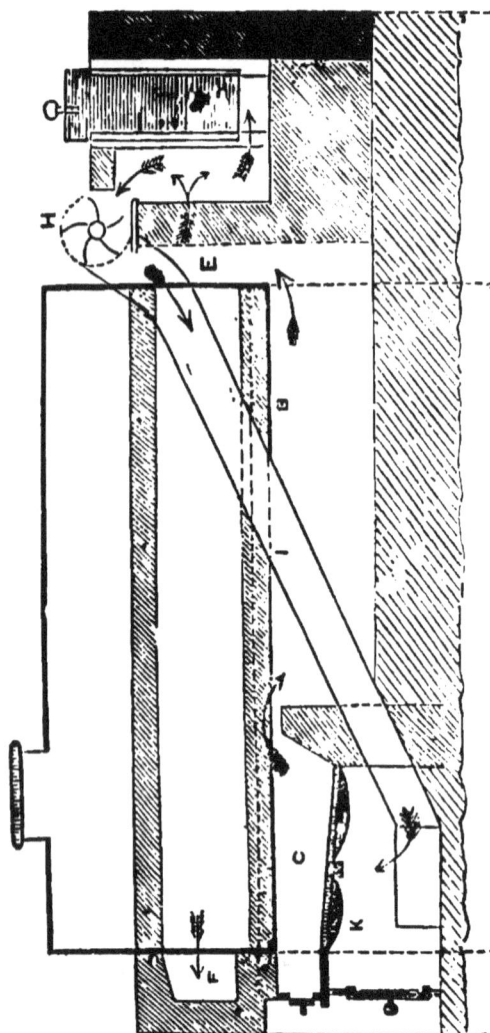

Fig. 59.—Blast of Smoke and Air.

Figs. 58 and 59.·This plan, as in Fig. 58, with the sectional view, 59, is also taken from Mr. West's Summary, and is here introduced with the view of further pointing to the hot air, and "smoke-burning" fallacy. The following is the description given by Mr. West: "The smoke, after having passed along the flues marked F, is intended to be caught by the fan H, before reaching the damper G, and, along with a sufficient quantity of atmospheric air, is propelled along the return flue I, into the enclosed ash-pit K, where it is again forced through the fire-grate C." It is not necessary to add any comment on what is so wholly opposed to chemistry and nature.

In the case of boilers already constructed, it may be asked how they should be altered so as to admit the required supply of air. In *land boilers,* where the furnace doors are set in brick, they may easily be enlarged, and at a small cost, to allow space for the requisite number of orifices, the aggregate area of which should average five to six square inches for each square foot of grate-bar furnace, according to the description of fuel.

In *marine boilers,* however, the enlargement of the door end is troublesome. Where sufficient space cannot be obtained, it will be advisable, in addition to as many half-inch orifices as can be inserted in the back plate of the close door box, or in the neighbourhood of the door, to introduce the ordinary perforated air-plate, as already shown in Fig. 41. This was the mode successfully adopted, in the present year, in the mail steam-packet, the "*Llewellyn.*" The boilers being new, and the maker not having allowed space sufficient for door-frame plates of the required size, the deficiency was supplied through the ordinary perforated box in the bridge.

The boilers previously in this vessel were remarkable for the continuous volume of dense smoke: the new boiler, independently of the absence of smoke, supplies more steam with a less consumption of coal. The contrast between the two modes of constructing furnaces, is well exemplified in the following extract from the report of Mr. Joseph Clarke, the Engineer of the Dublin Company, to whom this vessel belongs.*

In illustration of the alteration which should be made in marine boilers, Fig. 60 represents the usual mode of contracting the door end to the mere size of the door frame, as at *a*. Fig. 61 represents the mode of enlarging the opening, both at the sides and above the doorway at *b*, to allow of the introduction of a sufficient number of half-inch apertures, as shown in Fig. 62. It is here worthy of note, that as the ordinary mode of constructing the door end of marine boilers is difficult and expensive, as shown in Fig. 60, the mode shown in Fig. 61 is so much more simple as to cover all the outlay for the air boxes shown in the next figures.

Fig. 62 represents one of the modes adopted where the boiler has been *originally constructed to admit the required number of orifices.* This has been in successful operation for

---

* "The Holyhead mail steam packet, " *Llewellyn*," having now been at work three months with new boilers, I have to transmit you the result of their performances. This vessel has two boilers ; one before, and the other abaft the engines. Their construction are precisely the same : each having six furnaces. Both have all their furnace fittings exactly the same. In order to put the smoke-prevention principle in contrast with the ordinary mode, the *fore boiler* was allowed to remain as it came from the maker, while the *after one* had the door frames of each of the furnaces (which are made with box mouth pieces) perforated with 149 holes, each $\frac{7}{8}$ inch diameter, to admit the air. These not being sufficient, the perforated plate behind the bridge was added, in which there were 321 holes—in all, 470 holes; the gross area of which is equal to about 5 square inches for each square foot of fire grate. The result is, that the fore boiler gives out a continuous volume of dense smoke, and the after one none whatever. It is quite remarkable to see the steam blowing off from both boilers, and smoke only from one. I know nothing that could be more demonstrative of a principle than the contrast between the two boilers in this vessel. It attracted the attention of the passengers, and I resolved, therefore, on leaving the two sets of furnaces as they are for some time longer, to afford the public the opportunity of seeing that smoke prevention is practicable. When the vessel can be spared, it is my intention to make the furnaces of both boilers alike."

some years, and without requiring any repairs. In this plan it will be seen that air boxes are introduced at the sides and above the doors. The air entering to the upper box at

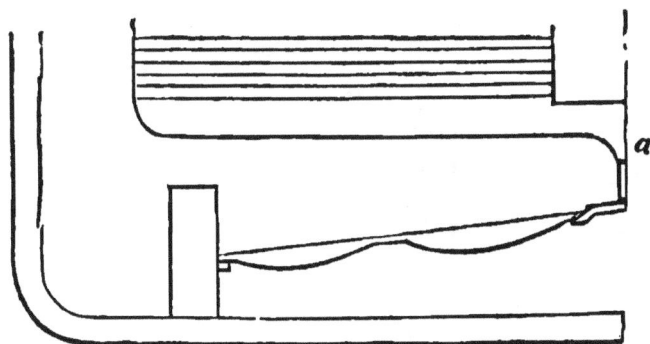

Fig. 60.—Common Furnace.

a, and to the side boxes at *b b*. (The left representing an outside, and the right an inside view of the orifices.) In the centre is a sliding plate P, by which, alternately, the right

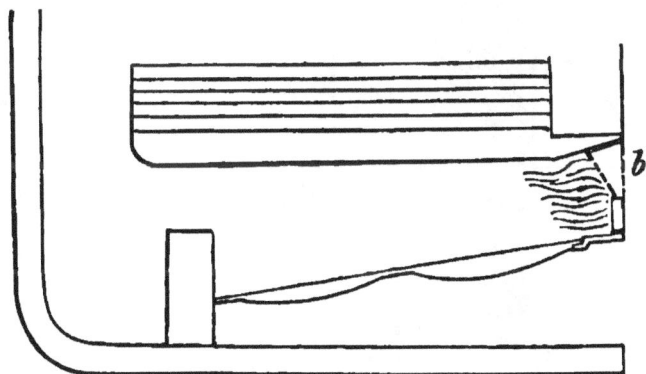

Fig. 61.—Argand Furnace.

or left hand upper orifices may be closed, when either furnaces are about to be charged.

As much stress has been laid on the value of having skilful firemen, it is important to show in what their real duties consist. The annexed figures will explain the differ-once in effect between the right and the wrong mode of charging a furnace.

Fig. 63 represents the proper mode of keeping a uniform

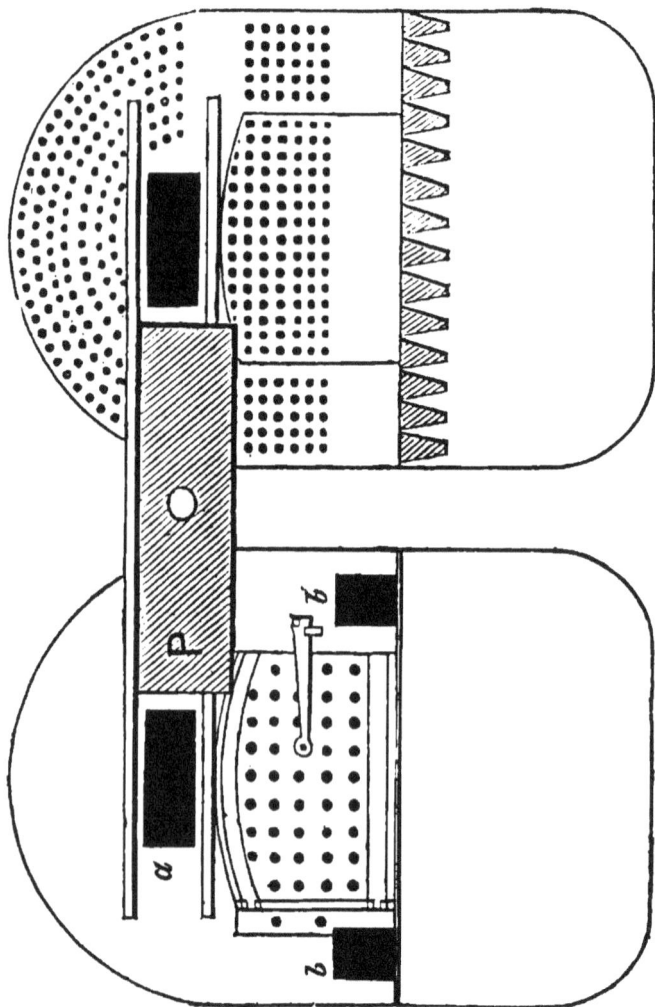

Fig. 62.—Argand Furnace.

depth of coal on the grate-bars :—the result of which will be, a uniform generation of gas throughout the charge, and a uniform temperature in the flues.

Fig. 63a represents the ordinary mode of feeding marine furnaces: charging the *front half* as high, and as near the door, as possible, leaving the bridge end comparatively bare. The result necessarily is, that more air obtains access

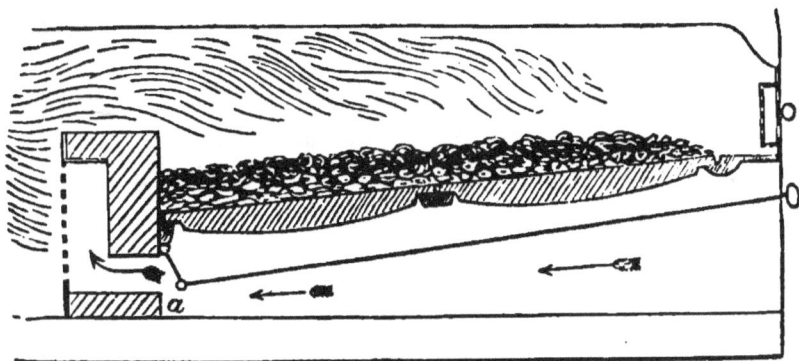

Fig. 63.—Proper Firing.

through the uncovered bars than could be required; thus defeating all efforts at introducing the proper quantity in the proper manner.

Fig. 63a.—Improper Firing.

Where the air is properly introduced, the duties of firemen are all contained in the following instructions :—

1st. Begin to charge the furnace at the bridge-end, and keep firing to within a few inches of the dead-plate.

2nd.   Never allow the fire to be so low, before a fresh charge is thrown in, that there shall not be at least four to five inches deep of clear, incandescent fuel on the bars, and equally spread over the whole.

3rd.   Keep the bars constantly and equally covered, particularly at the sides and bridge-end, where the fuel burns away most rapidly.

4th.   If the fuel burns unequally, or into holes, it must be levelled, and the vacant spaces must be filled.

5th.   The large coals must be broken into pieces not bigger than a man's fist.

6th.   Where the ash-pit is shallow, it must be more frequently cleared out.   A body of hot cinders overheat and burn the bars.

One important advantage arising from the control of the quantity of air is, that it enables the engineer to shorten the length of the grate by bricking over the after end of the bars, seeing that an unnecessary length merely gives the means of letting an improper supply of air pass in through the uncovered bars.

The facility with which the stoker is enabled to counteract the best arrangements, naturally suggests the advantage of *mechanical feeders*.   Here is a direction in which mechanical skill may usefully be employed :—the basis of success, however, should be the sustaining at all times the uniform and sufficient depth of fuel on the bars.

The plans of *Brunton's* revolving grate, *Jukes's* moving bars, or *Stanley's* self-feeding apparatus, need not here be described.*   There is in these no pretension beyond what they can perform ; each acts the part intended, and, where-

* Stanley's apparatus was early applied on board the Dublin Steam Company's vessel, the "*Liverpool.*"   Independent of its inconvenient bulk, it was wholly defective, when applied to large furnaces, requiring the most active firing, and the irregular demand for steam incidental to marine boilers.

over there is room for their introduction, and that the uniform amount of heat produced by these means falls in with the requirements of the steam engine and the manufacturer, these will answer the desired purpose.

We must here observe that these plans are inapplicable to marine furnaces, or where large quantities of steam, and active and irregular firing, are required.

The simple operation in these is, the keeping continuously a *thin stratum of fuel on the bars*, and, consequently, an abundant supply, and even an excess of air, through it, to the gases generated in small quantities over every part of the fuel. Neither must we be led to suppose, that they effect a more economical use of the fuel.

In an inquiry on the subject at the Society of Arts, much stress was laid on the annual saving by the use of the moving bars, at a large establishment in London. It appeared, however, that the saving arose, not from any more economic use of the fuel, or the generation of more heat, or by a more perfect combustion, but merely from the circumstance, that the mode of feeding the furnace, and keeping continuously a thin stratum of fuel on the grate, enabled the proprietor to use an inferior description of coal.

Although the combustion of the gases in locomotive boilers does not come within the scope of these remarks, the peculiarities of the boiler, as shown in Fig. 64, are so illustrative of the principle of admitting the air through numerous orifices, that it here merits attention.

Fig. 64. This plan of boiler is the invention of Mr. Dewrance, when Engineer on the Liverpool and Manchester Railway Company, and was adopted in their locomotive, the "*Condor.*" By this arrangement he was enabled to use *coal* instead of coke, and with entire success. It will here be seen that the air enters from a separate passage to a number of vertical perforated tubes, from which it passes to the gas, in a large mixing or combustion chamber, through numerous small orifices. The result is, immediate diffusion and combustion. The deflecting plate, to a certain extent, counter-

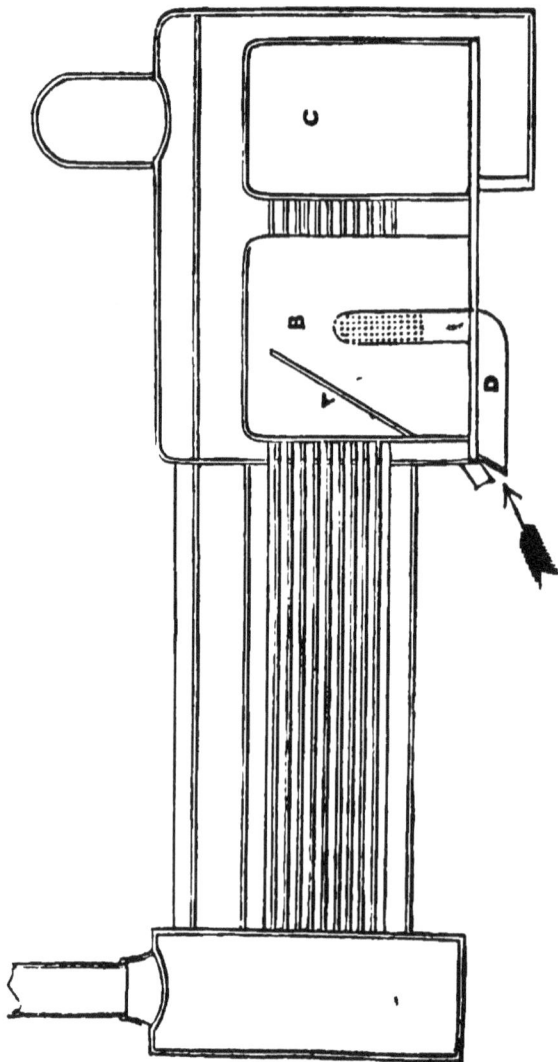

Fig. 64.—A· grand Furnace in a Locomotive Boiler.

acts the short run, or distance to the tubes. A, deflecting plate ; B, combustion chamber; C, common coal fire ; D, cold air passage.

In concluding these observations on the various modes of introducing air to the furnaces, it is only necessary to add, that all that manufacturers have to do is, to imitate, as near as possible, the principle of the common Argand gas burner. Let them introduce *the air by numerous small orifices to the gas, in the furnace, as the gas is introduced by small orifices to the air in the burner.* Let them begin by having as many half-inch or even three-quarter inch orifices, with inch spaces, drilled in the door and door frame, as possible. If the furnace be large, and the door-plate frame is not sufficient for the introduction of the required number of holes, let them introduce the perforated plate in the bridge, as shown in Fig. 41.

CHAPTER X.

ON PROVIDING ADEQUATE INTERNAL SURFACE FOR TRANSMITTING THE HEAT TO THE WATER FOR EVAPORATION.

ON this head, marine-boiler makers content themselves with calculating the gross internal superficies; and having provided a given number of square yards of so-called *heating surface*, they consider they have done all that is necessary for providing an *adequate supply of steam*.

As to general efficiency, the *flue system* is capable of supplying all that can be required, while it is free from the anomalies incidental to the multi-tubular plan. When larger quantities of steam are required for larger engines, this can be best obtained, not by additional tiers of tubes, but by extending the areas and length of run; thus increasing the number of units of *time*, *distance*, and *surface*, along which the heat-transmitting influence may be exerted.

As to the importance of *time and distance*, in connection with surface, it is only necessary to point to the *length of the flame*, in ordinary boilers, that being an unmistakeable evidence of the *duration of the process* of the combustion of *the carbon;* and which process cannot be interfered with, unless by the loss of that heat which would have attended its completion.

Again, in addition to the heat obtained by direct radiation from the flame, we have to consider that large quantity which would have been given out by the gases, *if their combustion had been completed.* It may here be observed that it is the obtaining the service of the heated products by an adequate run of flue, with sufficient time and surface, that characterises the *Cornish boilers.* In these, the main feature consists in generating, by *slow combustion,* no more heat than can be taken up, and transmitted to the water. In this respect, then, it is the direct reverse of the tubular system. In the former, there is slow combustion,—a continuous small development of combustible gas,—a long run, —abundant absorbing surface,—a moderate rate of current,— free access of the water to the flues,—and sufficient time to enable the surface plate to do its duty;—added to the adoption of every possible means of preventing the loss of heat, externally, by clothing the outside of the boiler.

In the marine *tubular boiler,* on the other hand, everything is the reverse. There is the most rapid combustion,—the largest and most irregular development of gas,—a rapid current,—a short run,—a restricted and imperfect circulation of the water,—and a total *inadequacy of time* for the transmitting and absorbing processes, with a great waste of heat by radiation from the boiler.

Another serious evil of this tubular system, and its short run, which carries the heat away so rapidly, is, *the over-heated state of the funnel and steam-chest;* and the consequent danger to the part of the vessel in their immediate contiguity.

CHAPTER XI.

OF FLAME, AND THE TEMPERATURE REQUIRED FOR
ITS PRODUCTION AND CONTINUANCE, AND ITS
MANAGEMENT IN THE FURNACES AND FLUES.

THAT a high temperature must, unintermittingly, be maintained in the chamber part of the furnace, will at once be
understood, when we consider that flame, continuous though
it appears to be, is but a rapid succession of electric explosions of atoms, or groups of atoms, of one of the constituents
of the gas—the hydrogen with oxygen; and as rapidly as
their respective atoms obtain access and contact with each
other; the second constituent—the carbon—taking no part
in such explosions. Whatever, therefore, interrupts this
succession (that is, allows the explosion of one group to be
terminated before another is ready, and within the range of
its required temperature), virtually causes the flame to cease:
in ordinary language, *puts it out.*

Again, if by any *cooling agency* we reduce the temperature below that of accension, or kindling, the effect is the
same: *the succession is broken*, and the continuousness of the
flame ceases; as when we blow strongly on the flame of a
candle, by which we so cool down the atoms of gas that
they become *too cold for ignition*, and pass away in a
grey-coloured vapour; but which, by contact with a
lighted taper, may again be ignited, and the succession
restored.

Thus we see there are two modes by which flame may be
interrupted, that is, extinguished; both of which are momen-

tarily in operation in our furnaces. 1st. By the want of successive mixture or groupings of air and gas. 2nd. When the gas is reduced in temperature by cooling agencies, as will be shown hereafter.

The two essentials of combustion are laid down by Sir H. Davy, viz.—*temperature and contact;* he considers the management or treatment of the flame, and the means by which it may be effected or extinguished. He states, that on mixing one part of *carbonic acid* with seven parts of *the mixture of gas and air;* or one part of nitrogen with six parts of the mixture, their *powers of explosion were destroyed,* —that is, *ignition was prevented.*

Again he observes : " If combustible matter requires a high temperature for its combustion, it will be *easily extinguished* by rarefaction or by *cooling agencies,* whether of solid substances, or of incombustible gases."

On examination of what passes in furnaces *using* coal, we see the direct connection between its effect, and what Sir H. Davy so clearly points out as the means of *extinguishing the flame.* On looking into a *flue boiler* from the back end, a body of flame will be seen flashing along from the bridge, and if air be properly introduced, extending a distance of 20 to 30 feet. This is the appearance which has to be sustained *until the process of combustion be completed,* if we would have the full measure of heat developed.

On the other hand, looking into a *tubular* boiler, across the smoke-box, the light of the flame may be seen through the tubes ; but, on entering their orifices, or at a short distance within them, it will appear to be suddenly cut short and extinguished, and converted into smoke.

The distance, then, to which flame will penetrate tubes, *before being extinguished,* will depend on the rapidity of the current,—the size of the orifices,—and the quantity and character of the gaseous products, *entering in company and in contact with it.* These products are—

F

From the coke . . carbonic acid and nitrogen.
From the gas   . . carbonic acid, nitrogen, and steam.

Here we have the very incombustible gases referred to by Sir. H. Davy,—not even in small, but in very large quantities,—forced into the most intimate possible mixture with the flame. The result necessarily must be, the reduction of its temperature, and consequent extinguishment.

Under the circumstances of an ordinary *flue* boiler, if the flue be of sufficient area, the products of combustion *separate themselves*, as seen in the flame of a candle, and as will hereafter be shown. So in the flue, the hottest portion, and the flame itself, will take the upper part, thus avoiding that unnatural mixture with its own incombustible products— carbonic acid, nitrogen, and steam; but which in the tubular system are again forced into contact with the flame from which they had separated themselves.

That the temperature *within the tubes* will be reduced below that required for continuous ignition, may be tested by looking into them through apertures across the smokebox end, or by introducing shavings or paper fixed to the end of an iron rod. In most cases (unless when the fuel on the bars is clear) the paper may be passed in and withdrawn, blackened with soot, or unscorched, according to the state of the furnace, indicating the low temperature within the tubes, and their utter uselessness *as steam generators*.

The inference which this inquiry leads to as regards the high temperature required—1st, for the ignition, and 2ndly, for the sustained existence of flame—is, that the tubular system is chemically, mechanically, and practically a destroyer of both.

CHAPTER XII.

OF THE CIRCULATION OF WATER IN THE BOILER.

THIS important branch of the subject—promoting circulation in the water in evaporative vessels—appears to have hitherto received but little attention ; yet promoting circulation is virtually promoting evaporation. Mr. Perkins proved by numerous experiments how much evaporation was increased by an unembarrassed action of the ascending and descending currents of the water : since then, no further effort has been made in that direction. If sufficient space be allowed for the action, the ascending and descending currents will of themselves take such directions as are most favourable for their respective function, as in Fig. 65, where an ascending current is seen in the centre, and a descending one on the sides.

Dr. Ure observes, " When the bottom of a vessel containing water is exposed to heat, the lowest stratum becomes specifically lighter, and is *forced upwards by the superior gravity of the superincumbent colder and heavier particles.*" Here we have the correct theory of circulation.

Fig. 65.—Circulation of Water.

So far as regards the motion in water, *previous to ebullition,*

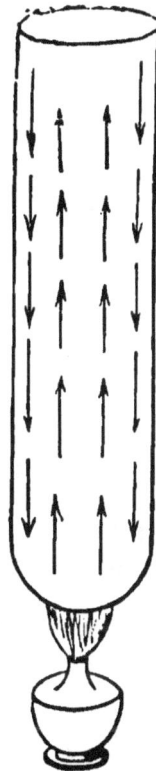

F 2

it has been commented on by all writers on the subject. The act of boiling, however, creates a species of currents of an entirely different and important character. These have not received due attention, yet they are the most important, inasmuch as they influence not only the amount of evaporation, but, as will be shown, the *durability* of the boiler itself.

With reference to the movements among the particles in water, it is a mistake to suppose they will descend in the same *vertical lines* in which they had ascended, as a shower of rain would through the opposing atmosphere. Such a direction would be impracticable on account of the resistance of the ascending currents of both steam and water, caused by ebullition. This may be illustrated by the annexed drawings. Fig. 66 represents a supposititious case of the particles of water on reaching the surface, turning and descending in the same vertical lines in which they had ascended. Fig. 67 represents the ascending particles of water *flowing along the surface to the coolest and least obstructed part for their descending course.* This is what takes place in all boilers.

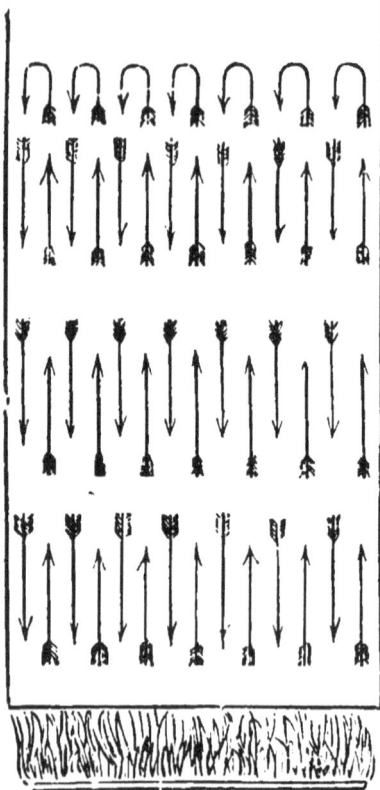

Fig. 66.—Circulation of Water.

When heat is first applied to water, the uniformity of the

motion is the mere result of *diminished specific gravity*, that being then the *sole motive power*. *After ebullition*, however, a new state of things is created. The columns of rising steam obtain great physical power, violently and mechanically forcing upwards the water which comes in their way. Vertical streams are thus induced, *putting in motion a body of water far greater than would be required for merely taking the place of that which had been converted into steam.* Now, as bodies or streams of water, commensurate with these continuously forced upwards, must necessarily return to prevent there being a vacant space, it is for these *returning or downward currents*, of what may be called *surplus water*, that we are called on to provide both space and facility.

The difference in the character of the currents *before* and *after* ebullition are shown in the annexed figures. These may be

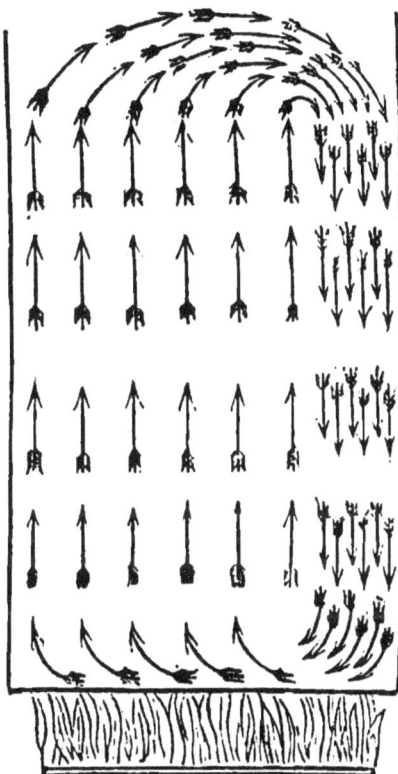

Fig. 67.—Circulation of Water.

well observed in a glass vessel, of the shape here indicated, and about 4 or 5 inches wide, suspended over the flame of an Argand burner. Fig. 68 represents the uniform motion which takes place *before* ebullition. Fig. 69 represents the water *after* ebullition in its descending and revolving currents, *forcing the rising columns of steam aside*

from their vertical course, as marked by the arrows. These
motions, which are not perceptible if the water be free from
foreign matter, will be seen on throwing in a great number of
small bits of paper, so as to occupy all parts of the water.
The entire mass will then be exhibited in violent and re-

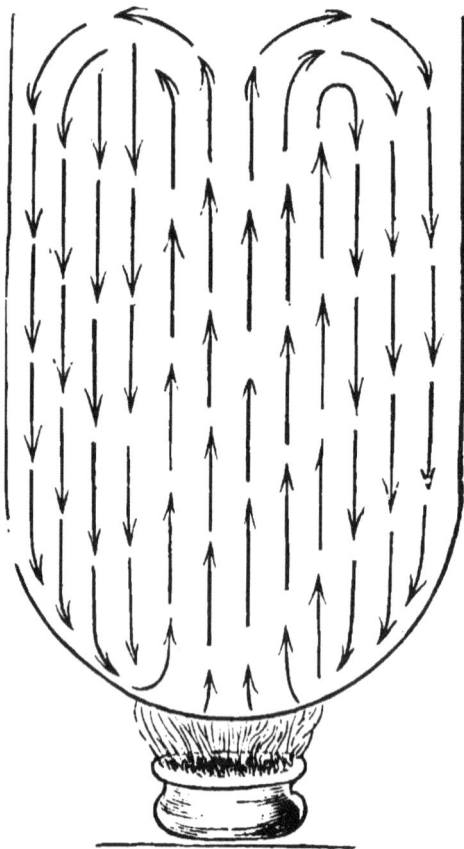

Fig. 68.—Ebullition.

volving currents—the ascending steam occupying one side,
and the descending body of water rapidly descending in some
other part, but manifestly *occupying a much larger area of the
vessel than the ascending portion.*

So great is the ascensional energy and velocity of the

rising steam, and the extra water forced before it, that numerous globules are borne along by the current, and carried even downwards. These may be observed at A, Fig. 69, in their slow oscillating motion, struggling to return upwards through and against the force of the descending

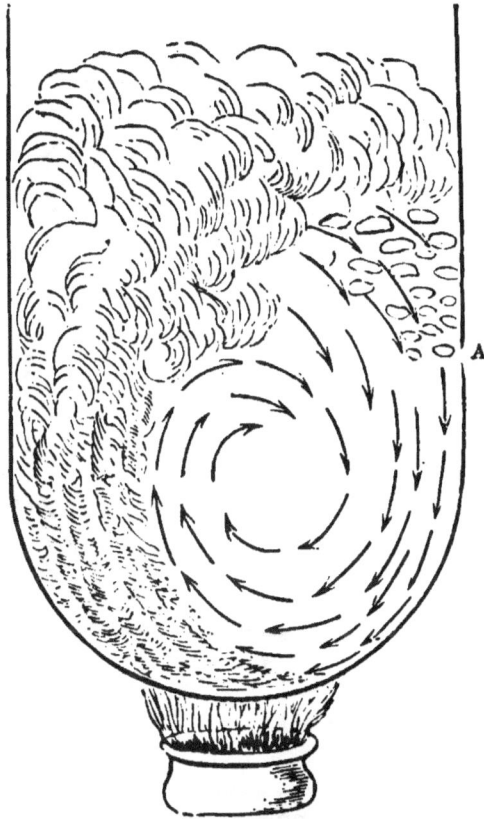

Fig. 69.—Ebullition.

water. These movements are highly instructive, and should be well examined, since, without an accurate knowledge of them, we cannot have a right conception of what is required for giving a due circulation to the water, and arranging the flues and water spaces in boilers to enable those motions to be completed.

The influence exercised by the descending body of water was strikingly illustrated in an experimental tin boiler, 12 inches long, with a single flue running horizontally through

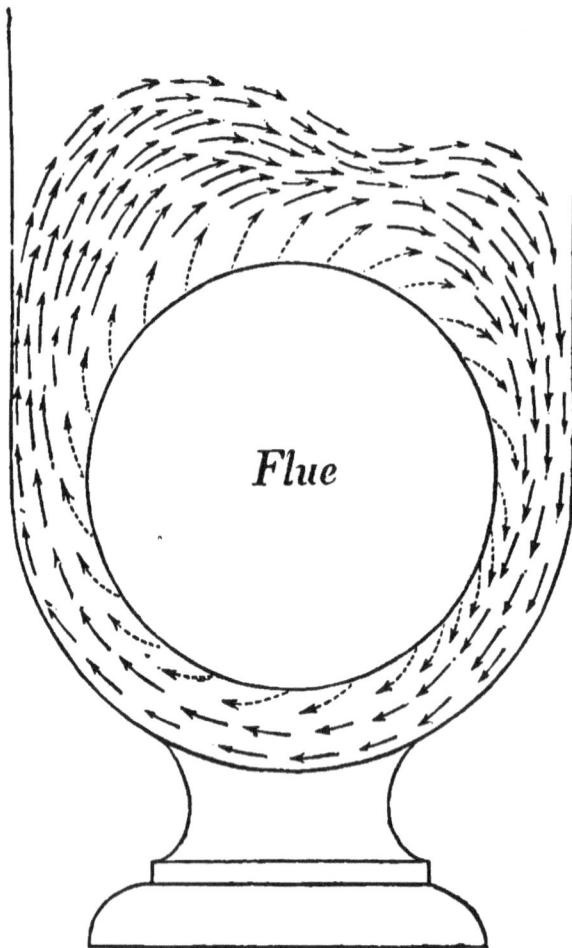

Fig. 70.—Ebullition.

it, the water being heated by the flame of a large laboratory gas lamp.   The boiler being open at the top, the movements of the steam and water were thus ascertained, as shown in Fig. 70.   So soon as the ebullition became strong, the water spaces round the flue appeared insufficient to allow the steam to ascend, and the water to descend, equally on both sides. The consequence was, that much of the water forced up by the steam on the one side, was carried over by its violence, and descended on the other, thus making a circular course round the flue, and forcibly carrying along with it much of the steam that came in its way. This circular motion is shown by the dotted arrows representing *the steam*, and the plain arrows, *the water*.

With the view of observing the injurious consequences of restricted water-ways, a very useful class of observations may be made by using a tall narrow glass as in Fig. 71, attached to a tin or iron vessel, with a flat bottom, to receive the heat from an Argand burner, or spirit lamp.

Here the *descending water* is so obstructed by the joint columns of *ascending steam and water*, that both are thrown into great confusion:—their respective currents continually changing sides, and the progress of evaporation considerably delayed. We here obtain a clear practical view of what must take place between the flues or tubes of boilers, with their usually *restricted water-ways*.

The violence and intermittent action which ensues where separate channels or sufficient space are not available, will be well illustrated in the following experiment: Fig. 72 represents two long glasses, each 2 inches wide by 18 inches long, A and B connected by means of a tin apparatus C and D, at the top and bottom,

Fig. 71.—Ebullition.

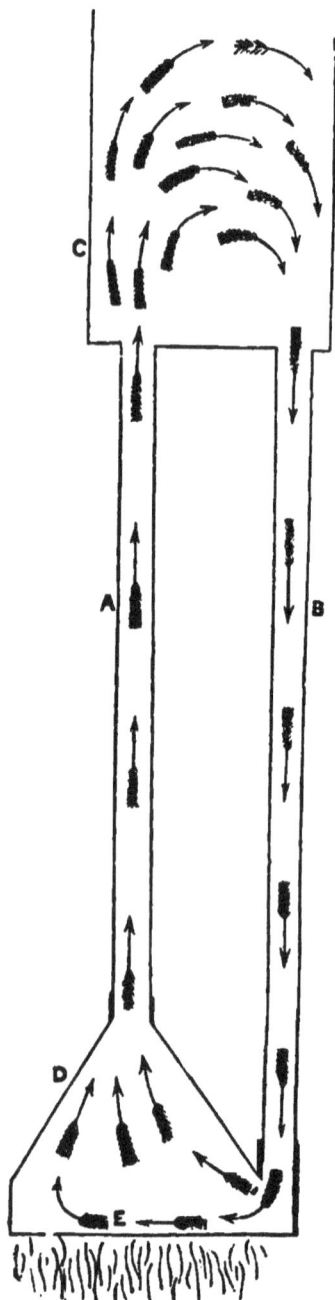

Fig. 72.—Ebullition.

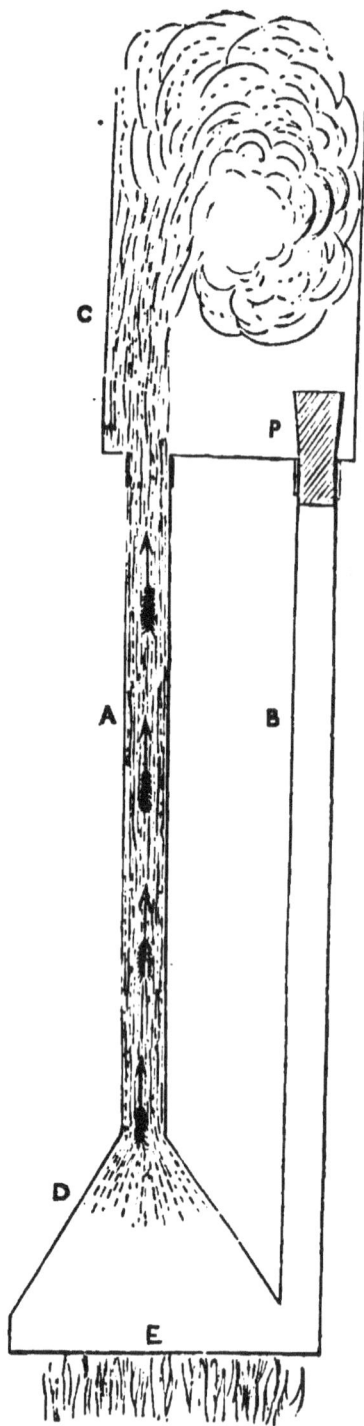

Fig. 73.—Ebullition.

leaving the communication open above and below; the
whole being suspended over a fire, or circular series of gas
jets producing a strong heat. On the heat being applied,
a current of mixed steam and water will be seen ascending
in one glass, and descending in the other, as indicated by
the arrows. There being here no confusion or collision,
a state of things will be produced highly favourable to
the generation of steam ; the colder water finding easy
and continued access to the heated bottom of the vessel
at E.

If, however, the communication between the two glasses
be cut off by inserting a cork or plug in one of the glasses,
as seen at P, Fig. 73, the circulation in glass B will be
suspended, and the glass A will then have the double duty
to perform of allowing the rising steam to reach the surface,
and the descending water to reach the bottom at E. The
previous uniform generation of steam will then be succeeded
by an intermittent action, explosive violence alternating
with comparative calm and inaction, clearly indicating that
the latter is only the interval of accumulating force to be
discharged by the former. The *rationale* of this intermittent action is, that the water being obstructed in its
descent, the steam is necessarily delayed or accumulated in
the lower chamber, and only discharged at intervals. The
motions exhibited in these intermittent changes are little
understood, and have not been examined either scientifically
or practically ; yet this branch of hydrostatics merits the
most serious investigation in connection with the construction of large boilers.

Again, this accumulated steam getting sudden vent is
discharged with great violence, literally emptying both the
glasses and lower chamber. An equally violent, but more
sudden, reaction of course follows, and a large body of
colder water as suddenly rushes down to fill the space
vacated. An interval will then necessarily be required to

raise the temperature of this large supply of colder water, and restore the previous state of ebullition.

Here, then, we have a natural and physical cause for the intermittent action on the small scale which takes place in boilers on the large scale, where free circulation is impeded by the want of adequate space. Here also may be seen the true source of *priming* in boilers where the act of ebullition is violent.

CHAPTER XIII.

ON THE CIRCULATION OF THE WATER IN RELATION TO EVAPORATION, AND ITS INFLUENCE ON THE TRANSMISSION OF HEAT.

WITH reference to the currents in water caused by the application of heat, the first point for consideration practically is, the direction in which the atoms of water approach the plates where they are respectively to receive heat.

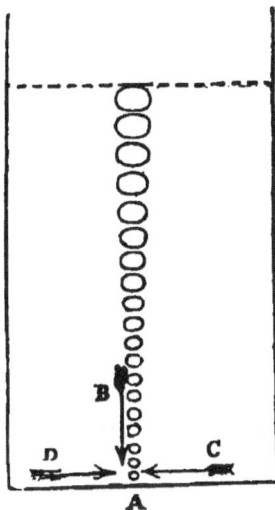

Fig. 74.—Steam rising.   Fig. 75.—Steam rising.

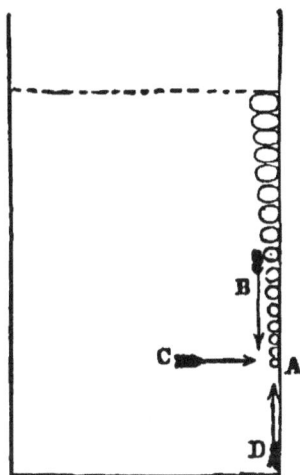

Remembering that before any particle of water can leave the heated plate, and rise to the surface, *as steam*, it must be "*forced upwards by some colder and heavier particles.*"

Unless, therefore, a due succession of such particles be enabled to take their places as rapidly as others have received the heat due to their vaporization, the plate itself cannot be relieved of its heat, and consequently, evaporation must be retarded.

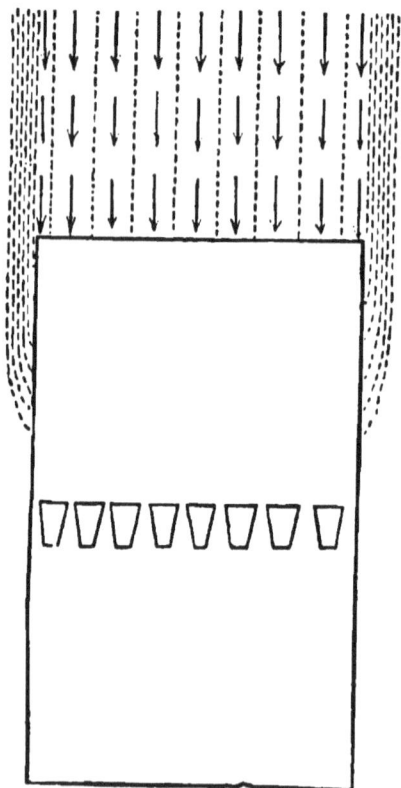

Fig. 76.—Circulation of Steam and Water.

To illustrate the direction in which the particles successively approach the heated plate, Fig. 74 represents a vessel of water—the heat being applied *from beneath*. The question here is, whether the colder atoms which are to take the places of the atoms of vapour, generated at the point A, will approach that point in the direction of the arrows

B, C, or D. From what has just been shown, it is manifest they cannot arrive in the *downward direction* of B, and must necessarily come in that of either C or D.

Again, suppose the heat be applied *literally, as to the side plates of a furnace*. The question then will be, whether the colder atoms will approach the point A, as in Fig. 75,

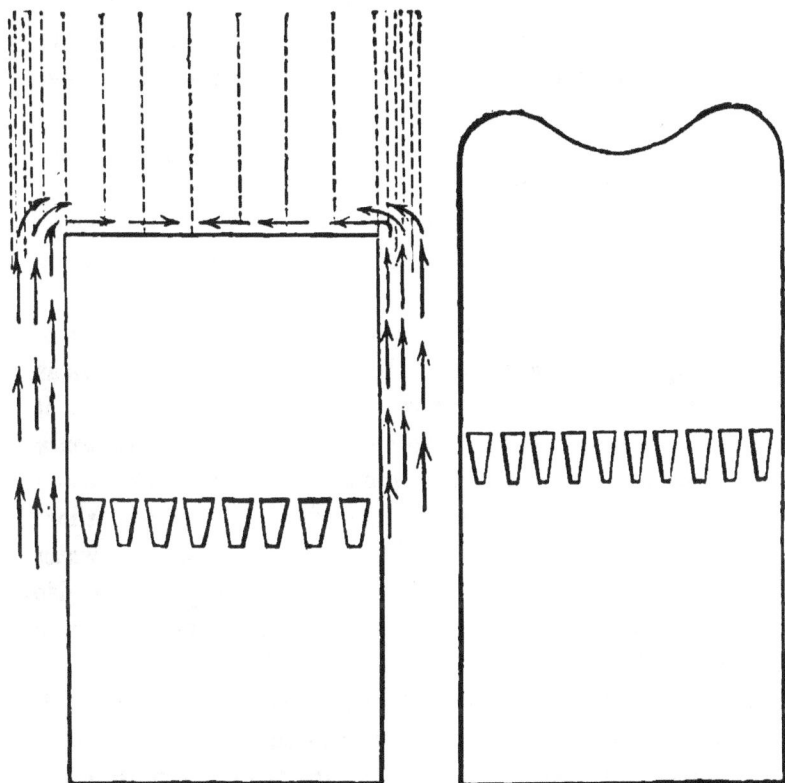

Fig. 77.—Circulation of Steam and Water.   Fig. 78.—Over-heated Furnace-crown.

in the direction of the arrows B, C, or D. For the same reasons it will be seen that it must be in that of either C or D. This is manifestly in favour of *vertical*, rather than *horizontal* surfaces, as practically has been proved in the case of vertical tubes. This advantage may be accounted for by the fact, that the currents of the atoms of vapour, and that of the water about to be converted into vapour, will then be running *in one and the same direction*, and consequently, without obstruction or collision.

Now as vaporization does not depend on the quantity of heat *applied to* the plate, but on the quantity *taken from it*, the amount of evaporation will be determined by the rapidity with which the colder atoms of water obtain access to the heated plate. Hence we see how more important it is to study the means of giving that access of the *water to the one side* of the plate, than of *heat to the other*.

The annexed figures will illustrate, practically, the direction in which the colder water obtains access to the *sides* and *crown plates* of a furnace :—the arrows representing the atoms of water, and the dotted lines those of the rising steam.

An important question is here raised,—do the atoms of water approach the plate in the direction of the arrows as in Fig. 76 or Fig. 77 ? Everything goes to show that it must be as in the latter. We here then find that the *crown* or *horizontal* plates can be supplied from the *vertical water spaces alone*: and hence the importance of making those channels so large, and conveniently arranged, that the water may reach them in full current and quantity. This also accounts for the fact that the crown plates are the most liable to injury from being over-heated. The side plates next the fuel on the bars, becoming over-heated and burnt, we shall see, is referable to a different cause.

The crown plates of a furnace are then the most liable to injury from the double cause of being exposed to the greatest heat, by direct impact of the flame, and from the water having greater difficulty of access to them, as shown in Fig. 76. This causes them frequently to bulge under the pressure of steam, as in Fig. 78. This bulging on one occasion occurred in the first voyage the vessel had made; nevertheless, no inconvenience resulted from it ;—the iron having been of good quality, the boiler remained in a state of perfect efficiency for many years; and when ultimately broken up, the bulged part was as sound and thick as it ever had been.

On this it may be observed, that if there be any obstruction or sluggishness in the water reaching the crown plates (on its rising from the vertical water-ways), the violence of

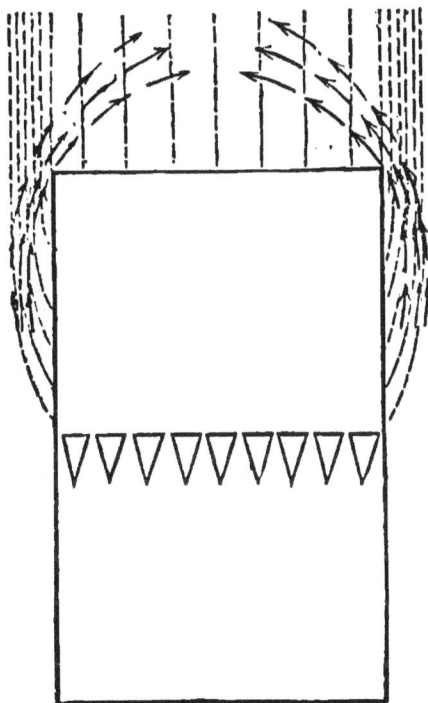

Fig. 79.—Circulation of Steam and Water.

the upward current of steam will carry it in the direction of the arrows in Fig. 79;—thus leaving the centre of the crown plate in *contact with steam rather than water*, and by which it necessarily becomes over-heated.

Looking to the direction of the water on passing from the vertical side spaces to the crown plates, the shape, as in Fig. 80, would appear most favourable for aiding the access of the water—(the arrows show the direction of the water, and the dotted lines that of the steam).   The cylindrical shape, perhaps, offers the most advantages as regards direction of the water, and the resisting power of the plate.

The inferences to be drawn from these facts are,—*First,*

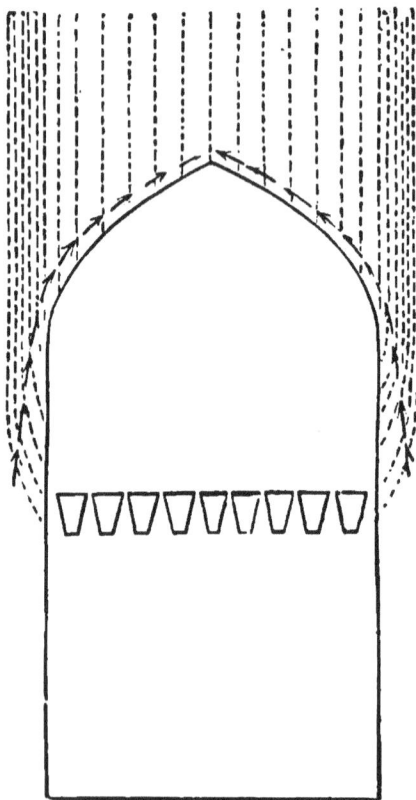

Fig. 80.—Circulation of Steam and Water.

that every possible facility should be provided for enabling the steam to reach the surface of the water without loss of time or temperature.   *Secondly,* that ample space should be given, at *the ends or sides, or both,* for the large *returning body of water,* which had been forced upwards by the violence of the ascending columns of steam.   *Thirdly,* that similar and adequate means should be provided to enable the water *to spread along the bottom*, in its course to supply the numerous vertical spaces between the furnaces.

It is more important that we investigate *the absorbent power of the recipient, than that of the plate.* All that the plate requires is, that its *transmitting power*—its proper function, *be brought into action.* On this head, M. Péclet correctly observes, that, "under ordinary circumstances, the quantity of heat which a metal plate has the power of transmitting, is far greater than what it is really called on to transmit!" To expect, then, that the plate will exercise a greater power of *transmitting* heat to some other body, than that body possesses for *receiving and absorbing it*, would be a physical absurdity.

It is manifest that as the heat absorbing power of *steam* is so inferior to that of *water*, a much less quantity will be taken up, in given times, by the former than by the latter. Again, since no more heat can be *transmitted through the plate from the one side than can be taken from it by the recipient* on the other—the *transmitting* power of any given surface must be absolutely dependent on the *absorbing* power of that recipient, whether it be oil, water, steam, or air. If, from any circumstance, the water be prevented or delayed *in gaining access to the plate*, the steam will in a like degree be obstructed *in leaving it;* the presence of the former being the very means by which the latter will be effected.

CHAPTER XIV.

OF THE CIRCULATION OF THE WATER IN RELATION TO THE DURABILITY OF THE PLATES.

HAVING considered the circulation of the water as regards evaporation; distinguishing the separate functions of the heat *transmitter*, and the heat *recipient*, we have now to examine its bearing on the durability of the plates.

In the first marine boilers, the flues were made *deep with narrow water spaces*, usually about four inches wide. The object, then, was to combine the two essentials—adequate length of run in the flue, with sufficient heating surface. Among the disadvantages of this arrangement of the flues was, the recurrence of injury to the plates in the region of the furnaces by becoming *over-heated*. Although this evil of over-heating continues to be experienced, the direct cause of it has remained without due inquiry. Its recurrence was usually attributed to neglect on the part of the fireman, or the want of sufficient water in the boiler ; hence more importance was attributed to the necessity of having "careful and experienced stokers," than to the remedying the de-\ fective construction of the boiler itself.

If the water indicated a level *above the flues*, all was considered right, and no thought was given to the possibility of its being deficient *below or around them*. Yet experience has shown, that although everything indicated a proper height of water in the boiler, the plates, particularly these connected with the furnaces, were, nevertheless, subject to be over-heated and injured. In such cases, if the iron was laminated, or otherwise of inferior quality, it became cracked, or burned into holes. This state of things was strikingly illustrated in the boilers of the "*Great Liverpool*" steamship, on her first voyage to New York, in 1842. The engineer, observing the side plates of the furnaces constantly giving way—some bulging and others cracked and leaking, and even burnt into holes, although there was always a *sufficient height* of water in the boiler, suspected something had interfered to keep the water from the plates, and with the view of testing it, introduced an inch iron pipe from the front into the water space between two of the furnaces. This at once brought the source of the evil to notice; for although the glass water-gauge always indicated a sufficient *height of water within*, yet nothing issued through the pipe but *steam*, so long as the boiler was in full action.

This fact unmistakably showed that the over-heating of the plates was unconnected with the duties of the fireman, and was the result of *insufficient circulation*, depriving the deep narrow flue-spaces of an adequate supply of water. There was then, manifestly, no remedy for this continually recurring evil, and a new boiler became necessary.

The important point then for inquiry is, Why was the plate thus left in contact with steam instead of water? There was here no apparent impediment to the steam rising to the surface. There was, however, as we shall see, extraordinary difficulty of access to the water to *dislodge the steam*; and, as Dr. Ure observes, it must remain, until "*forced upwards by colder and heavier atoms of water.*"

As the details of this boiler of the "*Liverpool*" will afford opportunities for comment, on what is practically necessary for promoting circulation, they are here annexed.

Fig. 81 is a plan of the boiler, showing the ten furnaces, and the narrow water-ways separating the series of narrow flues.

Fig. 82 is a section from A to B, showing the water
spaces of 5 feet deep by 4 inches wide, and the direction

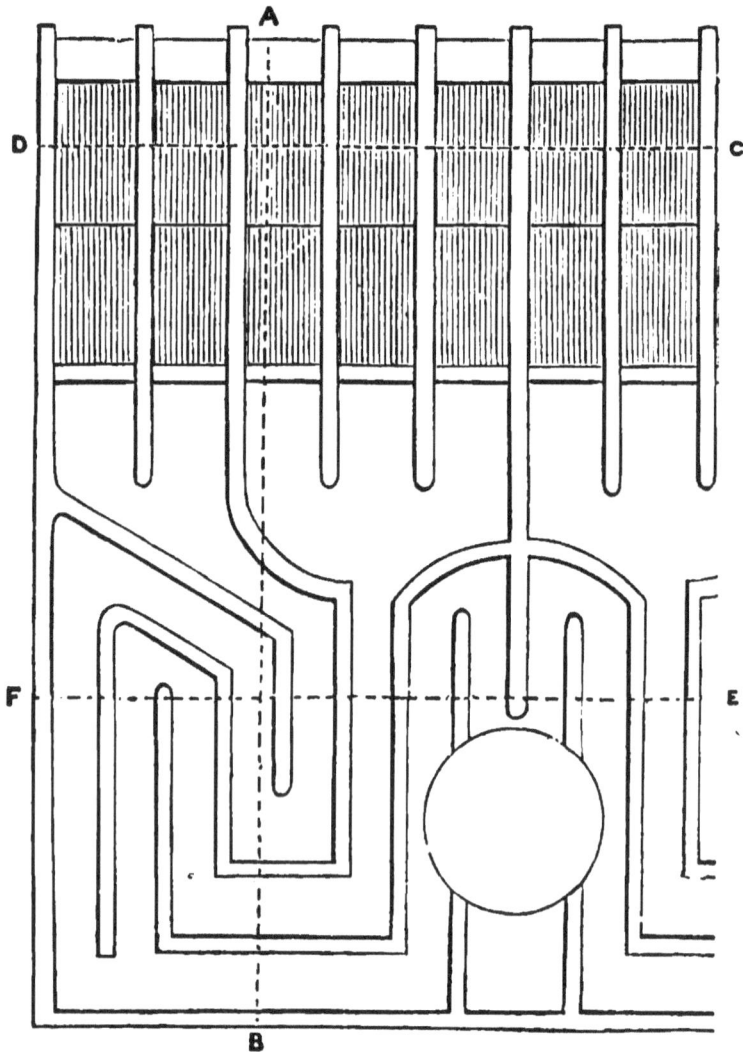

Fig. 81.—Boiler of the "*Liverpool*."

in which the water approached the side and crown plates
of the furnaces.   The *bottom horizontal* water-space is here

seen, of *but 5 inches deep*, and from which *all the vertical spaces were to be supplied*.   This bottom space was also found to be much obstructed by sediment and other deposit.

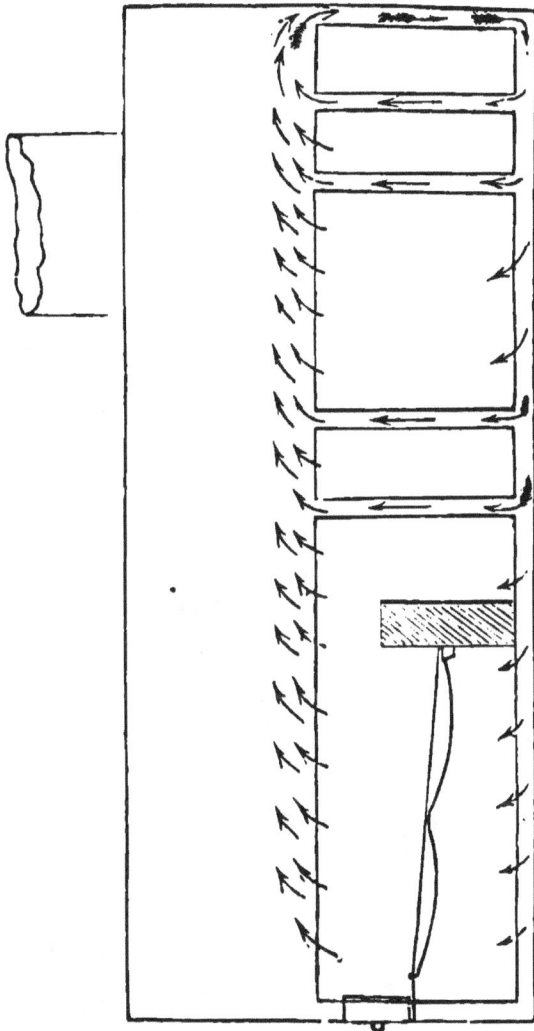

Fig. 82.—Boiler of the "*Liverpool*."   Section through ᴀ ʙ.

Here we see the direction the water had to take in its *downward* course was through one narrow four-inch space ; and then to be distributed over the bottom area of above

500 square feet in its way to the vertical spaces between the furnaces and flues.

Fig. 83 is a section across the furnace end of the boiler,

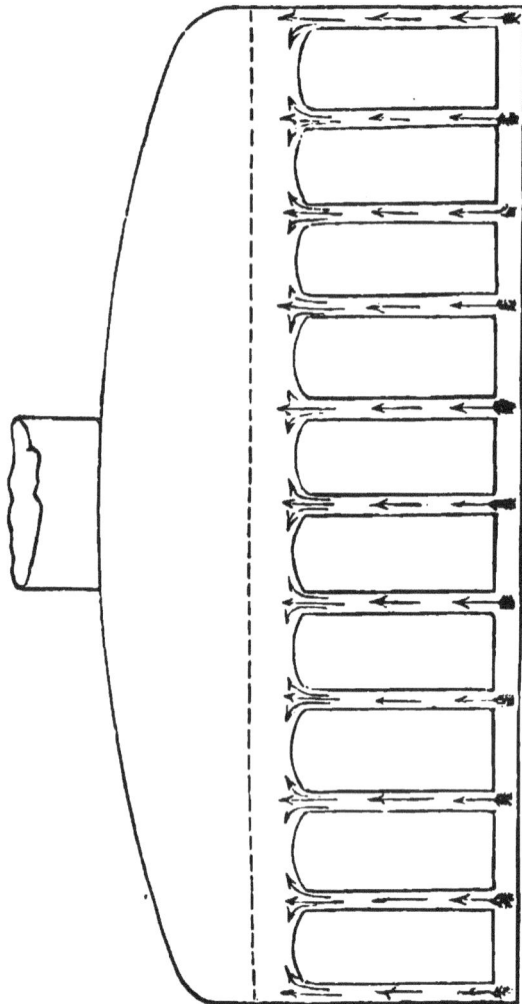

Fig. 83.—Boiler of the "*Liverpool*." Section through C D.

from C to D, shewing the eleven narrow water spaces, and into one of which the trial-pipe was introduced, as already mentioned.

Fig. 84 is a cross section of the after-part of the boiler, from E to F, showing the sixteen water spaces between the flues.

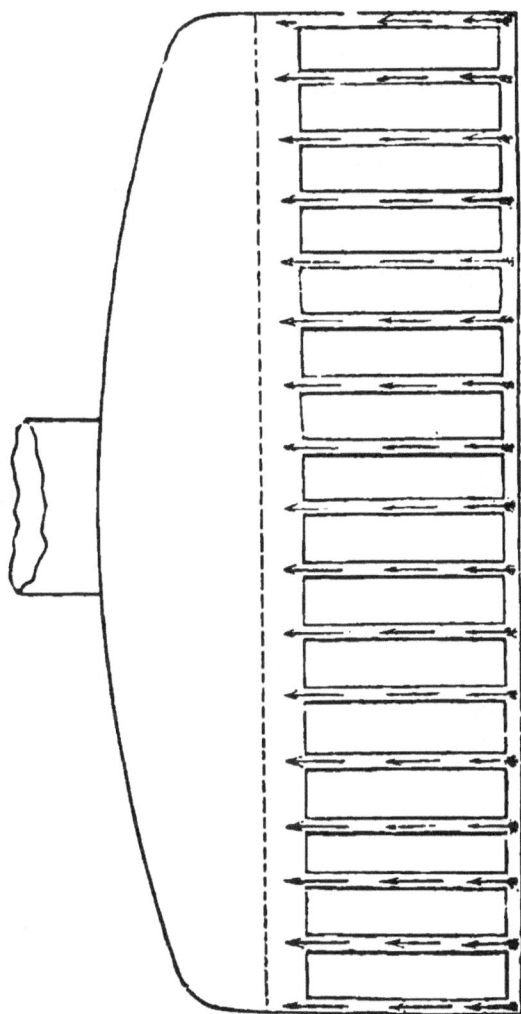

Fig. 84.—Boiler of the "Liverpool." Section through E F.

It may here be observed, that the side plates of the ten furnaces, which, as being the hottest, required the largest supply of water, were, necessarily, the worst supplied, being

G

the *farthest from the narrow downward current at the back end.* It is here manifest that the furnace side-plates could not have been adequately supplied with water, to take the place of the great volume of steam generated from their surfaces, and, consequently, that *the steam must have been retained in contact with them.* It can no longer, then, be a matter of surprise that the sides of the furnaces became over-heated and injured.

In this boiler, the ten furnaces being all at one end, the water taking the coolest place for its descent, would naturally flow along the surface from front to rear, as shown by the arrows, Fig. 82. (The construction of this boiler, as regards what takes place *within the flues,* will be further noticed in the Chapter on "*Draught.*")

That the transmitting plate cannot be unduly heated or destroyed, where the recipient of the heat is *water,* may be tested in many ways. Water may be boiled in an egg shell, or in a vessel, the bottom of which, though made of *card paper,* will not be injured. That the temperature of an iron vessel of even half an inch in thickness, containing boiling water, cannot be much, if any, above that of the water, may be tested by applying the fingers to it, immediately on being removed from even an "intense fire." In fact, the temperature will rather appear to *increase after its removal from the fire.* The reason is obvious :—the heat being so rapidly taken up and absorbed by the water, *its current through the plate* must necessarily be equally rapid. On removal from the flame, however, the *exterior surface,* then receiving no further supply or increment of heat, is instantly reduced in temperature ; the current through the plate then *becomes reversed, and passes from the inside to the outside.* The water side being absolutely the hottest, a new equilibrium is established, equal to that of the water at 212°. For the purpose of putting this to the severest test, an iron vessel with a flat bottom and half-inch thick plate was placed over a furnace, expressly constructed, so that it might be exposed to a very great heat from a coke fire, urged by a strong blast: the rapidity of the ebullition was extreme. It was so arranged, that it could be suddenly removed from the furnace, and the temperature of the bottom instantly ascertained by the touch of the fingers. This was done repeatedly, yet, on all occasions, it scarcely appeared to have the temperature of boiling water.

A conclusive illustration of the fact that the plates are in no way affected by thickness or temperature, is afforded by finding that, in breaking up old boilers, those parts which were exposed to the most intense heat and direct impact with the flame, have continued sound, and wholly undeteriorated, when the water had free access to them.

CHAPTER XV.

OF THE DRAUGHT.

THE draught, or current of air passing through a furnace, is occasioned by the difference in weight between the column of air within the chimney, and that of an external column of the same proportions,—the "ascent of the internal heated air," as Dr. Ure observes, "depending on the diminution of its specific gravity,—the amount of unbalanced weight being the effective cause of the draught." Since, then, this levity of the inside air is the result of increased temperature, the question here for consideration is, how that temperature may be obtained with the least expenditure of fuel?

In marine boilers, numerous cases of deficiency of draught will be found to arise from an injudicious arrangement of the flues, and the conflicting currents of the heated products within them.

M. Péclet observes, "Where several tubes or flues open into one common flue, the currents are continued beyond their orifices, and by their mutual action affect or modify their respective forces. If, for example, two flues, A and B (see Fig. 85) enter the common flue C, by orifices *opposite each other*, the influence of their currents on each other will be *nil, if they have equal rapidity;* because the whole will pass as if they had struck against a plane fixed between them. If, however, the currents be unequal, *that which has the greatest rapidity will reduce the speed of the other*, and more or less have the effect of closing the orifice

through which the latter flowed. So many proofs of this,"
he adds, " may be adduced, as to put the fact beyond doubt."
" These streams of air," he continues, " in this respect, act

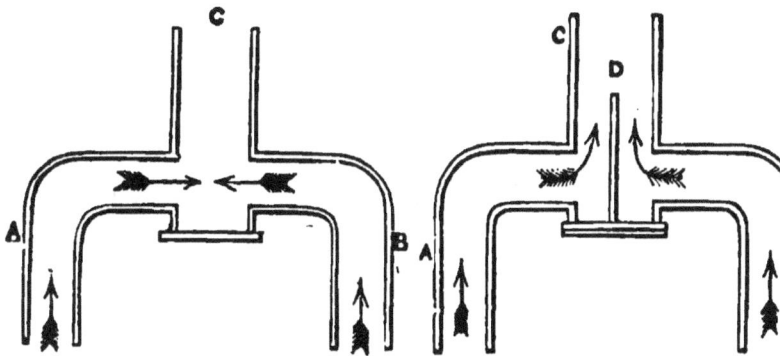

Figs. 85 and 86.—Draught in Flues.

on each other as streams of water. It is already known by
the experiments of *Savart*, that where two streams of water,

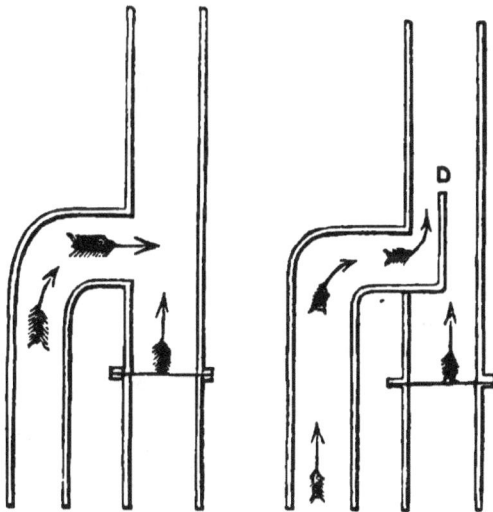

Figs. 87 and 88.—Draught in Flues.

of the same sectional area, act in opposite directions, and
that one of them has even but a little more speed than the
other, the latter is pushed back, and the influence is felt up

to its source. The result of this collision in the flue may be avoided by the Diaphragm D, Fig. 86."

Again, " Phenomena of the same kind will be produced

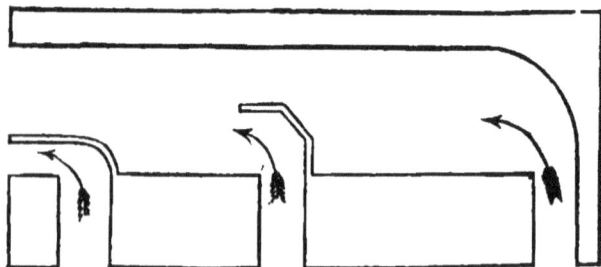

Fig. 89.—Draught in Flues.

where the courses of two flues are at *right angles to each other*, as in Fig. 87. These effects may also be avoided

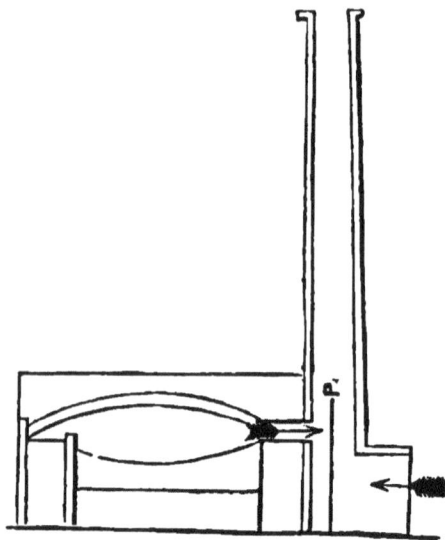

Fig. 90.—Draught in a Chimney.

by the Diaphragm D, Fig. 88." This also is of frequent occurrence, and seriously affects the general draught, as will hereafter be shown.

Again, " Where the chimney or flue is common to several furnaces, the arrangement should be such that the streams, or currents of heated gas, should not interfere with each other. Fig. 89 represents the arrangement that should be adopted in such cases." It is needless to observe how frequent this state of things occurs, and how little attention is given to it.

" So, where a current of hot products issues horizontally into a chimney, it may happen that its draught would be *entirely destroyed*, if the rapidity of such current was con-

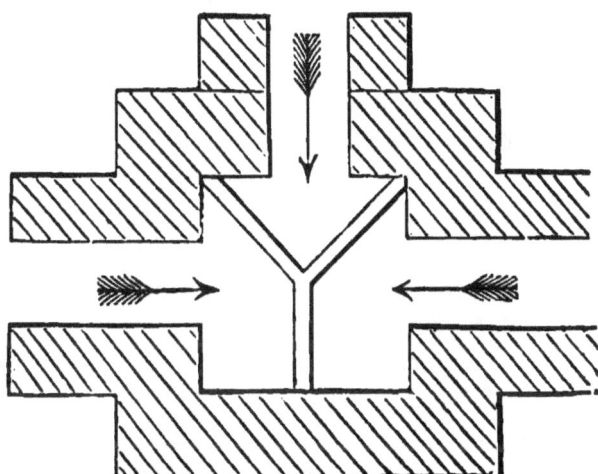

Fig. 91.—Draught in a Chimney.

siderable, as it would then have the effect of shutting tho chimney like a damper." He then describes what occurred at a soda manufactory, with a chimney for general draught, which he had to construct, and which was also connected with a flue from another apparatus, as shown in Fig. 90. In this case, "the current from the one flue completely neutralised that of the other." This he remedied by tho partition P.

Again, where three flues enter a funnel from *three different points*, it is evident, he observes, that " the diaphragms should be so placed as to leave each current an adequate section of the chimney," as in Fig. 91. The circumstance here referred to may be found to exist in almost all marine boilers. Rarely, however, is the interposition of these diaphragms thought of, yet numerous instances of the derangement of the draught, particularly of the wing boilers, must be within the knowledge of all engineers.

Let us now apply these judicious practical observations. The first boilers of the "*Great Britain,*" screw steamer, are in point. The arrangement of these boilers have already been noticed with reference to their impeding the due *circulation of the water.* We have now to consider them in respect to their influence on *the draught.*

In these boilers, attention was given, almost exclusively, to two objects :—providing the largest possible amount of *fire-grate areas,* and the largest aggregate of internal *heating surface.* As to the former, almost the entire area of these large boilers may be compared to an aggregate of furnaces. Nevertheless, there was no command of steam, and the engineer stated, that the wing boilers were unequal in draught to the centre ones. The deficiency of draught in the furnaces of the side boilers will easily be accounted for on examining the plan of the upper tier of flues, and the numerous collisions where the heated products from twenty-four large furnaces entered the funnels, as shown in Fig. 92.

The flues from the four furnaces of each wing boiler are here made to enter one common cross flue—each thwarting the current of the preceding one. No. 1 being checked by No. 2—which crosses it at right angles—which, in its turn, was checked by No. 3, and so on—the same mal-arrangement taking place in each of the sixteen flues of the four wing boilers. These, it will be seen, are the direct cases adduced by M. Péclet, where the products and current from one flue act as a damper on the *draught of its preceding one.*

Again, the joint products of the four flues of each wing
boiler are made to enter the funnel by a single opening,
which is not only at *right angles* with the flues from the
four centre boilers, but *directly opposite to those of the wing*

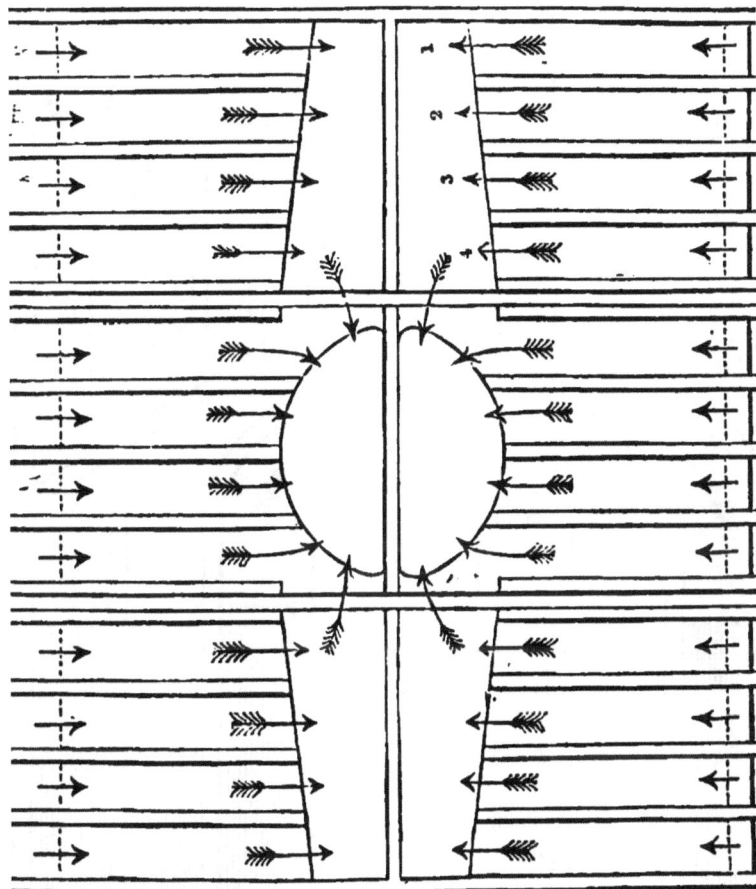

Fig. 92.—First Boilers of the " *Great Britain.*"

boilers on the other side. Thus the flues of no less than
eight furnaces, all entering by a single opening, are brought
into direct collision with those of the other four, and in
the most certain way to affect the draught of all. Here

we have a combination of the evils referred to by M. Péclet.

The case of the boilers of the "*Great Liverpool,*" is a still

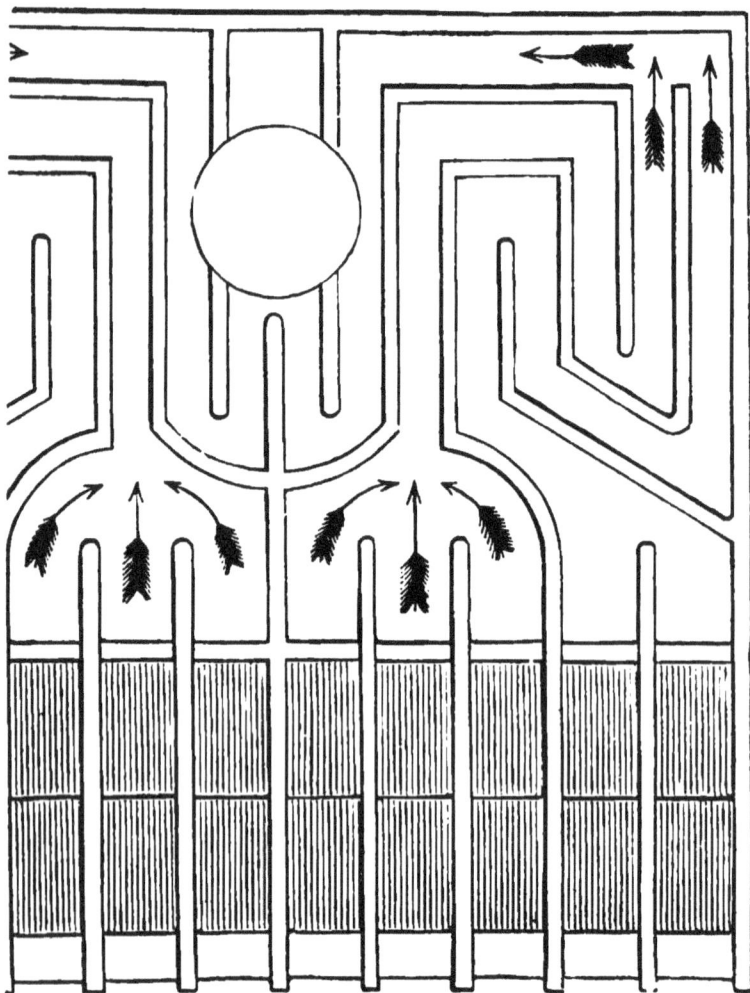

Fig. 93.—Boiler of the "*Great Liverpool.*"

greater violation of the rules which should regulate the draught. Here, there being but a single tier of flues, the required aggregate of *heating internal surface* was obtained by the labyrinth of windings shown in Fig. 93.

In the first place, the flames from the three centre furnaces of each half of the boiler are forced into a single flue of but 13¾ inches wide, as shown by the arrows. Again, the gaseous products of each set of three centre furnaces, and which are necessarily the more powerful, are made to enter the single back flue at *right angles*, and across the current of products from the two wing furnaces, as shown by the enlarged view in Fig. 94. It is scarcely possible to

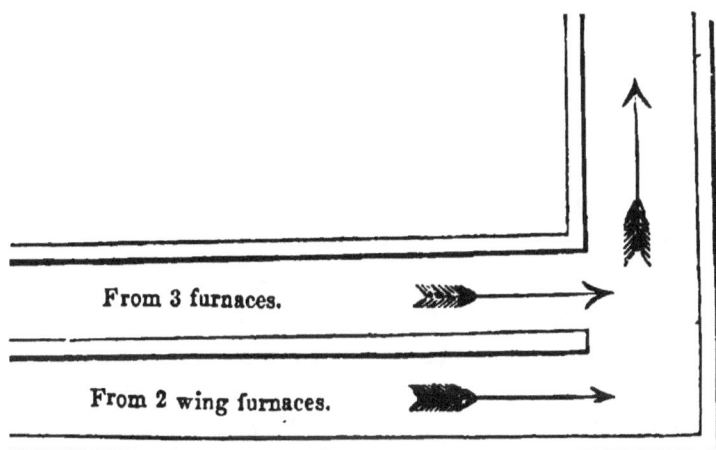

From 3 furnaces.

From 2 wing furnaces.

Fig. 94.—Boiler of the " *Great Liverpool.*"

conceive a more direct case of collision, or a more effectual damper by the hotter and larger current from three furnaces, on the smaller current than the two wing furnaces. In these boilers, it is manifest that nine-tenths of the steam was produced by the plates in connection with *the furnaces alone*, and by a system of continued forcing; the long run of flues being filled with dense black smoke.

A considerable improvement was effected by constructing furnaces *in pairs*. This had the important advantage of rendering any interference with the supply of air unneces-

sary, by giving uniformity to the quantity of gas passing
from the bridge to the flue; since, by firing the two furnaces
alternately, the supply of gas is equalised on entering the flue
from the bridges.

This plan had the disadvantage of *the split flue* at the
back end, where, as M. Péclet observes, *the hotter or
stronger current will always neutralise the other.* In practice, a strong or hot current of gaseous products will not,
voluntarily, divide itself, to meet the arrangements of the
flues : the whole, or nearly so, will pass either to the one
or the other, in proportion to the temperature then in the
flue, or to the length of the course each has to run to the
funnel.

The plan in Fig. 95 was adopted with the view of
dividing the gaseous products, and thus spreading the
heat along a double surface. This, however, was quite
defective, in as much as a gaseous stream cannot be induced
to divide itself contrary to the laws governing the currents
of fluids; the hottest and shortest course being always
taken by the gaseous products.

The plan of a land boiler, as in Fig. 96, is that of a
still more objectionable effort to divide the current into
two smaller flues, with the view of increasing the internal
surface. Here, the flue, after passing under the cylindrical
boiler, and returning through a central flue, is expected to
divide itself into two streams, one to pass on each side of
the boiler, on their way to the chimney. This is the case
referred to by M. Péclet. A commission, he observes, from
the *Société Industrielle* of the Grand Duchy of Hesse, made
a series of experiments to determine the influence of the
circulation of the products of combustion round boilers.
By these it was proved, that the flue passing round the
boiler had a considerable effect on the amount of evaporation. It was also established, that if the products pass
simultaneously by the two side flues, they will *not distribute*

themselves equally, and will only pass by that which represents
the least resistance.

On the *external* circumstances that influence the draught,
M. Péclet adduces many proofs of the importance of avoiding

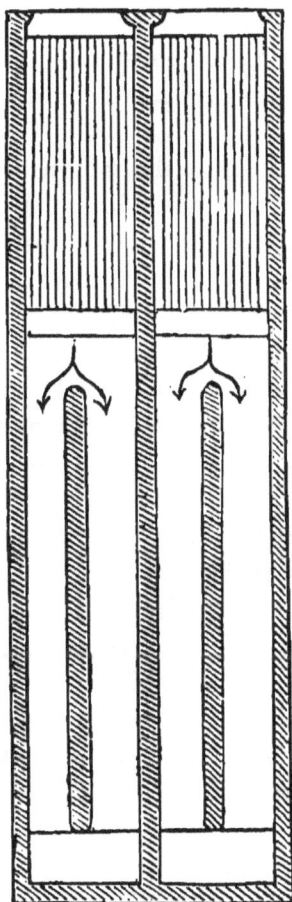

Fig. 95.—Split Draught.

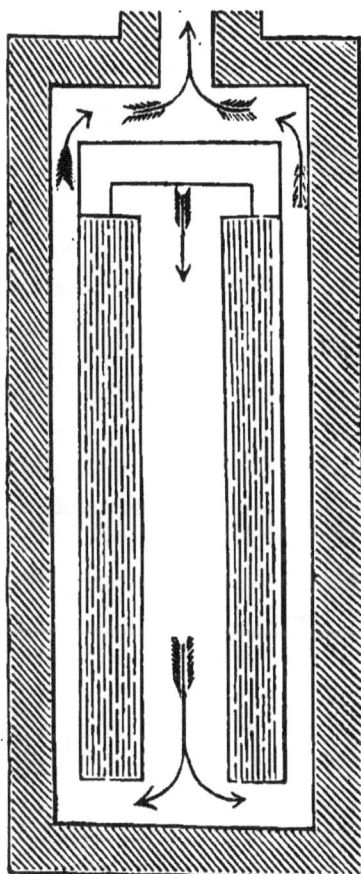

Fig. 96.—Split Draught.

any interference with the introduction of the air, by reason
of the *direction of the wind outside the building.* This is a
circumstance which has excited no attention from our
engineers. In marine boilers, placed low down in the

vessel, the direction of the wind, with reference to that of
the current entering the furnace, has often a considerable
effect on the efficiency of the combustion. So, if the wind
is opposed to the motion of the vessel or the reverse. The
importance of this consideration is exemplified where the
vessel contains two boilers, having their furnaces facing
different ways. In such case, according to the direction of
the wind, or the motion of the vessel, one boiler will have a
better draught than the other.

In a discussion at the Society of Arts, on the subject of
the prevention of smoke, Mr. D. K. Clark " testified to the
advantage of a rapid, or rather, intense draught, in per-
fecting combustion and extinguishing smoke." " This," he
observed, "was the panacea he constantly held forth for the
universal prevention of smoke in large furnaces." Mr.
Clark's views are, unquestionably, well founded ; but the
practical difficulty lies in the obtaining this " intense
draught," or an adequacy of draught for even imperfect
combustion, in many marine boilers.

The absolute *command of draught* for the generation of
the required quantity of steam, to enable the engines to
work to their full power being then so essential, it becomes
a question whether *other means than the natural draught*
should not be resorted to ; since, independently of the uncertainty in the amount of draught, and the consequent irre-
gularity in the working effect of the engines, the cost of
sustaining that draught may be so much in excess of what
an *artificial draught* would be.

This branch of the subject has excited little attention in
this country. M. Péclet has investigated it with his usual
care, and his results are worthy of record. " Where the
draught," he observes, " is created by the expenditure of
fuel and heat, *the expense exceeds one-fourth of the combustible used.* If we have not the means of otherwise
employing that heat, the natural draught, by the chimney,

is then admissible. If, however, that heat may be made
available for the purposes of evaporation, and if a draught,
mechanically obtained, would cost less, it would then be
more advantageous to use it."

Among other proofs, he gives two instances which

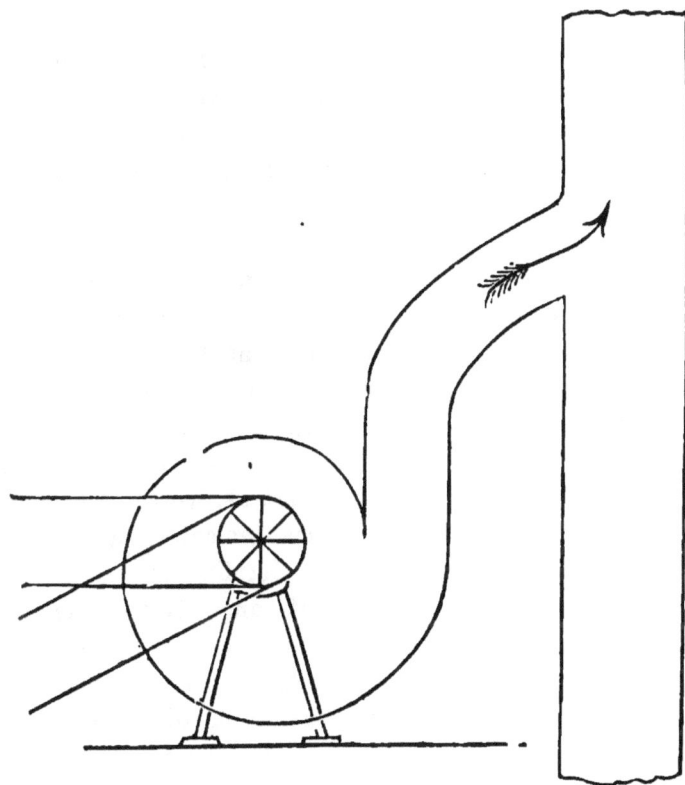

Fig. 97.—Mechanical Draught.

establish the fact. One, the hot baths on the Seine at Paris,
the result of which was, that what was effected by the
labour of one man alone, when the draught was *mechanically*
produced, cost the value of seventeen men's labour when
produced by the *natural draught* from the heat of the
furnace.

The second case was that of a large brewery, where the power employed was that of 200 horses. In this instance, *a ventilator which employed the power of but six horses, was sufficient to produce a draught equal to that of 50 horses, obtained by means of the natural draught of the chimney.*

He then proceeds to consider the relative merits of the several descriptions of ventilators ; and comes to the conclusion, that the rotary fan with plain wings, but with the eccentric motion is to be preferred. The mode recommended for its application is given in Fig. 97, where, by means of *an exhausting fan,* the heated products were directed into the ordinary chimney shaft.

This opens a new field of inquiry, which is far more important than any question arising from the mere relative cost of the *natural* or *mechanical* draught, the best constructed boilers not unfrequently being insufficient, from the mere circumstance of a deficient draught.

Impressed with the importance of the subject, several experiments by means of a fan apparatus, worked by a small steam engine, were made. The arrangement, as described in Fig. 98, also afforded the means of deciding many important points connected with the length of run,—the working temperature in the flue, and that of the escaping products at the chimney. In this Fig. A represents the boiler, 15 feet long, with an upper returning flue, having its own chimney A' furnished with a thermometer C, and a damper D, Houldsworth's pyrometer P being connected with the return flue farthest from the furnace. Two sight apertures were introduced,—the one at S, opposite the furnace, to observe the action of the air, introduced through the door and air-box above it, as already described ; the other S' looking into the upper flue.

To test the practicability of converting the great heat which escaped by the chimney, an auxiliary boiler B was

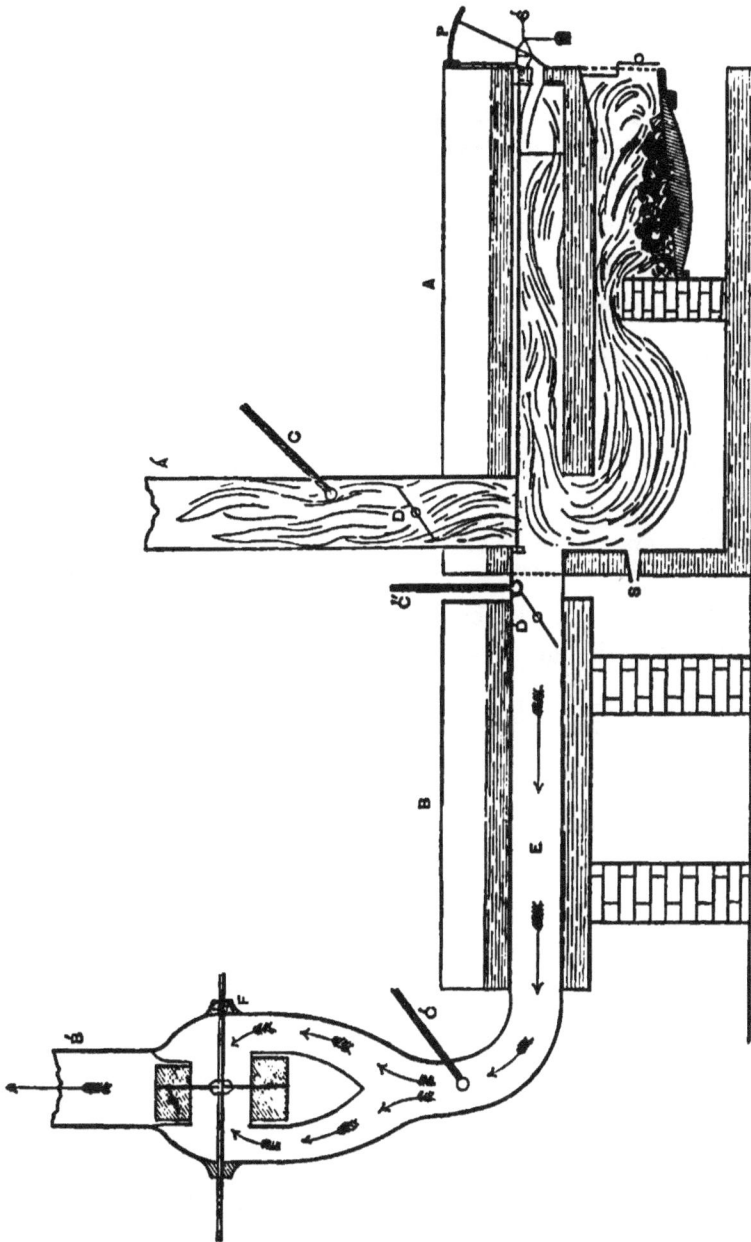

Fig. 96.—Mechanical Draught *versus* Ordinary Draught.

attached, by means of a continuation flue E, and furnished with a separate chimney B, with an exhausting fan F, to produce an increased draught. This auxiliary boiler had two thermometers, to ascertain the temperature of the escaping products, C' and C'', and a damper D', so that the two boilers might be used separately or conjointly.

Experiments with the auxiliary boiler and exhausting fan draught :—

Experiments.	Coal used per hour.	Water evaporated per hour.	Water evaporated per lb. of Coal.	Pyrometer heat in flue.	Temperature of heat escaping.
1 With fan draught.	265 lbs.	2454 lbs.	9·26 lbs.	1025°	650°
2 With ordinary } chimney draught. }	215 lbs.	1552 lbs.	7·21 lbs.	725°	410°

The effect produced by the fan draught was thus not only to increase the evaporative power of the boiler, *within the hour*, from 1552 to 2454 lbs. of water, but to increase the evaporative effect from *each pound of coal used*, from 7·21 lbs. to 9·26 lbs.

CHAPTER XVI.

OF THE TUBULAR SYSTEM AS APPLIED TO MARINE, LAND, AND LOCOMOTIVE BOILERS, IN REFERENCE TO THE CIRCULATION OF THE WATER AND THE PROCESS OF COMBUSTION.

HAVING considered these subjects in reference to *flue* boilers, we have now to examine them in connection with the *tubular* system. The annexed views of a locomotive and a tubular boiler will enable us to appreciate their respective peculiarities.

In the locomotive, Fig. 99, the furnace compartment, called the fire-box, is placed at one end of a long boiler, and so apart from the tubular compartment, that, as regards the objects of circulation and evaporation, they may be considered in the light of separate boilers.

In the marine boiler, as in Figs. 100, on the contrary, the tubes, placed directly over the furnaces, become enveloped in the atmosphere of steam generated, and rising from the latter. This is virtually the placing one boiler over another :—a tubular over a flue boiler, and within the same shell. By this arrangement, the steam generated from the furnace department (and which is necessarily the largest in quantity) cannot reach the surface without passing through the numerous close sets of tubes, and the steam and water which surround them.

So also of the water; it can neither ascend or descend without first working its way through the intricate mazes presented by the tubes:—no more certain method, therefore, could have been devised for producing a mischievous inter-

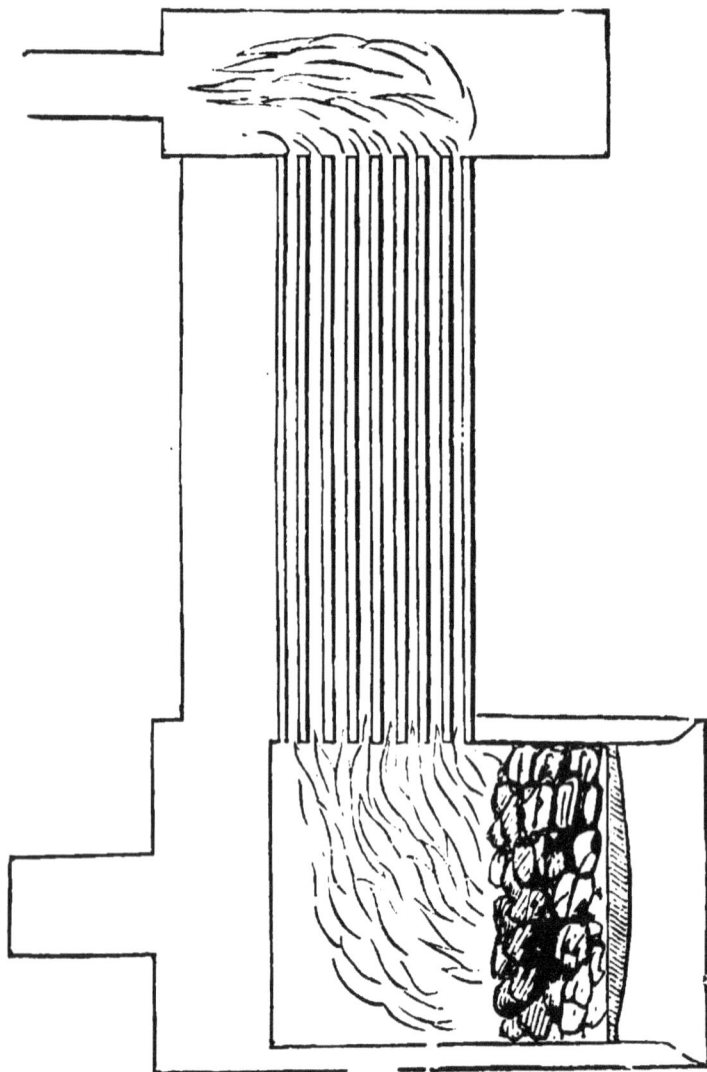

Fig. 99.—Locomotive Boiler.

ference with the respective functions, both of the water and the steam. Here, then, is another violation of the principle, that "no part of the heating surface should be so situated that the steam may not readily rise from it, and escape to the surface of the water."

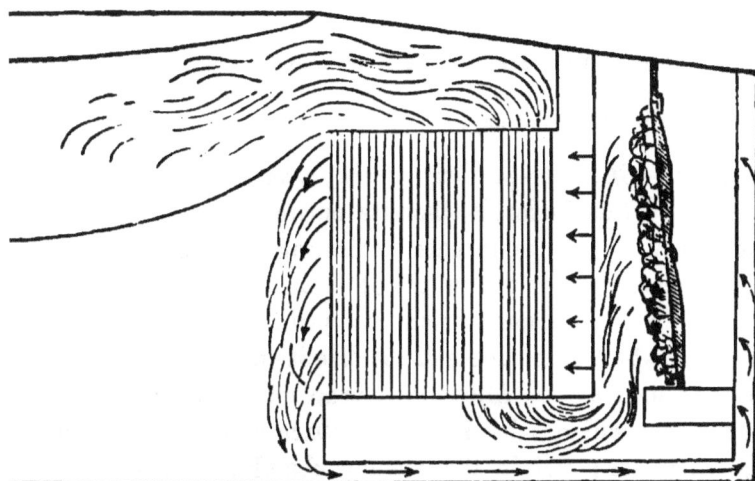

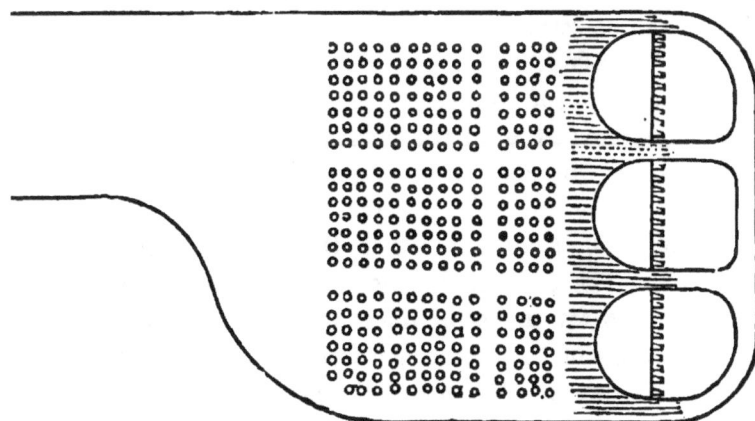

Fig. 100.—Marine Boiler.

In proceeding to consider the marine tubular boiler, it will be well to take one of modern construction, and examine how far its arrangements are in accordance with the operations of nature, and particularly as to the following points, viz. :—

1st.   The proportions of the furnace.
2nd.   The distance the flame and heat have to travel.
3rd.   The time available for giving out heat.
4th.   The admission of the air.
5th.   The circulation of the steam and water.
Cth.   The effects of the tubes, as heating surfaces.
7th.   The cconomy.

First.—*Of the proportions of the furnaces.* A remarkable feature of this boiler is, that its entire area may be considered as one large furnace, indicating the dependence on this department for the supply of steam. As to its several parts, there is disproportion everywhere. The chamber above the fuel is not one-half the size it should be in proportion to its length. Allowing a proper body of fuel on the bars, there will not be 12 inches space between it and the crown plates ; yet in this shallow chamber, and under the influence of a rapid current through it, all the operations of gas making, gas heating, mixing with the air, and combustion, are to be carried on.

So of the ash-pit ; although above 9 feet long, it has but an average depth of 18 inches. Thus the spaces, both above and below the fuel, have the character of long narrow passages, causing injudicious longitudinal currents, and counteracting all the processes of nature.

Second.—Of *the distance* which the flame and heat have to travel. The deficiency is here remarkable, there being, in fact, but a few inches of *lineal run* between the fuel and the first range of the tubes ; say from A to B. It may then be asked, what is to become of the great body of flame and half-burned gases rising from the coal on an area of 20 square feet, the bars of each furnace being 2 feet 6 inches wide, by 8 feet in length ; the six furnaces of each boiler thus presenting an aggregate bar surface of 120 square feet.

The inference is, that the tubular boiler (accompanied by

a short run, with a large area of fire-grate, and using bituminous coal) is wholly incompatible with perfect combus-

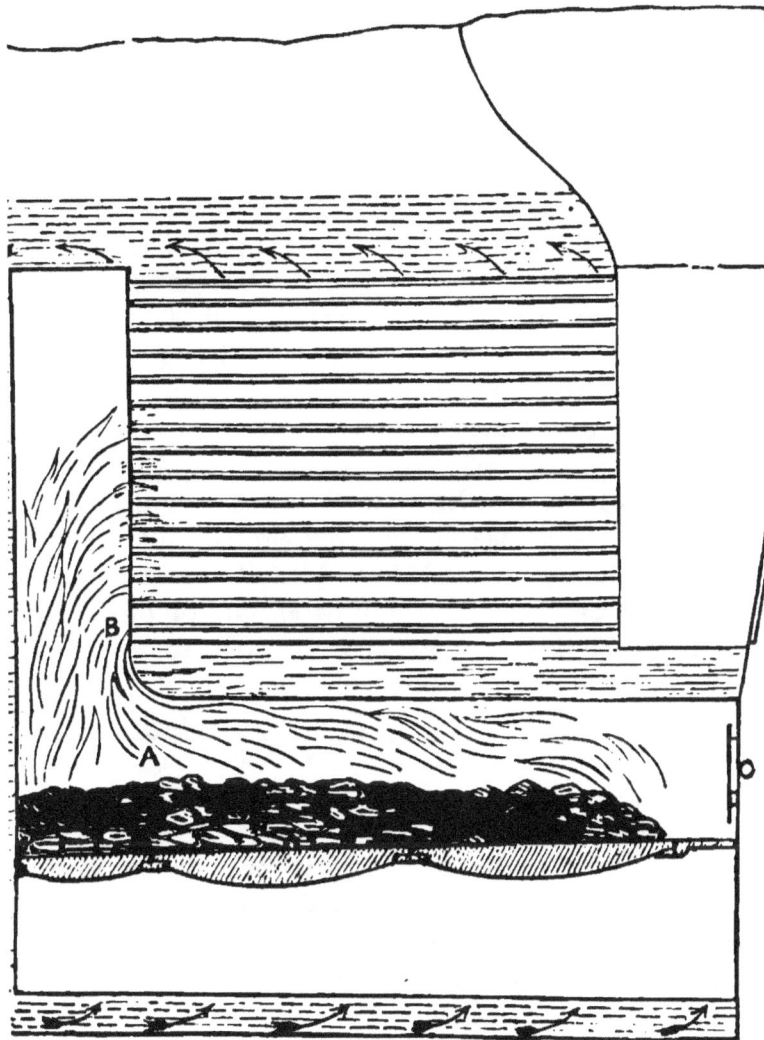

Fig. 101.—Marine Boiler.

tion, and obtaining the due measure of heat from the fuel
and flame.

Third.—*Of the time* allowed for giving out heat. *The*

question of time and distance being so directly connected, whatever influences the one must have a corresponding effect on the other. When, however, we consider that the time or interval that can elapse between the passing the flame from the furnace, and its reaching the tubes, can be but a *small fraction of a second*, such must be wholly insufficient for giving out the heat from so enormous a body of flame.

Fourth.—Of the *admission of air.* No provision whatever is here made, apart from that by the ash-pit. The manifest error of such an arrangement will be best understood by pointing to the utter impossibility of the 300,000 cubic feet of air required for the gas, together with the 600,000 cubic feet for the coke from the *three tons of coal hourly consumed* on the bars, to find access through the body of coal resting on them.

Fifth.—Of the *circulation of the water.* In this boiler nothing has been done in aid of the circulation, but much, on the contrary, to embarrass both its ascending and descending currents. This is even aggravated by the circumstance of the water space, at the back end, which in *flue boilers*, as being the *coldest*, and farthest from the furnaces, and most *favourable for the descending water, is here the hottest;* the great body of flame being projected directly against it at C, and therefore most unfavourable for that descending current.

Sixth.—Of *the tubes, as heating surface.* In the *marine tubular* boiler, where coal is used, and no transparent gaseous products exist, but on the contrary, where there must be always a large volume of fuliginous gas and flame, the tube system is wholly inapplicable.

Seventh.—Of *economy.* In the use of *fuel* in this boiler, *economy is out of the question.* The great supply of steam being generated from the radiated heat and impact of the flame in the furnaces, the system of *forcing the fires becomes an absolute essential;* the greater the weight of coal consumed

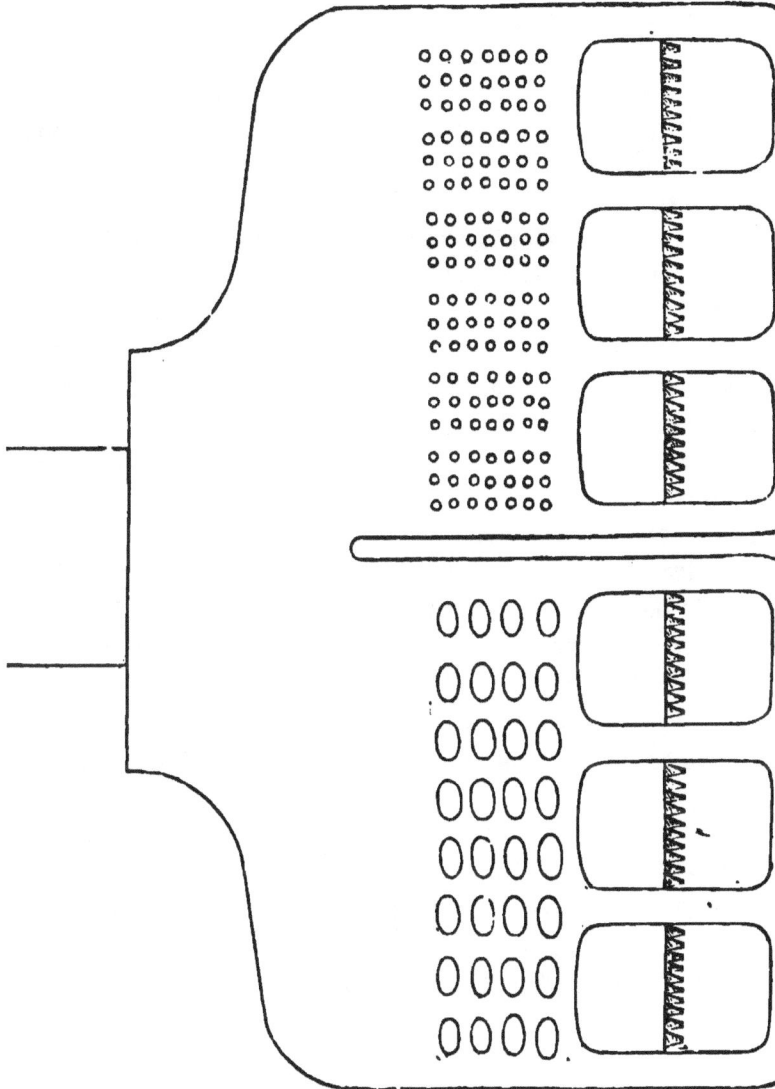

Fig. 102.—Boiler of the "*Leeds.*"

in given times, the greater being the amount of available flame produced.

H

A further disadvantage of the marine tubular boiler is, that there is no place for the deposit of soot, sand, ashes, and other matter which accompanies the use of coal; or the scale or sediment which is found both inside the tubes as well as outside of them, in the narrow water-spaces. The result is, that all this matter accumulates in the tubes and bottom of the smoke-box, requiring constant attention for its removal.

A practical illustration of the disadvantages of small tubes was afforded in the boiler of the steam-vessel, the *Leeds*, in which, for the express purpose of deciding the question, one-half the boiler was constructed with tubes of 3 inches diameter; the other having enlarged tubes of 7 by 5 inches, as shown in Fig. 102.

The result was, that for years no repairs whatever were required in the latter, while the former was a continued source of annoyance and expense; besides that, it was less effective as a steam generator. Many of the small tubes had to be renewed; the water spaces were liable to be filled with incrustation; and the face-plate, in which the tubes were inserted, required to be drawn in, and the tubes again riveted; innumerable patches and additional bolts were from time to time introduced to secure the back face-plate, and keep it in its place.

Here also was a practical confirmation of the fact, that the mere circumstance of having a larger aggregate surface had no effect in producing increased evaporation, the *aggregate surface of the small tubes*, in the one-half of the boiler, being *double that of the larger ones* in the other half.

CHAPTER XVII.

ON THE USE OF HEATED AIR, AND ITS SUPPOSED VALUE IN THE FURNACES OF BOILERS.

Among the devices by which the public have been led astray, may be mentioned, the use of hollow bars, supplemental flues, calorific plates, self-acting valves, double grates, heated tubes, and such like contrivances.

When these so-called inventions came to be examined, it was found that they were incapable of imparting any sensible degree of heat to the great volume of air required. That they were, in fact, but so many proofs of the ingenuity of their respective advocates, and of the ease with which the public may be imposed on ; and that the announcement of a scheme for consuming, or preventing smoke, by the use of hot-air was a mere professional and *ad captandum* averment, based on no principle, justified by no proofs, and supported by no chemical or practical authority.

The idea that there was some undefined value in the use of hot-air, originated in the hot-blast system in the manufacture of iron. The principle or process by which iron may be melted has, however, so little relation to that by which the combustion of the coal gas in furnaces is effected, that no analogy whatever exists between them.

To show in a still stronger point of view the deception practised, either on themselves or on others, it may be observed that it is not to the *coke* or incandescent part of the fuel on the bars that these patentees would apply the

H 2

hot air, (as is done in the iron furnaces,) but to *the gases* in the furnace chamber, where the great disproportion between the relative bulk of the air required, and the gas, is already so obstructive of rapid union and combustion, and one of the great difficulties to be encountered.

With reference to the use of hot-air in boiler furnaces, no inquiry appears to have been made, either as to the temperature to which its advocates would raise it; or even whether, by any of their plans, it would be heated at all. Still there was something so plausible in the enunciation of a plan "for consuming smoke by means of hot air," that it was listened to by many who had no means of investigating its supposed merits, or detecting its fallacy.

Fig. 103.—Air at 32° = 36 grains = 28 nitrogen, 8 oxygen.

The first question for inquiry here is, what would be the effect of heating the air before it would be introduced into the furnace? *Chemically*, no change whatever is effected. *Mechanically*, however, an important change takes place, namely, that its already unwieldy volume is still further increased. Thus, if a cubic foot of air be heated one additional degree, its bulk will be increased $\frac{1}{480}$ part; consequently, if heated by an addition of 480 degrees, its bulk will be doubled.

Let us then see if any effect be produced on its *constituents* by this enlargement of its volume. Let Fig. 103 represent a body of air at the temperature of 32°, and weighing 36 grains, viz., '28 grains of nitrogen, and 8 grains of oxygen; these being the proportions as they exist in the atmosphere.

Again, let Fig. 104 represent *the same weight of air*, heated to the temperature of 82 + 480 = 512°; its bulk being then doubled. Nevertheless, there are still but the same relative weights, viz., 28 grains of nitrogen and 8 grains of oxygen, *and no more.*

Now, as the efficiency of the air in producing combustion and generating heat is not in the proportion of the *bulk*, but of *the weight of oxygen* it contains, nothing has been gained by such increased temperature; while this great practical disadvantage has been incurred—that *double the volume* of air must be introduced into the furnace; and, of

Fig. 104 —Air at 32° = 36 grains = 28 nitrogen, 8 oxygen.

course, double the draught must be obtained before the same quantity of gas can be consumed.

The practical inconvenience of enlarging the volume of the air by heating it is easily illustrated; for if the oxygen of 300,000 cubic feet of air, at atmospheric temperature, be required for combustion of one ton of coal, it would require that of 600,000 cubic feet if raised to 512°—a volume which no *natural draught* would be equal to.

Sir H. Davy says: "By *heating strongly gases* that burn with difficulty, the continued inflammation becomes easy." Thus, as they are more easily inflamed when hot than cold, we have this testimony in favour of heating *the gas* rather than *the air*. With reference to heating the air, and thus expanding it, Sir H. Davy does not appear to have attempted it; but he has done what was more to the point—he tried the effect of *condensing* it. Professor Brande says : " Sir H. Davy found considerable difficulty in making the experiments with precision ; but he ascertained that both the light and heat of the flames of sulphur and hydrogen were *increased in air condensed four times.*" This is decisive against *heating the air*, and in favour rather of *condensing* it.

CHAPTER XVIII.

ON THE INFLUENCE OF THE WATER GENERATED IN FURNACES FROM THE COMBUSTION OF THE HYDROGEN OF THE GAS.

THE fact of the great quantity of water produced admits of no doubt. Bituminous coal, we have seen, contains from 5 to 6 per cent. of hydrogen, and as each pound of hydrogen, in combustion, combines with 8 pounds of atmospheric oxygen, the product is 9 pounds of water. Each hundred-weight of coal, then, containing on an average 5½ pounds of hydrogen, the product will be nearly 50 pounds of water. Thus the gas from each ton weight of coal will produce about half a ton weight of water, *in the form of steam.*

When the coal gas is generated in the furnace, the first operation towards its combustion is the union of its hydrogen with the oxygen from the air, forming water. This chemical union, as already shown, produces that intense heat which raises the other constituent—the carbon, to the temperature of incandescence in the form of bright visible flame. It is this heat which, on being applied to some solid body, as charcoal, or lime, produces the extraordinary luminosity exhibited in the oxy-hydrogen microscope.

By the apparatus shown in the annexed Figure 105, this water may not only be condensed, but collected. It consists of a tin vessel, A, about four feet long, filled with cold water:—the flame of a large gas burner B, and the heated products passing through the flue C, slightly inclined from c to d, to favour the escape of the condensed water of combustion.

The flame and other products of the gas being directed through the flue, the steam will be condensed within it, and

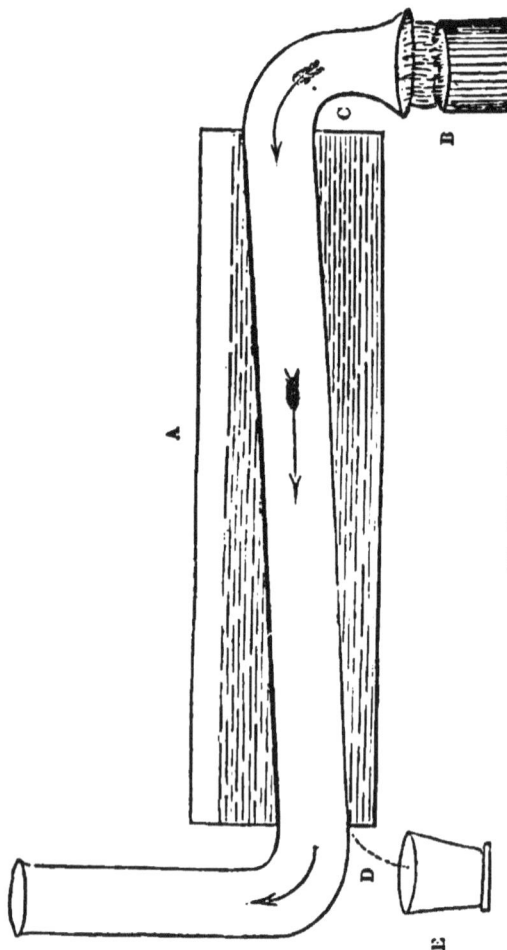

Fig. 105.—Condenser.

the water will continue dropping from the lower end into the vessel E, as long as the flue remains sufficiently cold:— the other products passing off by the funnel, and at a very low temperature.

In boilers on the *flue* system where there is sufficient room, this steam produces no injurious effect, by reason of its having space to *separate itself* from the flame, as rapidly as it is generated.

In the tubular system, however, the injurious influence of this mass of steam is serious and palpable.

CHAPTER XIX.

ON INCREASING THE HEAT-TRANSMITTING POWER OF THE INTERIOR PLATE SURFACE OF BOILERS.

UNDER the most favourable circumstances of boilers, a larger portion of heat will be lost than would be required for merely producing the necessary draught. Experiments were, therefore, made with the view of counteracting this great waste of heat, and which established the fact that it was possible, to a certain extent, to increase the quantity of heat transmitted by any given surface of plate. It is true a plate, 10 feet by 10, equal to 100 square feet, presents the same amount of surface area as one of 100 feet long by one foot wide. As a steam generator, however, the effect would be very different :—the *lineal run* or distance travelled over being as 10 to 1, and occupying *ten seconds* of time in the first, and but *one second* in the other.

When the gaseous products of combustion are carried through flues or tubes, this lineal current passes *at right angles to the line of transmission* of heat through the plate. If, however, we heat one end of a rod of iron, a large conducting power is brought into action, the heat passing *longitudinally* along its fibres. Now this is the power that has been here rendered available.

Independently of the *conducting* power which a metallic pin or rod may have, it possesses then a *receiving* power, greater than is due to its mere diameter.

Suppose an iron or copper pin of half an inch diameter,

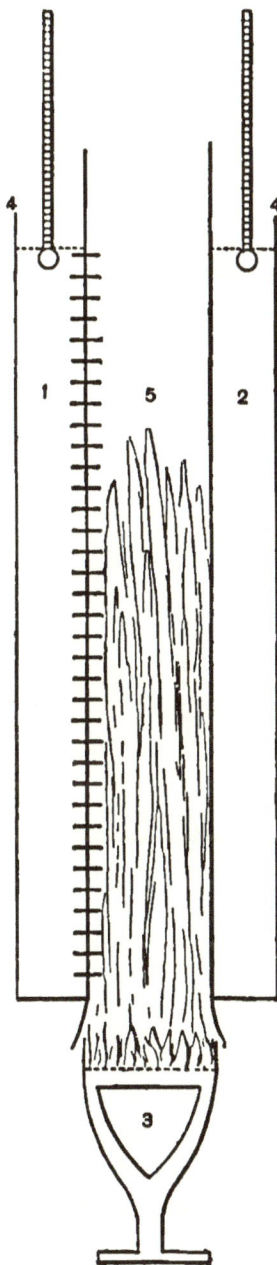

inserted in a plate, and projecting into the flue *three inches beyond its surface, and across the current of the heated products.* In such cases, the portion of such plate occupied by it will be equal to a disk of but half an inch in diameter, while the pin itself will present a heat-*receiving surface* of $4\frac{1}{2}$ inches. By this means we obtain an effective heat-*receiving* surface, nine times greater than the area of the plate which the pin occupied. If, then, a series of these conductors be inserted in the flue and furnace plates, there will be an increased effect from the circumstance of the current of the heated products *being directed against them,* instead of *passing along the surface* of the plate.

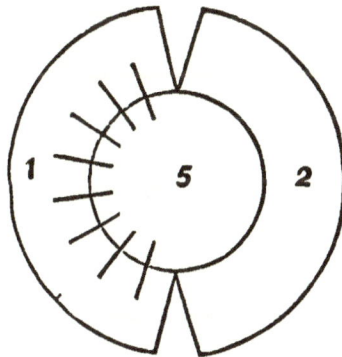

Fig. 107.—Conductor Pins.

Fig. 106.—Conductor Pins.

The popular impression that the three-legged pot boiled sooner than the one without legs, though it

passed as a fable, was, nevertheless, a true one,—the legs
acting the part of heat-conductors. This was tested by
having a large pitch-pot constructed with twenty legs instead
of three—the bottom being thus furnished with so many

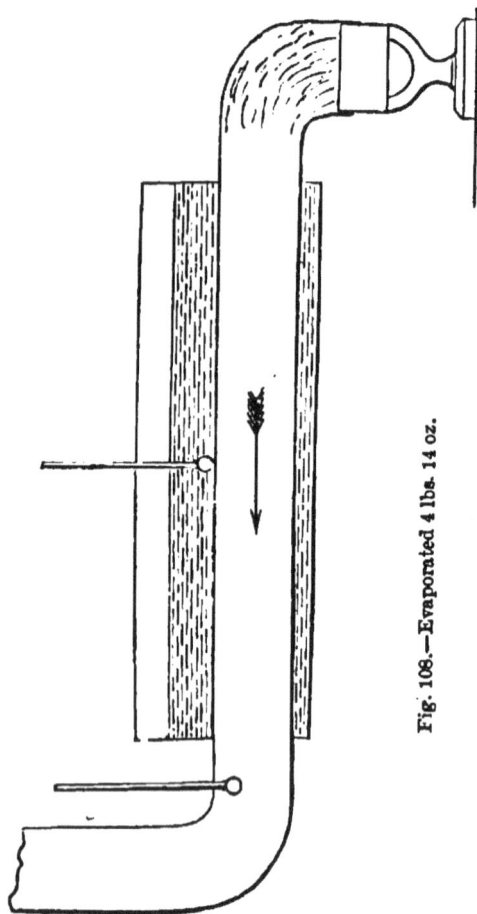

Fig. 108.—Evaporated 4 lbs. 14 oz.

projecting conductors, each six inches long.   The result was,
that pitch or water was more rapidly boiled in this than in
one of the ordinary kind.   The principle of projecting heat
conductors thus was shown to be entitled to attention, and
practically available.

The following experiment is illustrative of the increased evaporative effect produced by the conductor pins. In Fig. 106, 1 and 2 are tin vessels containing the same quantity of cold water, 1 being furnished with the con-

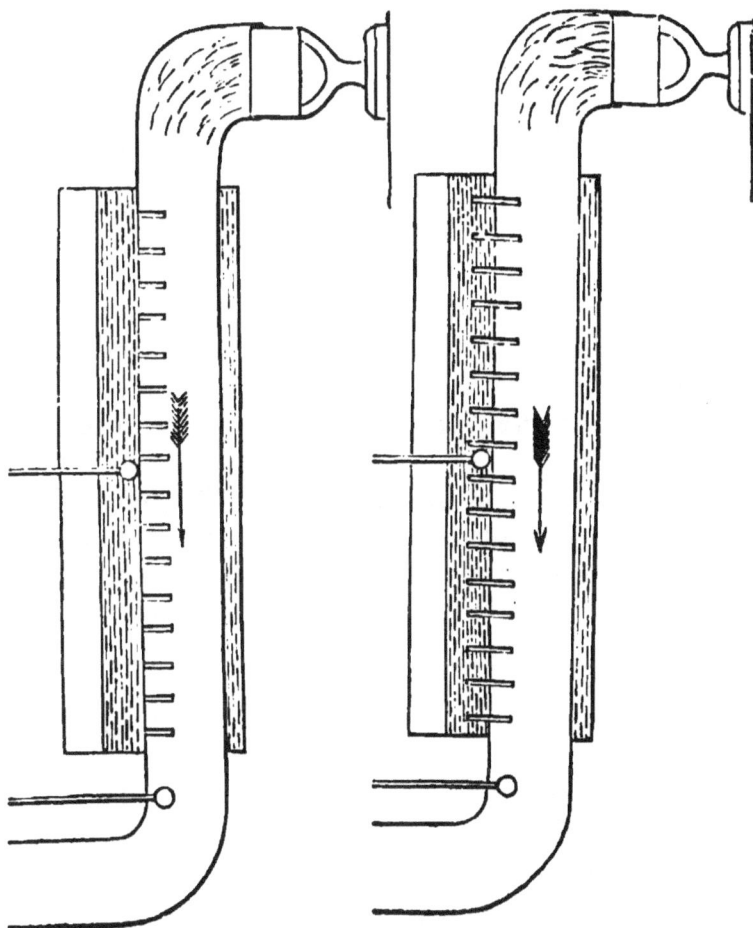

Fig. 109.—Evaporated 7 lbs. 14 oz.   Fig. 110.—Evaporated 8 lbs. 5 oz.

ductors made of ¼ copper wire, and two left plain; a thermometer, 4, being suspended in each: 5 is a vertical flue, through which the products from the flame of a large gas-burner passed. Fig. 107 is a sectional view of the same. The temperature in both vessels was taken in two minutes' time. The following are the progressive rates at which the water was raised to the boiling point.

Pan with Conductors.			Pan without Conductors.				
Initial Temperature, 61°			Initial Temperature, 61°			16 . 163°	
After 2 minutes,	75°		After 2 minutes,	70°		18 . 171°	
„ 4	„	. 95°	„ 4	„	. 82°	20 . 181°	
„ 6	„	. 124°	„ 6	„	. 101°	22 . 188°	
„ 8	„	. 161°	„ 8	„	. 118°	24 . 196°	
„ 10	„	. 177°	„ 10	„	. 130°	26 . 203°	
„ 12	„	. 201°	„ 12	„	. 146°	28 . 210°	
„ 13	„	. 212°	„ 14	„	. 156°	29 . 212°	

Thus it appears that the water in the pan with the heat-conductors, was raised to the temperature of 212° in 13 minutes, while that in the plain pan required 29 minutes.

The following experiment shows still further the value of assisting the evaporative power by the aid of these conductors.

Three tin boilers, as in Figs. 108, 109, 110, were placed in connection with a large laboratory gas burner. In each was put 22 lbs. of water; 30 cubic feet of gas were consumed in each experiment, in two hours and forty minutes. The result was as follows :—

		lbs.	oz.	
Fig. 108, without conductors, evaporated		4	14	of water.
109, with conductors on one side only		7	14	„
110, with conductors projecting on both sides	. . . .	8	5	„

The quantity of gas consumed was the same in both cases,—the heat generated was the same,—the area of flue plate was the same,—the difference in effect was therefore alone produced by the greater quantity of *heat transmitted to the water, longitudinally, through the conductors.*

In this case, the heat conveyed to the water, and that escaping by the funnel, showed that where the waste heat was greatest, the evaporative power was necessarily the least.

Fig. 108. PAN WITHOUT CONDUCTORS.

Gas consumed.	Heat of Water.	Heat escaping.
. . . .	58	62
5 feet	120	382
10	152	390
15	162	395
20	164	396
25	166	402
30	166	406
	988	2432

Evaporated 4 lbs. 14 ounces.

Fig. 109. PAN WITH CONDUCTORS PROJECTING ON ONE SIDE.

Gas consumed.	Heat of Water.	Heat escaping.
. . . .	58	62
5 feet	143	257
10	160	280
15	172	385
20	178	392
25	186	300
30	188	320
	1085	1996

Evaporated 7 lbs. 13 ounces.

Fig. 110. PAN WITH CONDUCTORS PROJECTING ON BOTH SIDES.

Gas consumed.	Heat of Water.	Heat escaping.
. . . .	58	62
5 feet	152	248
10	174	273
15	178	276
20	182	278
25	186	282
30	188	284
	1110	1703

Evaporated 8 lbs. 5 ounces.

The comparison of the three pans then stands thus :—

		Heat retained.	Heat lost.
1	Pans without conductors	988	2432
2	„ single conductors	1085	1996
3	„ double conductors	1110	1703

Conductor pins were applied to the boilers of a six-horse engine. The result was, that each inch deep of water, which previously required twenty-eight minutes to evaporate, was, by means of the conductors, done in twenty-one minutes.*

Encouraged by these results, conductor pins have been introduced into many marine and land boilers with unquestionable success. After many years of observation as to their durability, the conclusion is, that a projection into the flues of three inches is the most advisable. If longer, they will burn away to about that length.

Supposing the conductor to be made of half-inch rods, and inserted at intervals of three inches, the strength of the plate has been tested, and found to be rather improved, the conductor pins apparently acting the part of floor bridging, and giving increased stiffness.

Supposing the area of the flue to be two feet square, then the introduction of the pin conductors may be as shown in the annexed Fig. 111.†

As illustrative of the mode and extent to which the system has been practically applied, Fig. 112 and Fig. 118

* Dr. Ure, impressed with the same view, made some experiments with corrugated plates. The effect, he observes (see his Dictionary of Arts), was remarkable : the water evaporated when the current of heated products passed across the corrugations, and, as it were, striking against them; being so much greater than when it ran in the same direction. On the same principle, the heat transmitted was increased when the current of the products was intercepted by the conductors.

† The principle of these conductor pins has been adopted in sugar boiler pans, and other descriptions of evaporative vessels ; and would no doubt be applicable to the operations of brewing and distilling.

represent a plan and section of the boilers of " *The Royal William*." These boilers were in constant use for nine years, and with the most perfect success as regards economy of fuel,—freedom from the smoke nuisance,—evaporative power, and durability ; the number of conductor pins was 4859.

The plan of the boiler described as Lamb and Summers' patent, may here be given, inasmuch as one of its peculiarities is connected with the use of the heat conductors, and precisely corresponding with the description given by M.

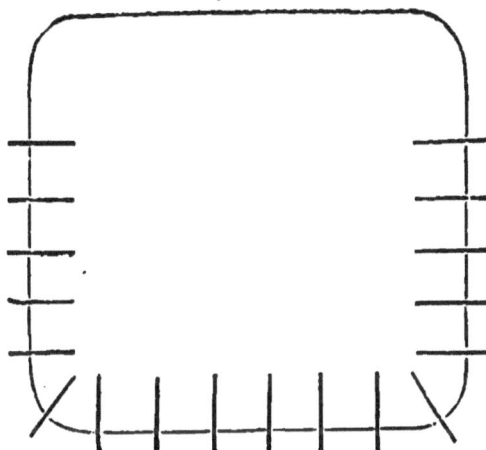

Fig. 111.—Conductor Pins.

Péclet,—the flue being "*traversed by metallic bars*," and which here act the double purpose of *stays* (as in locomotive boilers) and *heat conductors*.

In the boilers of the Peninsular and Oriental Steam Company's Ship *Pacha*, as in Fig. 114, the stays which act the important part of double heat conductors are of ⅞th-inch iron : of these there are 1920, and as they act effectually on the water spaces on each side, do the duty of 3840 most effective heat conductors.

The Patentees state that the superiority of this plan over

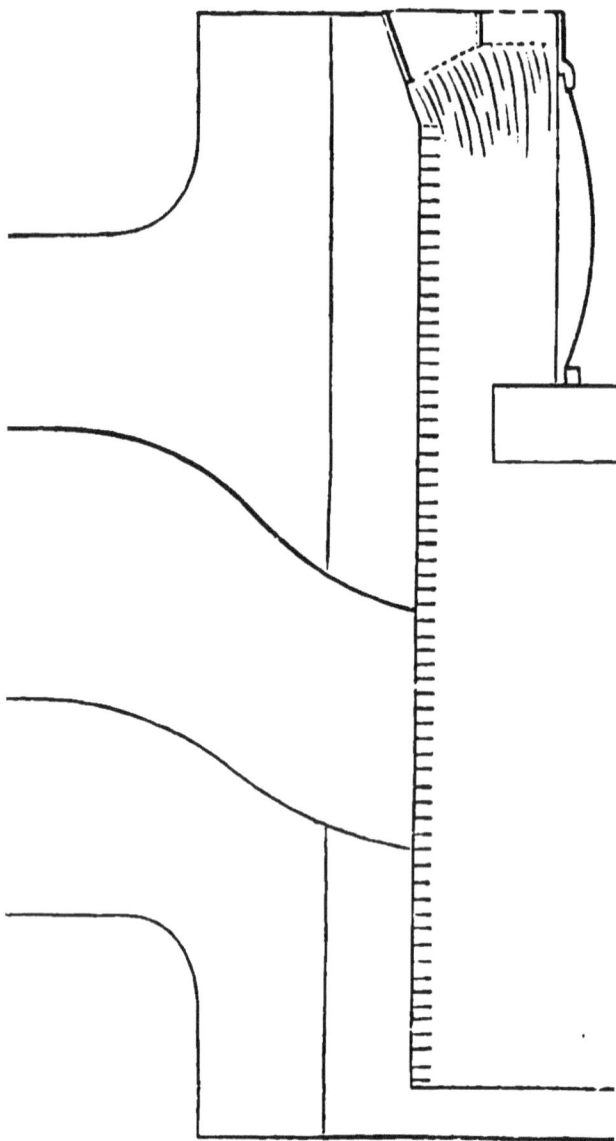

Fig. 112.—Boiler of the "*Royal William.*"

the common tubular "consists in the facility for cleaning;

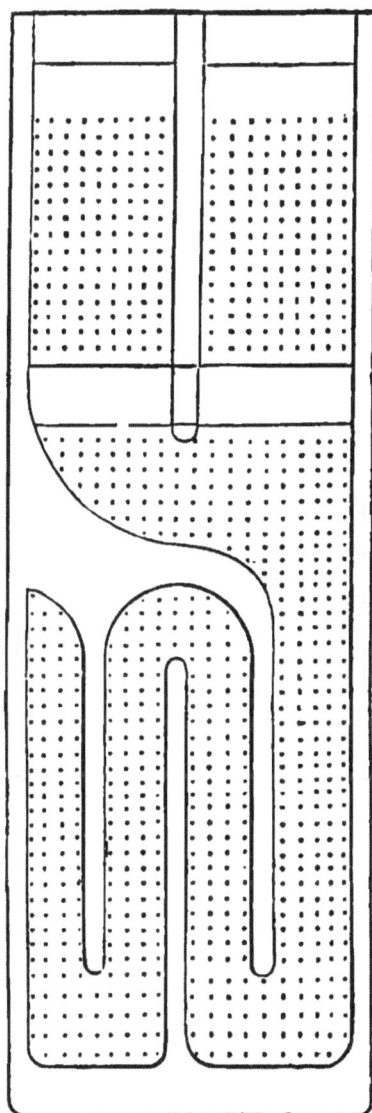

Fig. 113.—Boiler of the " *Royal William.*"

that is, for the removal of the scale or deposit which takes
place so largely in the boilers of sea-going vessels. The

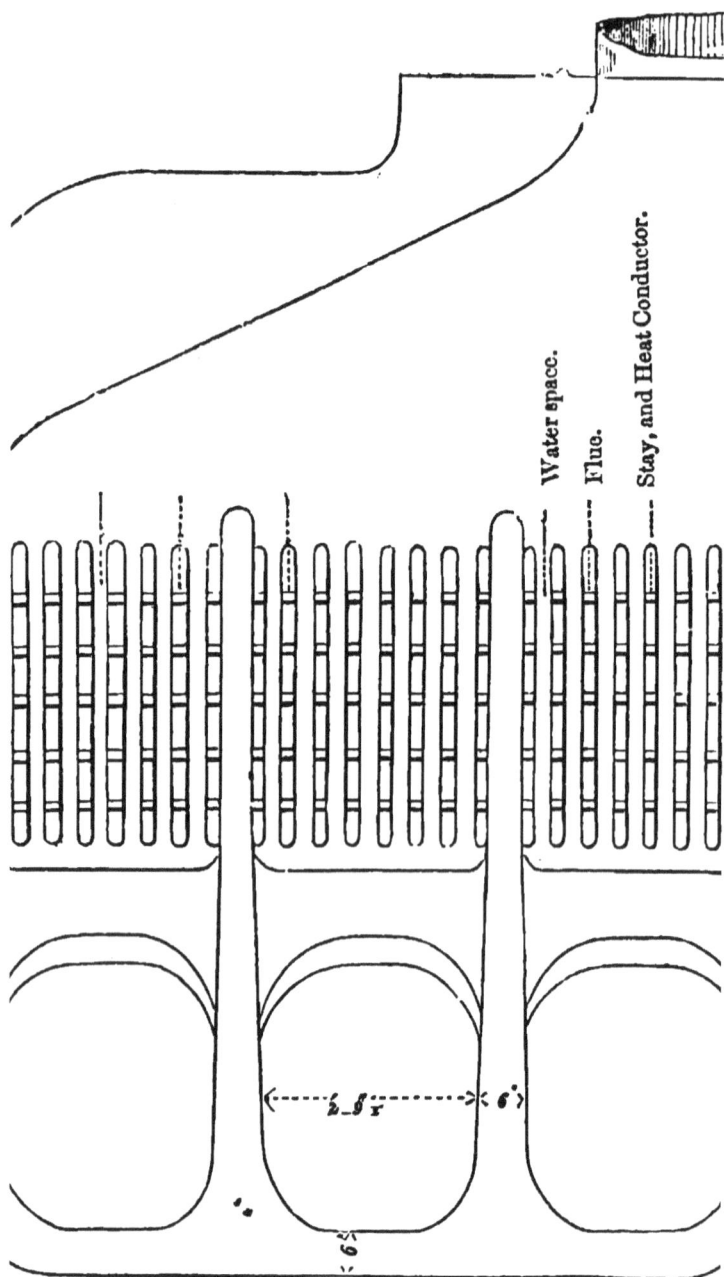

Fig. 114.—Lamb and Summers' Boiler.

vertical water spaces of these boilers afford an easy means of cleaning the sides of the flues, and so enable the water to come in contact with the iron flue. That in tubular boilers the horizontal position of the boiler spaces between the tubes renders it an impossibility to clean them; the consequence of which is, that a constant succession of deposit takes place. The flues of boilers which had been constantly at work for several years, presented no appearance of deterioration."

The principle of these heat conductors is too self-evident to avoid adoption hereafter in all descriptions of vessels where heat has to be communicated, or abstracted.

CHAPTER XX.

ON THE GENERATION AND CHARACTERISTICS OF SMOKE.

So much has been said and credited on the subject of the *burning* and *combustion* and even *consumption*, of smoke ; and it has been so often asked, *What is smoke?* that the subject cannot here be dismissed without comment.

When we see a dark yellow vapour rising from heated coal, as at the mouth of a retort, or from a furnace, or domestic fire, after fresh coal has been thrown on, this colour is not occasioned by the presence of *carbon*, but is caused by the sulphur, tar, or earthly impurities which might happen to be in the coal. All these are subsequently separated from the carburetted hydrogen in the purifying process—the gas remaining transparent—so minute are the several atoms of the carbon, and so diffused are they when in connection with the hydrogen. That the *solid carbon is there*, notwithstanding this transparency, is proved by its subsequent liberation; as when a polished body is thrust into the flame of a candle or gas jet, and brought out with a deposit of the carbon on it. Carbon, in fact, when in chemical union with gaseous matter, is always invisible and intangible. The following experiment will sufficiently illustrate these facts, exhibiting both the *gas* and the *smoke* in their separate states of existence, and with their separate characteristics.

Fig. 115 represents a tin vessel *a*, capable of holding a

quart measure; in it was placed some small coal, resin, and tar, to produce a quick and large development of gas. The

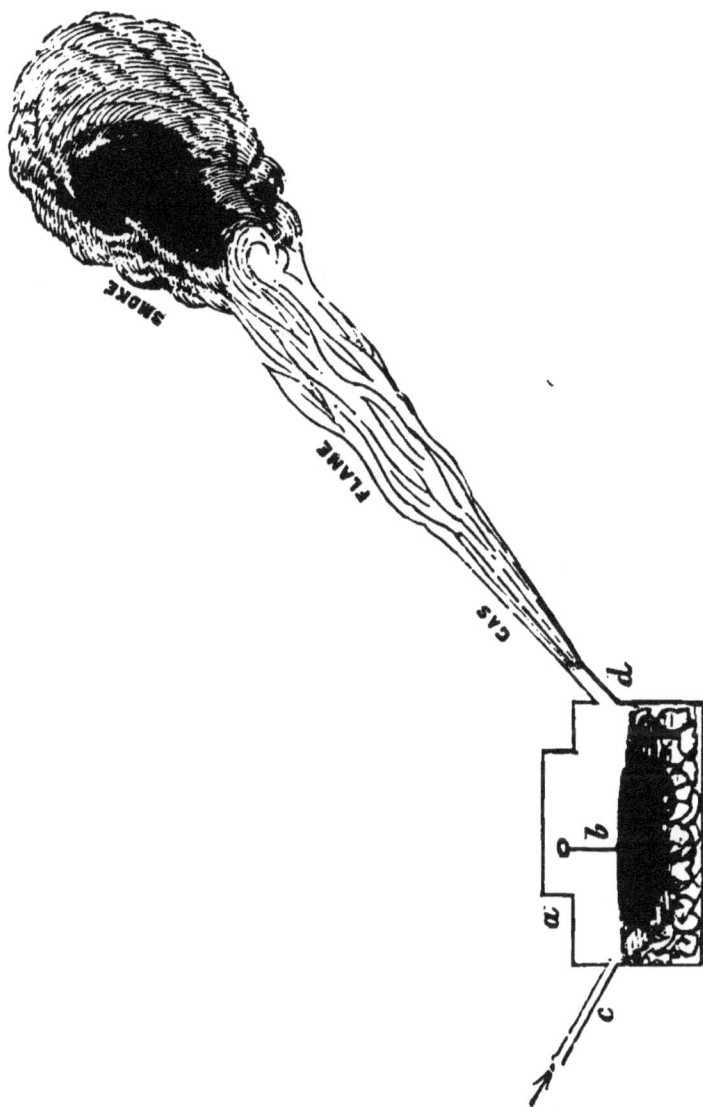

Fig. 115.—Gas, Flame, Smoke.

lid being removed, an iron, *b*, made red-hot, was introduced, and the vessel again close covered. A small tube is then inserted at c, to be blown into, as into a blowpipe, to expel the gas in a stream.

By blowing through this tube, a copious volume of the gas will issue from the nozzle d. That the carbon in this gas is inaccessible, is proved by presenting a sheet of paper to the stream, and, although it may be slightly stained, if there be much tar present, no carbon will be deposited.

On this stream of gas, many inches long, being lighted, a lurid flame will be produced, but which, becoming cooled down before it can be sufficiently mixed with the air, produces a large volume of true smoke. Here, then, is exhibited *the gas, the flame,* and *the smoke,* at the same moment, and in succession, just as they are produced in the furnace,—the gas being converted into flame, and the flame into smoke.

It may be well here to notice an error with which we are generally impressed, namely, that the cloudy volume of smoke, as we see it issuing from a chimney, and filling a large space in the atmosphere, is formed of carbonaceous matter. This black cloud is merely the great mass of *steam,* or watery vapour, formed in the furnace, as already described, but *coloured by the carbon;* and when we consider, that no less than half a ton weight of water (*in the expanded form of steam*) is produced from every ton weight of bituminous coal consumed, we can easily account for the enormous volume and mass of this *blackened vapour* called smoke, as it appears to our vision, and the palpable error of supposing that this cloud of incombustible matter was capable of being consumed, or converted to the purposes of heat.

Were it not for this mass of steam the carbon would soon fall, as a cloud of black dust; but, being intimately and atomically mixed with the large volume of steam from the furnace, it is carried along by the atmosphere, only differing in colour, like the cloud of steam we see issuing from the chimney of a locomotive when in action.

CHAPTER XXI.

CONCLUDING REMARKS.

A CONSIDERATION of the nature of the products into which the combustible constituents of coal are converted in passing through the furnace and flues of a boiler, will enable us to correct many of the practical errors of the day, and ascertain the amount of useful effect produced, and waste incurred. These products are:

1st. Steam—highly rarefied, invisible, and incombustible.

2nd. Carbonic acid—invisible and incombustible.

3rd. Carbonic oxide—invisible, but combustible.

4th. Smoke—visible, partly combustible, and partly in-combustible.

The fourth—smoke—is formed from such portions of the hydrogen and carbon of the coal-gas as have not been supplied or combined with oxygen, and, consequently, have not been converted either into steam or carbonic acid.

The hydrogen so passing away is transparent and invisible; not so, however, the carbon, which on being so separated from the hydrogen, loses its gaseous character, and returns to its natural and elementary state of a black, pulverulent, and finely-divided body. As such, it becomes *visible*, and this it is which gives the dark colour to smoke.

Suppose the equivalent of air to be supplied in the proper manner to the gas, namely, by jets, for in this respect the operation is the same as if we were supplying gas to the air,

I

as in the Argand gas-lamp. In such case one-half of the

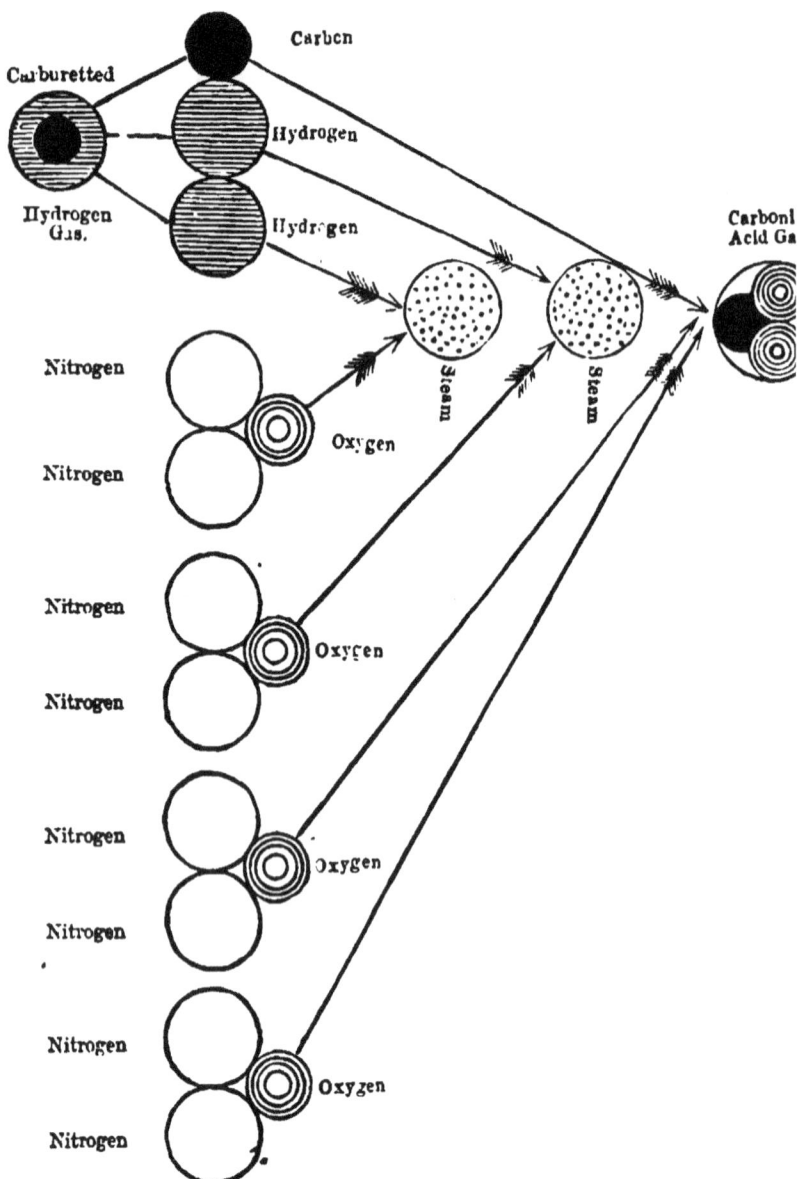

Fig. 116.—Combustion of Carburetted Hydrogen.

oxygen absorbed goes to form steam, by its union with the

hydrogen; while the other half forms carbonic acid, by its union with the carbon. Both constituents being thus supplied with their equivalent volumes of the supporter, the process would here be complete—perfect combustion would ensue, and no smoke be formed; the quantity of air employed being *ten times the volume of the gas consumed.* See Fig. 116.

Again, suppose that but one-half, or any other quantity, *less* than the saturating equivalent of air were supplied. In such case, the hydrogen, whose affinity for oxygen is so superior to that of carbon, would seize on the greater part of this limited supply: while the carbon, losing its connection with the hydrogen, and not being supplied with oxygen, would assume its original black, solid, pulverulent state, and become *true smoke.* The quantity of smoke then would be in proportion to the deficiency of air supplied.

But smoke may be caused by an *excess* as well as a *deficiency* in the supply of air. If the gas be injudiciously supplied with air, that is, by larger quantities or larger jets than their respective equivalent number of atoms can *immediately combine with,* as they come into contact, a *cooling effect* is necessarily produced instead of a *generation of heat.* The result of this would be, that, although the quantity of air might be correct, the second condition, the required temperature, would be sacrificed or impaired, the union with the oxygen of the air would not take place, and smoke would be formed.

And now as to the relative *quantities* of the several constituents of smoke: 1st, of the *invisible nitrogen.* As atmospheric air contains but 20 per cent. of oxygen, the remaining 80 per cent. being the *nitrogen,* passes away, invisible and uncombined. If, then, a ton of coal requires, absolutely, for its combustion the oxygen of 300,000 cubic feet of air, the 80 per cent., or 240,000 cubic feet of invisible and incombustible nitrogen, forms the first ingredient of this black cloud. 2nd, of the *invisible carbonic acid*. This portion of the cloud may be estimated as equal in volume to the 20 per cent. of oxygen which had effected the combustion of the carbon *both of the gas and the coke* of the coal. 3rd, of the *invisible steam* formed by the combustion of the hydrogen of the gas. In this will be found the great source of the prevailing misapprehension; yet no facts in chemistry are more accurately defined than those which belong to the formation, weight, and volume of the constituents of steam.

The next consideration is, as to the *value of the carbon* which produces the darkened colour of the smoke cloud. Now, the weight of this carbon, in a cubic foot of black smoke, is not equal to that of a *single grain*. Of the extraordinary light-absorbing property and colouring effect produced by the inappreciable myriads of atoms of this finely-divided carbon, forming part of the cloud of the steam alone, some idea may be formed by *artificially* mixing some of it when in the deposited state of soot with water. For this purpose, collect it on a metallic plate held over a candle or gas-jet, and touching the flame. Let a *single grain weight* of this soot be gradually and intimately mixed on a pallet, as a painter would, with a pallet-knife : first, with a few drops of gum-water, enlarging the quantity until it amounts to a spoonful. On this mixture being poured into a glass globe containing a gallon of water, the whole mass, on being stirred, will become opaque, and of the colour of *ink*. Here we have physical demonstration of the extraordinary colouring effect of the minutely divided carbon—*a single grain* weight being sufficient to give the dark colour to a gallon of water. Whatever then may be the quantity or number of its atoms, we see from the cloud of incombustible matter with which this carbon is so intimately associated, *as smoke*, that even attempting its separation and collection, indedendently of its combustion, borders on absurdity.

Finally, we see that the cause of the formation of smoke is, the absence of the proper supply of air to the combustion of the gas (the only combustible that it contains), *at the time when from its. high temperature of incandescence* it was best fitted to receive it.

APPENDIX.

EXTRACTS FROM THE SECOND REPORT OF MESSRS. LONGRIDGE, ARMSTRONG, AND RICHARDSON TO THE STEAM COAL COLLIERIES' ASSOCIATION, NEW-CASTLE-UPON-TYNE.

GENTLEMEN,—

In submitting to you our further Report upon the question which you have referred to our decision, we have to observe, that it would have been easy for us to have selected and submitted to trial certain of the competitors' plans, and to have reported to you on their comparative merits at a much earlier period. But such a course would neither have done justice to you nor to the important question which we had to decide, inasmuch as one of the principal conditions established for the competition was, that the plans submitted should not diminish the evaporative power of the boiler.

It was, therefore, our first object to ascertain this evaporative power as a standard of reference.

The boiler built for these experiments presented no peculiar features. The annexed drawing will show that it was the ordinary type of a marine multitubular boiler, such as is generally considered to present the greatest difficulty as regards the prevention of smoke.

It contained two furnaces, each three feet wide, and 135 tubes $5\frac{1}{2}$ feet long and three inches internal diameter, and had an aggregate heating surface of 749 square feet.

The heater, which was subsequently added, as mentioned in the ninth paragraph of our former Report, was used for the purpose of heating the feed water. It in no way altered the condition of the boiler, except by reducing the temperature of the escaping gases, and thereby, to some extent, diminishing the draught and rendering the prevention of smoke somewhat more difficult, whilst, at the same time, it slightly increased the evaporative effect by its additional absorbing surface.

This increase was, however, much less than might have been expected from the large absorbing surface of the heater, which contained 320 square feet; yet it was found that, when the products of combustion before entering the heater were at 600°, the passage through it did not reduce the temperature more than about 40° to 50°.

The whole of the experiments with the competitors' plans were made with the boiler after the heater was added, as also were those made previously for establishing the standard of reference.

We have established as the standard the means of a series of experiments during which the firing was conducted according to the ordinary system, every care, however, being taken to get the maximum of work out of the boiler by keeping the fire-grates clean and by frequent stoking. No air was admitted except through the fire-grates, and as a consequence much, and often a very dense smoke was evolved.

As the economic effect of the fuel increases when the ratio of the fire-grate surface to the absorbing surface is diminished, we have adopted two sizes of fire-grates, and consequently two standards of reference. With the larger fire-grate the amount of work done by the boiler per hour is greatest, but this is done at a relative loss of economic value of the fuel as compared with the smaller grate.

The one gives us the standard of maximum evaporative power of boiler,—the other the standard of maximum economic effect of the fuel.

The fire surfaces used for fixing those standards were 28½ and 19½ square feet respectively.

Each competitor was allowed to vary his fire-grate to meet these two standards, and in the tabulated forms hereinafter given, the results obtained are compared with the standards as well as with the maximum results which we have arrived at in our own experiments.

With these prefatory remarks we now proceed with our Report.

The total number of plans submitted to us was 103, which, upon examination, we found might be arranged in the following classes:—

1st Class.—Requiring no special apparatus, and depending upon the admission of *cold* air into the furnace or at the bridge.

2nd Class.—Requiring no special apparatus, and depending upon admission of *hot* air into the furnace or at the bridge.

3rd Class.—Requiring special adaptations of the furnace of more or less complexity, but yet applicable to the ordinary type of marine boiler. The most of this class admitting air above the fire-grate surface.

4th Class.—Requiring self-acting or mechanical apparatus for supplying the fuel.

5th Class.—The smoke burning systems, the principle of which is to pass the products of combustion through or over a mass of incandescent fuel. This class might be subdivided into two, in one of which the gases pass downward through a part of the fire-grate into a close ash-pit, and thence to the flame chamber or tubes, and in the other the gases, &c., from one furnace are passed into the ash-pit and upwards through the fire-grate of another furnace, and in which arrangement the process is alternated by a system of doors or dampers.

6th Class.—Proposing the admission of steam mixed with the air into the furnace as a means both of preventing smoke and increasing the evaporative effect of the fuel.

7th Class.—Such projects as are either impracticable or not applicable to the ordinary type of marine boilers, and consequently not in accordance with the established conditions.

The following table shows the number of plans sent in, arranged in the above classes :—

Class 1	.	.	.	.	.	.	9
„ 2	.	.	.	.	.	.	16
„ 3	.	.	.	.	.	.	15
„ 4	.	.	.	.	.	.	6
„ 5	.	.	.	.	.	.	12
„ 6	.	.	.	.	.	.	1
„ 7	.	.	.	.	.	.	44
							103

After full consideration we selected the following plans for trial at your expense :—

From Class 1.—Messrs. Hobson and Hopkinson, Huddersfield.
Mr. C. W. Williams, Liverpool.
Mr. B. Stoney, Dublin.

From Class 3.—Mr. Robson, of South Shields.

We did not feel ourselves justified in trying any of the other plans at your expense, but in acquainting the remaining competitors with our decision, we stated that we were ready to submit their plans also for trial if they desired it, in conformity with the fifth paragraph of the original advertisement. None of these parties, however, availed themselves of the opportunity thus given of testing their plans at their own expense.

The standard of reference alluded to in the 14th and 15th paragraphs of the present Report are as follows:—

	Fire-grate 28¼ Square Feet.		Fire-grate 19¼ Square Feet.	
	A.	B.	A.	B.
Economic value, or lbs. of water evaporated from 212° by 1 lb. of coal ..	9·41	11·15	10·06	12·58
Rate of combustion, or lbs. of coal burned per hour per square foot of fire-grate	21·15	19·00	21·00	17·25
Rate of evaporation per square foot of fire-grate per hour in cubic feet of water from 60°	2·62	2·93	2·909	2·995
Total evaporation per hour in cubic feet of water from 60°	74·80	79·12	56·01	57·78

The columns A contain the standards of reference alluded to, as above, whilst the columns B give the *mean of the best* results ob'ained by our own experiments *when making no smoke.*

The first plan submitted for trial was that of Mr. Robson, of Shields, which we selected as a type of several of the plans comprised in Class 3, and as in our opinion the most likely of its kind to prove successful. The principle of this plan is to divide the furnace into two fire-grates, the one at the back being shorter than the other, and placed at a lower level. This back grate is furnished with a regular door-frame and door, for the purpose of enabling the stoker to clean the bars and remove the clinker when required. This door is also provided with an aperture fitted with a throttle valve, and in the inside a distributing box perforated with half-inch holes, after the manner practised by Mr. Wye Williams. The front grate is like the ordinary fire-grate, but without any bridge. The mode of proceeding is to throw all the fresh coal upon the front grate, and to keep the back or lower grate supplied with cinders, or partially coked coal, which is pushed on to it from time to time from the upper or front grate. No air is admitted at the door of the upper grate, but the gases arising from it meet with the current of fresh air admitted

through the door of the lower grate, and in passing over the bright
fire upon it are to a greater or less degree consumed.

With respect to absence of smoke, we have to report that this plan
is only partially successful. It diminishes the amount of smoke considerably, but it requires careful and minute attention from the
stoker, otherwise a good deal of smoke at times appears, and particularly when fresh fuel is pushed forward from the upper to the lower
grate.

Mr. Robson's fire-grate surface was 32½ square feet.

As regards economic value of fuel and work done, the following was
the result :—

Economic value of fuel . . . . 10·70 lbs.
Rate of combustion . . . . . 15·50 „
Rate of evaporation per square foot per hour   2·14 cubic feet.
Total evaporation from 60° ditto . . 70·50   „

Comparing these results with the standard, we get

	Robson.	Standard.	More.	Less.
			Per Cent.	Per Cent.
Area of fire-grate	32·50	28·50	14·03	...
Economic value of fuel . .	10·70	9·41	13·7	...
Rate of combustion . .	15·52	21·15	...	26·7
Rate of evaporation . . .	2·14	2·62	...	18·4
Total evaporation. . . .	70·50	74·80	...	5·8

From this it appears that though there was an increase of economic
value of fuel to the extent of 13·7 per cent., there was a loss of work
done by the boiler to the extent of 5·8 per cent., and this, although
the fire-grate was greater by four square feet, or 14 per cent.

This result may be traced to the nature of the apparatus. Owing
to the large admission of air at the fire-door of the lower or back
grate requisite to prevent smoke, the fuel on the front grate burns
sluggishly, and hence the falling off in the rate of combustion and the
work done.

The heat in the back grate was very intense, but the generation of
heat being thus thrown nearer to the tubes, the effect of the absorbing
surface above the front grate was greatly impaired.

We think also that the very intense heat in the back grate would be more injurious to the boiler and the tubes than the more equally distributed temperature which results from the ordinary description of fire-grate.

Another objection to this system is the constant attention required from the stoker, to keep the fires in order, and the difficulty in removing the clinker from the back grate, where it tends to form in considerable quantity.

The next plan submitted to trial was that of Messrs. Hobson and Hopkinson, of Huddersfield. In this system air is admitted both at the doors and at the bridge. At the doors by means of vertical slits, which may be opened or shut at will by a sliding shutter, and at the bridge through apertures in hollow brick pillars placed immediately behind it. The entrance of the air to these pillars is regulated by throttle valves, worked by a lever in the ash-pit. There are also masses of brickwork placed in the flame-chamber, with the intention partly of deflecting the currents of gases, so as to ensure their mixture with the air, and partly to equalise the temperature.

As regards prevention of smoke, we have to report that this plan was very efficient, though in hard firing it required considerable attention from the stoker. Whilst burning about 15 lbs. of coal per square foot of grate per hour, *no smoke* was visible, even with ordinary firing, but when the quantity was increased to 21½ lbs. per square foot per hour, the fire required to be very carefully attended to, or smoke, though in no great quantity, began to appear.

Messrs. Hobson and Hopkinson's fire-grate surface was originally 27½ square feet, but this was subsequently reduced to 18¼ square feet.

As regards economic effect and work done, the following were the results :—

	Fire Grate, 27½ Sq. Feet.	Fire Grate, 18¼ Sq. Feet.
	lbs.	lbs.
Economic value of fuel	11·08	11·70
Rate of combustion	14·25	21·50
Rate of evaporation per square foot per hour from 60°	Cubic Feet. 2·18	Cubic Feet. 3·49
Total evaporation from 60°	60·03	63·62

Comparing these results with the standards, we get—

LARGE FIRE-GRATES.				
	Hobson and Hopkinson.	Standard.	More.	Less.
Area of fire-grate .	Feet. 27·5	Feet. 28·5	Per Cent. ...	Per Cent. 3·7
Economic value .	lbs. 11·08	lbs. 9·45	17·1	...
Rate of combustion	14·25	21·15	...	32·7
Rate of evaporation	Cubic Feet. 2·18	Cubic Feet. 2·62	...	16·8
Total evaporation .	60·03	74·80	...	19·8

SMALL FIRE-GRATES.				
	Hobson and Hopkinson.	Standard.	More.	Less.
Area of fire-grate .	Feet. 18·25	Feet. 19·25	Per Cent. ...	Per Cent. 5·2
Economic value .	lbs. 11·70	lbs. 10·06	16·3	...
Rate of combustion	21·50	21·00	2·3	...
Rate of evaporation	Cubic Feet. 3·49	Cubic Feet. 2·909	19·9	...
Total evap. from 60°	63·62	56·01	13·5	...

From these tables it appears that with the large fire-grate there was an increase of economic value of fuel, although less work was done: whilst with the small grates there was a decided increase both of economic value and of work. Had the fires been harder pushed with the large grate, we have no reason to doubt that, although the economic value would have been somewhat less, the work done would have been up to the standard.

The only objection to this system is that the brickwork is liable to crack and get out of repair; but we do not attach much importance to this, as we believe that the existence of this brickwork is of no consequence, and that the results obtained are due simply to the admission of air to the gases.

The system is applicable to all the usual forms of boilers, the combustion is very good, and, with moderate firing, it does not much depend upon the stoker, and we are therefore of opinion that it complies with all the prescribed conditions.

The next plan tried was that of Mr. C. Wye Williams, of Liverpool.

Mr. Williams' system, as is well known, consists in the admission of air at the furnace door, or at the bridge, or at both, by numerous small apertures, with the intention of diffusing it in streams and jets amongst the gases. In the plan adopted in the present instance, Mr. Williams introduces the air only at the front of the furnace, by means of cast iron casings, furnished on the outside with apertures provided with shutters, so as to vary the area at will, and perforated in the inside with a great number of half-inch holes. The mode of firing which Mr. Williams adopts merely consists in applying the fresh fuel alternately at opposite sides of the furnace, so as to leave one side bright whilst the other is black.

The original fire-grate proposed by Mr. Williams was 22 square feet, which was subsequently reduced to 18 square feet.

As regards economy of fuel and work done, the following were the results :—

	Fire Grate, 22 Sq. Feet.	Fire Grate, 18 Sq. Feet.
Economic value of fuel	lbs. 10·84	lbs. 11·30
Rate of combustion	26·98	27·36
Rate of evaporation	Cubic Feet. 4·04	Cubic Feet. 4·31
Total evaporation	88·96	76·92

Comparing these results with the standards, we get—

LARGE FIRE-GRATES.				
	Williams.	Standard.	More.	Less.
Area of fire-grate . .	Feet. 22·0 lbs.	Feet. 28·5 lbs.	Per Cent. ...	Per Cent. 24
Economic val. of fuel	10·84	9·45	11 5	...
Rate of combustion .	26·98	21·15	27·4	...
Rate of evaporation .	Cubic Feet. 4·04	Cubic Feet. 2·62	54·2	...
Total evaporation .	88·96	74·80	19	...

SMALL FIRE-GRATES.				
	Williams.	Standard.	More.	Less.
Area of fire-grate .	Feet. 18·00 lbs.	Feet. 19·25 lbs.	Per Cent. ...	Per Cent. 6·5
Economic value . .	11·30	10·06	12·3	...
Rate of combustion .	27·36	21·00	30·3	...
Rate of evaporation.	Cubic Feet. 4·31	Cubic Feet. 2·909	48·0	...
Total evaporation .	76·92	56·01	37·3	...

These results show a large increase above the standard in every respect.

The prevention of smoke was, we may say, practically perfect, whether the fuel burned was 15 lbs. or 27 lbs. per square foot per hour. Indeed, in one experiment, we burned the extraordinary quantity of 37½ lbs. of coal per square foot per hour upon a grate of 15½ square feet, giving a rate of evaporation of 5½ cubic feet of water per hour per square foot of fire-grate, without producing smoke.

No particular attention was required from the stoker; in fact, in this respect, the system leaves nothing to desire, and the actual labour is even less than that of the ordinary mode of firing.

Mr. Williams' system is applicable to all descriptions of marine boilers, and its extreme simplicity is a great point in its favour.

It fully complies with all the prescribed conditions.

The next and last plan submitted to trial was that of Mr. B. Stoney, of Dublin.

In principle, so far as regards the prevention of smoke by the admission of air through the doors, and at the front of the furnace, this plan is identical with that of Mr. Williams. Its peculiarity consists in the adoption of a shelf outside the boiler, forming, in fact, a continuation of the dead plate outwards. Upon this shelf the fresh charge of coals is laid in a large heap, about half of the heap being within the furnace, and the rest outside. The door is a sliding frame, which shuts down upon the top of this heap of coals, so that air is admitted through the body of the coals as well as through perforations in the front plate of the furnace. When the furnace requires fresh fuel, a portion of that forming the heap, and which, to some extent, has parted with its gases, is pushed forward and its place made up by fresh fuel laid on in front.

This plan did not succeed in preventing smoke, for whenever the coal was pushed forward upon the fire, dense smoke was evolved.

We regret that Mr. Stoney was not personally present to see the result, which we think would have entirely satisfied him that the method he proposed did not comply with this important condition. Under these circumstances, we did not proceed to determine the economic value of the fuel or work done by this system.

In the following tables the results in each case are compared with the standards, and also with those of our own experiments when making no smoke. The former marked A and the latter B.

	A. Standard	B. Our experiment	Robson.	Hobson and Hopkinson.	Williams
	sq. feet.	sq. feet.	sq. feet.	sq. feet.	sq. feet.
Area of grate, square feet	28½	28½	32½	27½	22
Economic value of fuel or water evaporated from 212° by 1 lb.. . . .	lbs. 9·41	lbs. 11·15	lbs. 10·27	lbs. 11·08	lbs. 10·84
Rate of combustion per square foot of grate per hour	21·15	19·00	15·52	14·25	26·98
Rate of evaporation per square foot of grate per hour from 69° . . .	c. feet. 2·62	c. feet. 2·93	c. feet. 2·14	c. feet. 2·18	c. feet 4·04
Total evaporation in cubic feet per hour from 60°	74·80	79·12	69·52	60·03	88·96

LARGE FIRE-GRATES.

SMALL FIRE-GRATES.					
	A. Standard	B. Our ex- periment	Robson.	Hobson and Hop- kinson.	Williams
Area of Grate	sq. feet. 19¼	sq. feet. 19¼		sq. feet. 18¼	sq. feet. 18
Economic value of fuel or water evaporated from 212° by 1 lb. of coal .	lbs. 10·06	lbs. 12·58	Small grate not tried.	lbs. 11·70	lbs. 11·30
Rate of combustion per square foot of grate per hour	21·00	17·25		21·50	27·36
Rate of evaporation per square foot of grate per hour	c. feet. 2·909	c. feet 2·995		c. feet. 3·49	c. feet. 4·31
Total evap. per hour . .	56·01	57·78		63·62	76·92

With the above results before us, we are unanimously of opinion
that Mr. Williams must be declared the successful competitor, and we
therefore award to him the premium of £500 which you offered by
your advertisement of 10th May, 1855.

It is true that in economic value of fuel the *tabulated* results of Mr.
Williams' trial are about 2 per cent. inferior to those of Messrs. Hobson
and Hopkinson, but on the other hand the amount of work done is
much greater.

By Mr. Williams' plan the quantity of water evaporated with a
22 feet grate was 48 per cent. greater than with the 27 feet grate used
in Messrs. Hobson and Hopkinson's case, and 20 per cent. more with
an 18 feet grate.

We should also mention that, in an experiment not tabulated, Mr.
Williams obtained an economic value of 11·70, and a total evaporation
of 61·59 cubic feet, with a 22 feet fire-grate, results which exceed those
of Messrs. Hobson and Hopkinson's experiments, with 27½ feet fire-
grate, and equal in economic value of fuel their results with 18 feet
fire-grate.

An important feature in Mr. Williams' system is that it may be
successfully applied under very varied circumstances. We have above
given results obtained with fire-grates of 22 square feet and 18 square
feet; but in order to test the matter still further, we reduced the fire-
grate to 15½ square feet, with the following result :—

Area of fire-grate   .   .   .   .   .   15½ sq. feet.
Economic value of fuel   .   .   .   .   10·66 lbs.

Rate of combustion per square foot of grate
per hour . . . . . . 37·4 lbs.
Rate of evaporation per square foot of grate
per hour . . . . . . 6·51 c. feet.
Total evaporation per hour . . . 85·30 „

The results which we ourselves attained exceed, in economic value
of fuel, all the results of the experiments made with the competitors'
plans. This was chiefly the case with the small fire-grates, and was
due in a great degree, if not altogether, to the smaller amount of fuel
burned per square foot of grate per hour.

The consequence of this was a more complete *absorption* of the heat
generated, so that the products of combustion escaped from the
chimney at a temperature lower by about 200° when we obtained our
best economic results, than they did during the trials of the competitors' plans. It must be remembered that this increase in the
economic value of the fuel is obtained at the expense of the work
done, but it is highly satisfactory to find that (as is shown in columns
A and B of the last tables), *the great increase in the economic value is also
accompanied with a decided increase in work done when perfect combustion
is attained and smoke prevented.*

Before concluding we might offer some further observations upon
the results we have obtained, and on various interesting and important
questions which have presented themselves during the course of our
inquiries, but to do so in a manner at all satisfactory would be impossible within the limits of a Report like the present.

We must, therefore, content ourselves with pointing out three chief
conclusions at which we have arrived, and which, we believe, will prove
of great advantage as well to your interests as to those of all connected with steam navigation.

*1st.—That by an easy method of firing, combined with a due admission
of air in front of the furnace, and a proper arrangement of fire-grate, the
emission of smoke may be effectually prevented in ordinary marine multitubular boilers whilst using the steam coals of the "Hartley District" of
Northumberland.*

*2nd.—That the prevention of smoke increases the economic value of the
fuel and the evaporative power of the boiler.*

*3rd.—That the coals from the Hartley District have an evaporative
power fully equal to the best Welsh steam coals, and that practically, as
regards steam navigation, they are decidedly superior.*

This last conclusion is contrary to the general opinion, which, based
upon the Reports presented to Government by Sir H. de la Beche and
Dr. Lyon Playfair, is strongly in favour of Welsh coal.

The effect of those Reports has been to do the Northumberland

coal-field an immense injury and we feel this so strongly that we beg
to lay before you a few observations on the subject in a short supplementary Report accompanying this.

We cannot conclude this Report without bringing to your notice the
services of Mr. William Reed, to whom we entrusted the practical
management of the long series of experiments which we deemed it
right to make.

To his intelligence and unwearied attention we are much indebted,
and we can only add that we have every reason to congratulate ourselves and you upon having had the benefit of his valuable assistance
throughout the whole of this long and important inquiry.

We have the honour to be, Gentlemen,
Your obedient Servants,

JAS. A. LONGRIDGE,
18, Abingdon Street, Westminster.

W. G. ARMSTRONG,
Newcastle-on-Tyne.

THOMAS RICHARDSON,
Newcastle-on-Tyne.

NEWCASTLE-ON-TYNE, 16th January, 1858.

PART II.

ON ECONOMY OF FUEL.

BY T. SYMES PRIDEAUX.

ON ECONOMY OF FUEL.

INTRODUCTION.

To produce rapid combustion and intense heat in a furnace, it is necessary for the fuel to be rapidly supplied with air. A current of air must in fact be kept constantly rushing through it, a desideratum which may be effected in two ways, either by allowing the products of combustion to pass into the chimney, at a sufficiently high temperature to produce by their rarefaction a partial vacuum adequate to cause the requisite current of air through the fuel by atmospheric pressure, or by compressing the air by some *mechanical appliance*, and forcing it through. Thus *sucking* the air through the fuel in the first instance, and *blowing* it through in the latter; producing the current in one case by a partial *vacuum behind*, in the other by a *compression before*, the fuel.

Now, however compatible with the objects sought to be attained, and allowable as a question of economy, may be the plan of keeping up the draught through a fire, solely by the instrumentality of a chimney, where *slow* combustion only is required, as in the case of the domestic grate, or the Cornish steam boiler; whenever, on the contrary, rapid combustion and intense heat are a desideratum, such a system can only be carried out, and the air discharged by the chimney at a temperature sufficiently high to produce the powerful draught required, by an enormous sacrifice of fuel.

The result of the employment of this wasteful and unscientific system is seen in the fact, that in steam engines

where an intense draught is employed, the consumption of
fuel is 25 per cent. more than that of those where the products of combustion are reduced to a comparatively low
temperature before they enter the stack. In fact, unless
working with compressed air be resorted to, we cannot have
multum in parvo in a steam engine—*great power in little
space*, without great waste of heat. To burn a large quantity of fuel in a small area, we must have a rapid rush of air
through the furnace,—to produce a rapid rush of air through
the furnace, we must have a powerful draught in the stack,
—to attain a powerful draught in the stack, we must discharge the products of combustion into the stack at a high
temperature,—and to discharge the products of combustion
into the stack at a high temperature, necessarily entails an
enormous waste of fuel—the actual, though very unsatisfac-
tory result at present. Enclosed in a vicious circle, each
step in the process is necessary and inevitable, as long as so
erroneous a path is pursued, and there is no avenue of
escape, but by having recourse to a blowing machine, and
producing the required current of air, with $\frac{1}{100}$ part the fuel
at present wasted in the chimney, in effecting less efficiently
the same object.

In no case is the wasteful result of trusting solely to
draught in the *stack*, to create draught through the *fuel*,
more striking, than in furnaces for the manufacture of iron.
Here, from the iron lying at the bottom of the furnace out
of the axis of the line of draught, and the impossibility of
causing the heat to circulate round it as in a steam boiler,
only a small fractional part of the heat generated enters the
iron. In fact, it is impossible to be otherwise, as the stack,
acting the part of a suction pump through an orifice 1¼ foot
in area, empties the furnace of its gaseous contents *twice in
a second*, keeping up at the same time a state of exhaustion,
which draws in cold atmospheric air at every crack and
cranny, and particularly in puddling furnaces at the working

hole ; oxydising and wasting, or, as the workmen say, *cutting* the iron. In the principal axis of the draught, the products of combustion dart in a straight course from fire bridge to flue bridge, at the rate of 30 feet per second, and can, as a matter of course, leave but a small portion of their heat behind them. If we contrast with such a state of things, that of a furnace *distended* with heat impelled by pressure from behind, and struggling to escape in all directions faster than a *comparatively* contracted neck will allow of, and suffering moreover no cold air to enter, it will be easily seen with which system the superiority lies.

Injudicious as the method of working furnaces on which I have animadverted must appear to my readers, yet they are, I suspect, scarcely prepared to learn the actual results : viz., that in a melting furnace, the amount of fuel consumed is adequate to produce twelve times as much heat, as would, if all absorbed by the metal, raise its temperature from 60° to the melting point ; whilst in puddling furnaces the disproportion is still greater, the heat generated being, if the fuel employed be taken as an index, 16-fold more than can be contained in the metal during any period of the operation.

Now I am not so visionary as to suppose that a furnace can be constructed so as to make all the heat generated enter the iron. It is impossible to isolate the body of heat contained in a furnace and prevent its loss by radiation, and where processes requiring intense heat are carried on, from the necessity that exists for keeping up a rapid supply of fresh heat to replenish this loss, and enable the requisite temperature to be maintained, discharging also at the same time from the other end of the furnace, the *slightly cooled*, though *still intensely hot* gaseous products—the consumption of fuel must ever be very great, in proportion to the actual amount of heat taken up. In short, we must never lose sight of the fact, that the quantity of *available* heat generated for any process, has no relation to the *whole* quantity generated, but is merely the amount of the *excess* of the temperature produced, above the temperature required. Fully weighing these considerations, however, I am nevertheless still sanguine enough to believe, that the results obtained in the puddling and heating furnaces, may be effected with a great diminution in the quantity of fuel now consumed ; and notwithstanding it ever must be necessary in order to keep up the requisite heat in the body of iron furnaces, to discharge the products of combustion whilst still at a very high temperature, yet at the same time, there is nothing to prevent our economising this heat, by applying it to various purposes for which its temperature, though no longer equal to the manufacture of iron, is amply sufficient.

The preceding observations must have made it apparent, that processes of manufacture requiring intense heat, and where, as a necessary corollary, the products of combustion must be discharged at a high temperature, can never be economically conducted, unless some employment be found for the waste heat so discharged. In an iron work, none can exist more advantageous and convenient than the generation of steam for the machinery employed.

A vast field for economy is here open, by the introduction of the *system of working with compressed air ;* for without this invaluable assistant, our power to employ the waste heat is limited to a comparatively narrow compass, inasmuch as precisely in the ratio that our arrangements for this purpose become more perfect, do we destroy the draught of the furnace. By the employment of compressed air, however, to furnish a current for the supply of our fires—and by this means *only*—do we possess the power of using the waste heat at pleasure, however high may be the temperature demanded, and however keen the draught through the fuel required.

I confess, I did at the outset delude myself so far as to

suppose that the notorious fact—that furnaces work better with a breeze blowing into the ash-pit, and also when the barometer is *high*, than when it is *low*—would have secured a favourable opinion on behalf of a proposal for merely further carrying out these conditions, and making them *constant* by *artificial* means. Even with those whose knowledge did not enable them to see very deeply into the matter, the analogy was, I thought, too obvious to be overlooked : observation has taught me, however, that where the mind possesses no fixed scientific principles for its guidance, every process is surrounded with mystery, and every proposal for change invested with doubt. Incapable of separating the *accidental* from the *necessary*, no *à priori* convictions are possible.

CHAPTER I.

ON THE BEST MEANS OF RENDERING COMBUSTION PERFECT.

Since combustion, in the *ordinary* acceptation of the word, is the only means had recourse to in the arts for the development of artificial heat, *perfect combustion* may, for our purpose, be defined to be—the combination of a combustible body with the largest measure of oxygen with which it is capable of uniting. In fact, for all practical purposes, the fuel or combustible body employed may be regarded as composed exclusively of carbon and hydrogen.

Assuming 100 lbs. of coal to contain 80 lbs. of carbon and 5 lbs. of hydrogen, since the oxygen is to the carbon, in carbonic acid, as 16 to 6, to effect perfect combustion, 80 lbs. of carbon will require 313⅓ lbs. = 2,527 cubic feet of oxygen, to furnish which, 967·26 lbs. = 12,635 cubic feet of atmospheric air will be required, air consisting of 1 volume of oxygen to 4 of nitrogen, or 8 parts by weight of the former, to 28 parts of the latter; and since oxygen is to hydrogen, in water, as 8 to 1, 5 lbs. of hydrogen will require 40 lbs. = 473 cubic feet of oxygen, or 181·5 lbs. = 2,365 cubic feet of atmospheric air.

967·26 lbs. + 181·5 = 1148·76 lbs. = 15,000 cubic feet of atmospheric air, required for the perfect combustion of 100 lbs. of coal.

And the product resulting will be : 2,527 cubic feet of

carbonic acid, 946 cubic feet of steam, and 12,000 cubic feet of uncombined nitrogen.

We thus perceive that each 1 lb. of coal requires 150 cubic feet of air for its perfect combustion, or, in other words, for the conversion of all its carbon into carbonic acid, and all its hydrogen into water.

It is commonly, but erroneously supposed, that when no smoke appears at the chimney top, combustion is perfect. Smoke, however, may be absent, and yet the carbon may only have united with 1 atom of oxygen, forming carbonic oxide (a colourless gas), instead of with 2 atoms, forming carbonic acid, and consequently have only performed half the duty, as a fuel, of which it was capable, whilst the loss of duty on the coal taken as a whole (supposing all its hydrogen to have become oxydised) will be upwards of 40 per cent.

Hydrogen, having a stronger affinity than carbon in the gaseous state, for oxygen, when the supply is short, still seizes on its equivalent, and leaves the carbon minus. Thus, when coal gas (carburetted hydrogen) is inflamed with an insufficient supply of air to effect the perfect combustion of both its constituents, the hydrogen is still converted into water, whilst the carbon, in different proportions according to the oxygen present, becomes—deposited in the form of soot —converted into carbonic oxide,—or partly into carbonic oxide and partly into carbonic acid.

Where a fresh supply of coal is put on a briskly-burning fire, the first thing which takes place is, that the coal softens and swells, attended with the evolution of a large quantity of carburetted hydrogen gas, requiring for its combustion a corresponding large supply* of atmospheric air. A furnace immediately after a fresh supply of fuel, requires more than double the quantity of air it did the instant before, whilst we

* 1 measure of carburetted hydrogen or coal gas, requires for its perfect combustion 10 measures of air.

have no contrivance for furnishing such a supply, although
without it, throughout the space of time during which rapid
gasification of the hydrogenous portion is going on, more
than half the fuel consumed is wasted, and passes off unburnt, becoming thereby not only totally unproductive in
itself, but absolutely an agent of evil, by robbing the furnace of the heat absorbed in its own volatilization.

Only two methods present themselves by which the supply of air and the wants of the furnace can be made to cor-
respond,—*either both must be made constant and regular*, or
*the fluctuations of one must be made to coincide with those of
the other.*

The contrivance by which I propose to achieve the great
desideratum sought, effects its object by admitting an increased supply of air at the periods of coaling. The stoker,
when he closes the furnace door after firing, will raise the
arm of a lever appended to it; this movement throws wide
open a sliding valve in the face of the door, which immediately commences closing slowly and automatically, by the
gravity of the lever, regulated and restrained by the motion
of a balance wheel connected with it by appropriate gearing,
and affords, during the progress of its descent, a *gradually
diminishing* supply of air to the fire, in harmony with the
gradually diminishing requirements of the fuel. The area of
the valve, and the period of time throughout which the act
of closing is to be prolonged, are of course questions of
detail to be determined by circumstances, and should be so
adjusted according to the nature of the coal, and the average
quantity supplied at one time, as entirely to prevent the
appearance of smoke. The door of the furnace should be
double, and the air should pass into the furnace through a
series of perforations in the inner plate. By this arrangement, three important points are secured : 1stly, the heating
of the air; 2ndly, its subdivision into minute jets; and
3rdly, the keeping of the outer surfaces of the furnace door

comparatively cool, and thereby both economising heat, and preventing its radiation outwardly to the annoyance of the attendants.

In ordinary furnaces, the air becomes heated in passing through the grate bars and the burning fuel. Where air is suffered to enter the furnace above the fuel, too much stress cannot be laid upon the necessity for either supplying it *hot*, or making a provision for subjecting it, together with the volatile products of the fuel, to an adequate amount of heat before they are allowed to escape by the flues. Little practical benefit is to be derived from leaving the doors of steamboiler furnaces ajar for a certain period after coaling. The fuel unquestionably demands a larger supply of air at this period, but the air being admitted *en masse*, and cold, and the heat of the furnace having been just previously lowered by the refrigerating effect of the fresh supply of coals and sudden development of gas, at the expense of the stock of heat previously existing, the result is, that the temperature produced by the union of the cold air with the gases from the fuel, is below the point required for inflammation.

In reverberatory furnaces for the manufacture of iron, on the contrary, the gases arising from the fuel in the grate, have to pass through the *intensely heated body, or working chamber of the furnace* in their way to the stack; and this cavity, which may be said to be an oven with the sides and roof at a white heat, although having another office, becomes in reality a provision for subjecting the gases and the air with which they are mingled to an adequate amount of heat, to insure their inflammation; and the result is, that if an adequate *quantity* of air be admitted, whether cold or hot, all the combustible matter is consumed, and smoke entirely prevented.

Many times have I stood before such a furnace just after a fresh supply of coals has been added, with my hand on the open door, and found that by regulating the width at which

I kept it open, I could exercise a perfect control over the action of the furnace. If shut, a dense black smoke would issue from the chimney; if opened a very little, the smoke would be slightly but perceptibly diminished; if opened a little more, the change would be denoted by a still further diminution of smoke; whilst an opening of a certain extent would cause the smoke to cease altogether. As the process of gasification progressed, and the demands of the fuel for air became less, the same effect would follow from a smaller and smaller orifice, till at length the door might be closed altogether without smoke resulting, the ordinary supply of air through the grate bar openings being adequate to the current demands of the fire. In furnaces of this description, beyond all doubt a self-closing valve, which should perform automatically, what I effected by watching and close supervision, would be a valuable acquisition and tend greatly to economy, whilst nothing can be more easy and simple than its application.

CHAPTER II.

ON CONTRIVANCES FOR THE EMPLOYMENT OF INFERIOR KINDS OF FUEL.

THE principal use to which the better description of slack are now applied, is the generation of steam in the steam boilers, used in the manufacturing processes carried on in the locality where it is raised.

Many attempts have been made to substitute slack, or even a mixture of slack and coal, for best coal in the manufacture of iron ; but in the ordinary reverberatory furnace this cannot be done, without a diminution in the activity of the combustion and the intensity of the heat generated, entailing a delay in the process, prejudicial to the yield of iron. The great difficulty in the way of attaining rapid combustion and intense heat from slack, even of the best quality, arises from the fact that its small size makes it lie so close together, as to form a mass almost impervious to air, and overlying and closing the grate bar openings, the passage of air into the furnace is in great measure cut off.

The remedies for this difficulty are, either the admission of air through *orifices in the sides of the furnace*, to compensate for the diminished quantity which enters through the grate bars, or the attainment of the same object by the use of *compressed air in a closed ash-pit.*

The first time I entered an iron work, I was struck with the enormous quantity of heat escaping to waste in all directions, to the annoyance of the workmen. Upon closer

inspection, I found there were various draught holes leading
to the space beneath the iron bottom of puddling furnaces,
for the express purpose of allowing the air to circulate freely
and carry off the heat from the iron bottom plates with
sufficient rapidity to prevent their melting; whilst conjointly
with this arrangement I saw that the fire grates of these
same furnaces were fed with a supply of cold air from the
atmosphere.   My first attempt to partially remedy this
waste was by constructing flues beneath the bottom plates,
so arranged as to convey the air, heated by passing over the
surface of these plates and through the bridges, into the side
flues for the supply of the air chamber of the improved furnace, instead of allowing these flues to draw their supply
direct from the atmosphere, as had previously been done.

The result achieved by this furnace worked with ordinary
coal was a saving of 17½ per cent. in weight of fuel, not as
compared with the *average* consumption, but when matched
against the *best* furnace in a work containing about forty.
These furnaces are constantly at work night and day, from
Monday morning to Saturday night, so that it is only once a
week that an opportunity for inspecting the interior occurs.
On getting inside the grate of the furnace after its first week's
work, I found that the small orifices round the sides com-
municating with the air chamber were either partially or entirely sealed up, by the melting and running of the face of
the bricks, and also by the adhesion of the clinkers or slag.

It occurred to me that the holes from the air chamber into
the grate might be kept permanently open, by having suitable
openings in the external wall of the furnace, leading into the
air chamber, by means of which a tool might be passed
through the orifices in the internal wall of the air chamber,
into the fire, once or twice a day, for the purpose of cleaning ;*

* In the first furnace which I constructed with these cleaning
openings, with a view to prevent the men destroying the brick-work
of the side of the grate, in passing the cleaning bar through the per-

such openings at other times being kept closed by doors, so as to prevent the entrance of cold air ; and I accordingly lost no time in carrying out my ideas, but erected a furnace on these principles. Its success was complete, though eventually abandoned from the opposition of the workmen, who set their faces against it the moment they found it was to be applied to burning an inferior description of coal, which they considered inimical to their interests.

forations, I had an iron plate cast, with perforations in it corresponding to those in the brick-work, to serve as a guide for the cleaning tool. In order to gain space for the air chamber, and at the same time to confine the exterior of the furnace as far as possible within its former dimensions, it was found necessary to build the side of the grate of only 2½ in. brick-work, the iron plate being relied upon to give the requisite strength. After being in use a short time, a considerable portion of one of the sides fell down, leaving the iron plate intact. To my surprise, instead of melting, it stood alone, and proved as efficient and durable as the other parts of the grate. Acting on the hint, I have since repeatedly made use of iron as a material for grate sides, where these are required with perforations, as more durable than brick-work, which is apt to be damaged by the passing of the cleaning tool.

CHAPTER III.

ON THE USE OF COMPRESSED AIR IN REVERBERATORY FURNACES.

ALTHOUGH a great saving had been effected by the furnace with perforated sides, with the improvement of cleaning doors for keeping the perforations open, I regarded it rather as an earnest of future advances; whilst a great quantity of heat escaped to waste, the principal supply of air entered under the ash-pit bars *cold*. As there was ample waste heat for the purpose, why should not the *whole* supply of air furnished to the fire be previously heated?

Whenever I took any steps to effect this object in the puddling furnace, however, I encountered the fact, that precisely as my arrangements for heating the air became more perfect, did I destroy the draught through the fuel, deaden the fire, and lessen the yield of iron. This unexpected result I attributed to the rarefaction of the air in the ash-pit.

I arrived at a clear perception, that to give the system of supplying the fuel with hot air the fullest development of which it was capable, it would be necessary to supply this air to the fire *under pressure*.

Being in an iron work one squally day, my attention was arrested by a volley of oaths from one of the puddlers, whose furnace, being situated at the extremity of a rank, was particularly exposed to the wind, which blew directly into his working hole. With two large pieces of sheet iron suspended behind him, in the vain hope of warding off his adversary, he was labouring away in a state of great exasperation, and apparently to but little purpose. The dull red colour of the iron showed the deficiency in temperature, and it was with difficulty it could be made to cohere into a mass. When the more violent gusts of wind came, so great was the quantity that entered the furnace, and so completely did it destroy the draught, that puffs of smoke were forced out at the ash-pit.

The workman himself was in much too ill a humour for conversation, but another puddler who was standing by, observing the interest with which I watched the proceedings, addressed me, saying, "You see, sir, how very badly this furnace works; the man can hardly make his iron at all. Well now, if the wind should change in the course of the day, and a nice steady breeze set into the ash-pit, this furnace will work as well as any furnace on the works; and this man's partner, who comes on to-night, will do his work with comfort in two hours less time, and perhaps make 1 cwt. more iron."

Now it immediately occurred to me that if the *accident* of the wind's blowing into the ash-pit made such a difference, the same difference ought to be created *artificially*, and made *constant*, and concomitantly with this idea, I recognized the fact, that with the immense quantity of waste heat escaping by the stack, no fuel would be required for a steam-engine to compress the air, and that therefore the expense of the system would be restricted to providing the machinery, and keeping it in repair.

In working with compressed air, the action of the furnace *is always under direct control*, and can be heightened or diminished at pleasure, and in such a case as the one just referred to, by turning on an adequate current of air, the power of the wind acting in the adverse direction might be defied, and none of it allowed to enter the furnace : with the result of a saving of iron, of fuel, of labour, and of time,

CHAPTER IV.

ON THE ECONOMY TO BE ATTAINED, BY INCREASING THE TEMPERATURE OF FURNACES.

THE existence of this very important means of effecting economic results seems to have escaped due recognition. The principles on which its value depends may be briefly explained as follows. It is only that portion of the heat of a furnace which is in *excess* over the temperature of the body to be heated which constitutes in reality *available heat*, conducive to the performance of its office. Thus the nearer the temperature of the body operated upon approaches the temperature of the furnace, the larger is the portion of the heat of the latter which becomes ineffective ; whilst the more the temperature of the furnace predominates, the larger becomes the proportion of serviceable heat, to the whole heat generated. Therefore, if the intensity of the heat of a furnace can be heightened, *its efficiency will be increased in the ratio of the accession made to its surplus heat* (or the degree by which it exceeds the temperature of the body to which heat is to be imparted); so that for processes requiring very high temperatures, and where the ordinary heat of the furnace is but slightly in excess, a very small increment of temperature will be productive of a great efficiency, and corresponding saving of fuel.

In Staffordshire, 24 cwt. of coal (long weight 120 lbs. to the cwt.) is consumed to produce from pig, 1 ton of puddled iron, the consumption being about 240 lbs. per hour, or

4 lbs. per minute. Now since oxygen is to carbon, in carbonic acid, as 8 to 3, and oxygen to hydrogen, in water, as
8 to 1, 4 lbs. of coal will require for perfect combustion,
$10\frac{2}{15}$ lbs. = 120 cubic feet of oxygen, and since oxygen is to
nitrogen (by volume), in atmospheric air, as 1 to 4 = 600
cubic feet of air; and the products resulting will be : 101·08
cubic feet of carbonic acid, 37·84 cubic feet of steam, and
480 cubic feet of uncombined nitrogen; total 618·92 cubic
feet of gas. Estimate the temperature at 3,108°, and multiply the volume of gas by 7 for expansion, and we find that
4,332·44 cubic feet of rarefied gases pass through the furnace
per minute ÷ 60 = 72·27 cubic feet per second. And,
finally, estimating the capacity of a puddling furnace between
the bridges at 36 cubic feet, we find, to our surprise, that it
is unquestionably filled and emptied twice in a second.

Now, if we estimate the temperature required for the
process of puddling at 3,000°, which must be a very near
approximation to the truth (since the melting point of pig-
iron is found by Daniel's pyrometer to be 2,786°), and
suppose that the products of combustion pass over the fire
bridge and enter the body of the furnace at a temperature of
3,300°, then it appears, that since, to maintain the necessary
temperature, the contents of the furnace have to be renewed
twice in a second, the loss of temperature, by radiation,
conduction, &c., is at the rate of 600° per second. If therefore we could succeed in increasing the temperature generated
by 300°, and thus raise the atmosphere produced by the
furnace to 3,600°, since we have found that the rate at which
the cooling process goes on is 600° per second, we should
only have to renew the contents of our furnace once in a
second, instead of twice, to maintain the requisite temperature
of 3,000°, and should effect a saving of 50 per cent.*

In mill or heating furnaces, used for the manufacture of

* M. Prideaux has omitted to make allowance for the increase of
loss of heat by radiation externally.—D. K. C.

iron, the result of heightening the temperature would be equally beneficial. When first charged with cold iron the heating goes on rapidly, *but the rate at which it proceeds constantly diminishes as the temperature of the iron approximates to that of the furnace*, and at last becomes very slow. It is during this latter period that the beneficial effect of a more intensely heated furnace becomes most conspicuous.

Now if we suppose the temperature of the iron to be 100° when in the furnace, the temperature generated by combustion to be 3,300°, and that half the *difference* (in excess) or surplus heat, when this surplus is 3,200°, enters the iron, we shall have at the commencement (3,300°—100° = 3,200° ÷ 2 =) 1,600°, within a fraction of *half* the whole heat generated, imparted to the iron. When, however, the iron has attained the temperature of 1,700°, and the surplus becomes halved, we shall have only 680° communicated to the iron, or a little more than a *fifth* of the whole. When the iron has reached the temperature of 2,500°, and the surplus heat becomes quartered, only 290 parts of heat out of 3,300 generated, or less than $\frac{1}{11}$, will be profitably applied. Whilst, if we suppose that iron requires to be raised to 3,200° to bring it to a welding heat, the state of things when it has attained the temperature of 3,100°, and is beginning to receive its last 100° of heat, will be as follows :—Out of 3,300 parts of heat generated, only 52, or less than $\frac{1}{64}$, will enter the iron, the remaining 3,248 parts, or more than $\frac{63}{64}$ of the whole, being wasted. Could we, however, succeed in increasing the temperature of our furnace by 400°, thus raising it to 3,700°, the state of things at the same point would be, that 206 parts of heat, or $\frac{1}{16}$ instead of only $\frac{1}{64}$ of the whole heat generated, would enter the iron ; thus exhibiting the fact of an addition of temperature to the amount of $\frac{1}{8}$, increasing, the velocity of heating, and consequently the efficiency of the furnace at this crisis, fourfold.

The following tables contrast the effects of two furnaces, one heated to 3,300°, the other to 3,700°, in imparting heat to iron at five different stages, whilst being raised to a welding heat: clearly showing, that although the rate at which heat is communicated rapidly diminishes as the temperature of the iron increases, *yet a very trifling addition to the temperature of the furnace* is capable of producing a great result in retarding the *rate* at which the diminution of heating proceeds, and consequently, in the economic working of the furnace. Column A gives the temperature of the furnace, B the temperature of the iron, C the difference in favour of the furnace, D the portion of this difference imparted, E the proportion of the whole heat generated imparted, F the velocity at which the heating proceeds, calling the smallest velocity 1 :—

```
     A.      B.        C.         D.        E.        F.
   3300°—  100°=3200°:...1600....·484....30·49
   3300°—1700°=1600°.... 680....·206....12·97
   3300°—2500°= 800°.... 290....·087.... 5·52
   3300°—2900°= 400°.... 123....·037.... 2·35
   3300°—3100°= 200°....  52....·015.... 1
   3700°—  100°=3600°....1871....·505....25·64
   3700°—1700°=2000°.... 911....·246....17·35
   3700°—2500°=1200°.... 485....·131.... 9·25
   3700°—2900°= 800°.... 290....·078.... 5·52
   3700°—3100°= 600°.... 206....·055.... 3·93
```

To facilitate the comparison, the results are arranged below in the parallel columns, the prefix + denoting those of the hotter furnace.

B.	C.	+C.	D.	+D.	E.	+E.	F.	+F.
100°	3200°	3600°	1600	1871	·484	·505	30·49	35·64
1700°	1600°	2000°	680	911	·206	·246	12·97	17·35
2500°	800°	1200°	290	485	·078	·131	5·52	9·25
2900°	400°	800°	123	290	·037	·078	2·35	5·52
3100°	200°	600°	52	206	·015	·055	1	3·93

The most efficient method that I am acquainted with, for increasing the temperature of reverberatory furnaces used for the manufacture of iron, is the use of compressed air, in a closed ash-pit, thus creating *artificially* the state of things present when a steady breeze sets into the ash-pit. It is a fact, that a furnace thus favourably situated will heat *three* charges of iron, whilst its opposite neighbour, in other respects its equal, is heating only *two ;* and the wonder is, that the hint thus conveyed has not been sooner acted upon, and that state of efficiency which now only has place with a few furnaces, and at uncertain intervals following the caprice of the wind, made the permanent condition of all.

Whatever method of working furnaces be employed, the construction of the envelope so as to prevent as much as possible the passage of heat, must ever be an unmixed good as far as economy of fuel is concerned, and the better the fire-brick material employed, the more completely can the principle be carried out, without neutralizing the economy attained in fuel, by the extra expense in repairs, resulting from the increased rapidity with which the bricks are burnt out. By constructing furnaces with hollow walls, and filling the intervening space with cinder dust, or some other incombustible bad conductor of heat, a great saving of fuel might be effected ; but to what extent this would be counter-balanced in an economical point of view, by the increased amount of materials and labour necessary to keep them in repair, with the quality of fire-brick generally in use at present, I am not prepared to say.

CHAPTER V.

ON FEEDING FURNACES WITH HOT AIR.

CONSIDERING the impossibility of admitting just the precise quantity of air to produce the state of theoretical perfection for the combustion of coal, the fact that coal contains a small quantity of nitrogen, and also that some heat passes into the sides of the grate, we may pretty safely assume that the heat generated in the body of the furnace seldom exceeds 3,300°, and is probably more frequently below than above this point. Nevertheless, as this number, which is certainly about what theoretical deductions would lead us to fix upon, coincides with the observation of Professor Daniel with his pyrometer, who found 3,300° to be the greatest heat of an air furnace, I shall adopt it as the basis of my calculation.

Now, as lead, the melting point of which is 600°, melts with extreme rapidity in the hot-air flues of a furnace, running like sealing-wax in the flame of a taper, we certainly shall not err on the side of excess in estimating the temperature of the air at 720°, and if we subtract 60° for the atmospheric temperature, we get an addition of heat of 660°, or $\frac{1}{5}$ of 3,300°, the whole temperature generated, showing a saving in fuel of 20 per cent.; and if, as is probably the case, a portion of the diminution in the heat between the theoretical number 4,047°, and that which experiment shows 3,300°, is to be attributed to a surplus quantity of air passing through the fire, and lowering the temperature of the furnace, then this per-centage of saving by the use of hot air should be increased.

My anticipations of an *increase* in *the temperature and heat-imparting powers* of reverberatory furnaces, by feeding them with hot air, were not realised. One cause of the impaired heat of reverberatory furnaces supplied with hot air must be sought in the fact that gaseous combustion being accelerated by it, a larger amount of chemical union takes place in the *grate*, and less in the *body* of the furnace, where it would be more beneficial.

The next point to which I shall call attention, as one of the causes of the diminished heat of reverberatory furnaces supplied with hot air, is, the greater state of expansion and tenuity, attained by the mixture of air and coal gas, *prior* to inflammation, and the consequent result, that in a given *space* there is a less *quantity* of heat developed by their union. There is, no doubt, more heat contained in the products of combustion from a given *weight* of coal gas and air, heated before ignition to 720°, than would be contained in the products of the combustion of the same quantities by weight, of gases ignited at the temperature of 60°. When we take given *measures*, however, instead of given *weights*, the case is reversed. Hence the flame resulting is characterised by comparative tenuity.

The third and last circumstance that I have to adduce in explanation of the effect of hot air in diminishing the heat of the working chamber of reverberatory furnaces is, that from the greater tenuity of the air, it does not produce the same attrition in its passage through the fuel, nor (owing to its less sudden expansion) the same amount of *quasi* detonation or percussion, at the instant of combustion. The disintegration of the carbon being thus retarded, at the same time that the escape of the hydrogen is facilitated by the heat, the result is, that the ordinary relation between the rate of consumption of these two constituents of the coal is deranged, and the gases passing from the fuel contain an undue proportion of hydrogen, whilst the carbon accumulates as cinders on the bars.

On this theory we should be led to expect a considerable difference in the adaptability of different varieties of coal for use with hot air, an inference fully carried out by practice. There is a great variation as to the increased ratio with which the cinders accumulate on the bars, according as the texture and composition of the coal render the disengagement of the carbon more or less facile, and there can be little doubt but that, *cæteris paribus*, the more the carbon of a fuel is disposed to assume the gaseous form, as with cannel coal—or the less the quantity of hydrogen present, as with anthracite—the greater will be the chance of being able to seize the economic advantages attendant upon the increase quantity of heat attainable by the use of hot air, without having this heat so diluted as to make the temperature inefficient.

It is, however, in its deficiency in heating power with reference to its *volume*, that we must seek for the cause of the diminution of temperature which occurs in practice, by substituting hydrogen for carburetted hydrogen. The difference between the quantity of air required by equal volumes of the two gases is so great, amounting to fourfold, that with furnace arrangements adapted for carburetted hydrogen, that is, calculated to allow ten volumes of air to pass with each volume of gas, it is next to impossible but that just in proportion as hydrogen is substituted for it, does the quantity of air passing become an injurious excess.

It thus appears that three circumstances,—1st, an increased proportion of the combustion being completed in the grate,—2ndly, the higher temperature and consequent rarefaction attained by the gases prior to ignition,—3rdly, the retarded disintegration of the carbon on the grate-bars (and the resulting diminished quantity of combustible matter passing in a given time, and undue preponderance of hydrogen) all combine to neutralise—where a very high temperature is demanded—the economy, which might otherwise be attained by feeding the fire of reverberatory furnaces with hot air,—by attaching to the practice the condition,—that just in proportion as it is carried out, does the flame in the working chamber become attenuated, and its power of rapidly imparting intense heat impaired.

CHAPTER VI.

ON THE MANUFACTURE OF IRON.

The requirements for making iron are, 1st, the requisite chemical ingredients in the requisite proportions ; 2ndly, sufficient heat :—fluidity, to allow of the free motion of all the particles of matter ; 3rdly, sufficient motion amongst the particles to ensure their adequate intermixture.

The circumstance that hot-blast furnaces work faster than cold-blast has been supposed to be quite conclusive as to a higher temperature being attained by them : a little examination will, however, show that the fact affords no foundation for such an inference. If we assume that the perfect fusion of the minerals of a blast-furnace requires a heat of 4,000°, that the temperature of the point of most intense heat in the vicinity of the *tuyéres* in each description of furnace is the same, viz., 5,000° ; since the minerals will in the hot-blast furnace approach this point at a temperature 560° higher than they will approach the same point in the cold-blast furnace, and consequently will require 560° less additional heat imparted to them, to raise them to. the temperature of 4,000°, their melting, will, as a matter of course, be more quickly performed in the former furnace than in the latter ; and thus this supposed decisive fact in favour of the higher temperature attained by the use of the hot-blast is satisfactorily explained without leaving any difficulty behind it.

In considering the manufacture of iron with reference to economy of fuel, the most striking thing which forces itself

upon the attention is the fact, that whilst the two great processes of iron making, viz., the separation of the iron from
the ore in the blast furnace, and its decarbonization, and
kneading to develop fibre, in the boiling furnace, each
requires for its performance, and *is best performed by*, one
of the two constituents into which coal is resolvable, taken
singly; viz., the revival of the iron, by the solid carbon or
coke, and the operation of boiling, or puddling, by the
gaseous carburetted hydrogen. Yet, instead of separating
the constituents of coal before using, and employing each for
the operation for which it is best suited, we use it in its
raw state for both operations, not only wasting one constituent in each process, but absolutely producing inferior
results. For, the gaseous portion of the coals used in the
blast-furnace,—which if separated and purified, would suffice
to make all the pig-iron produced by the furnace, into pud-
dled-iron, of a quality very superior to that made by raw
coal,—exercises in the blast furnace a deteriorating effect on
the quality of the iron produced. Thus, all the coal at
present used in puddling is wasted, with the result of producing an inferior description of iron.

My notions of attainable perfection demand that iron
should be manufactured as follows:—the coal should be
coked by the waste heat from the furnaces, and set apart
for use in the blast-furnaces, having previously had its pores
impregnated with a preparation composed with reference to
its power of combining, and forming an easily fusible cinder
with the ashes of the coke on the one hand, and the foreign
matter in the ore on the other.

The gaseous products of distillation after being purified
from sulphur, and also from phosphorus, and arsenic if
present, should be devoted to the process of puddling, for
which they are for many reasons much better adapted than
raw coal. In the first place, the dust and fine cinders carried
over from the grate, become mixed with, and injure the iron,

and the same may be said of various volatile products of the coal ; in the second place, the operation of puddling consists of three different stages, during each of which the iron requires a modification in its treatment, which the old puddling furnace is powerless to supply. In fact, no prompt nor *adequate* power of control is possessed by the workmen over it, even for simply exalting its temperature, and as far · as the scientific performance of its functions is concerned, it must be regarded as a barbarous apparatus which ought long since to have been superseded.

In the first stage—that of melting—an intense heat, with no excess of oxygen, will save both time and iron, although it is not so absolutely indispensable in this stage as in the third.

In the second step of the process—that of decarbonization, or refining—an excess of oxygen is wanted, and the addition of the vapour of water is also often beneficial, since it has the power, not only of accelerating the decarbonization of the metal, but also of carrying off large quantities of silicic acid in a volatile form,* the effect of which in increasing the *yield* (inasmuch as the silicic acid would otherwise form a silicate of iron, and carry its base into the cinder) is sufficiently obvious.

During the third stage of the process—that of kneading the decarbonized iron into a tenacious, fibrous mass—an excess of oxygen passing is productive of great loss, from the waste it occasions by oxydizing the pure malleable iron, now no longer covered by the cinder. From the circumstance, however, of a great heat being necessary (requiring a clear fire) during this stage, particularly at its termination, just before the iron is removed from the furnace to go to the squeezer, it is impossible to guard against a surplus of oxygen being present,—such is, in fact, always the case under

---

* An instructive example of the law, that non-volatile bodies may become volatile by mixture with volatile ones.

the ordinary system,—a considerable loss is unavoidable, and
may be considered a perfectly normal result of the operation
as carried on in the old furnaces.

Let us compare with this narration of the inefficiency and
unavoidable defects of the old furnace, the advantages placed
within our reach by the judicious application of gas.

Suppose a furnace constructed as follows : let the end of
the furnace be formed of an iron box, divided vertically (and
transversely as regards the body of the furnace) into three
divisions : let the outer one be connected with a reservoir of
gas under pressure, the middle one with a similar reservoir
of air ; let a series of tubes (with a slight downward inclination) lead from each of these chambers, through the inner
chamber, into the furnace ; let the inner chamber be supplied
with water, to prevent the melting of the tubes or *tuyères*.

The pipes leading from the main reservoirs of gas and air
to the chambers of the furnace should each be furnished with
a throttle valve, the handle of which should form the index
to a dial, graduated to 100° subdivisions ; and the size of the
pipes and the valves should be so arranged, that when the
indexes pointed to the same figures on the dial, the respective
quantities of air and gas passing should be the exact combining equivalents of each.

As it would be desirable to have a surplus heating power,
beyond that demanded for ordinary working, for extraordinary
occasions, such as getting up the heat after the furnace had
been allowed to get cold, &c., we will suppose, that upon
charging with fresh iron, the hands of both dials are placed
at 75, for the first stage of the operation. The inclination of
the *tuyères* driving the flame downwards directly on the iron,
it will be melted and brought to the proper point of fluidity,
in about half the time occupied by the present furnace, where
the axis of the draught being horizontal, and the iron lying
completely below it, by far the larger portion of flame darts
over it, and enters the flue without having at all impinged on

the iron. In fact, the difference in the mode of operation of the two furnaces, is the difference between placing a body at the point of the flame of a blowpipe, and placing it at the exterior of its under side.

The next stage of the operation—that of refining, or decarbonization—requires a surplus of oxygen to be present, to obtain which, the workman moves the index of the air-pipe, say to 90 (the index of the gas-pipe remaining stationary at 75), and turns on in addition, if thought desirable, a supply of steam into the air chamber, to pass, with the air, into the working chamber. That such a furnace would be a better instrument for refining iron than our present one cannot admit of a doubt, whilst it is equally unquestionable that it would do its duty in less time.

During the third stage of the operation, a very intense heat is required, particularly just at the latter end, whilst, at the same time, it is of vital importance—to prevent the waste of the now decarbonized and exposed iron—to ensure that a surplus of oxygen be not present. Let the workman set the index of the gas-pipe at 95 (the air-pipe index remaining, as before, at 90),—thus guarding against the possibility of any free oxygen reaching the iron through imperfect mixture, by supplying a slight excess of carburetted hydrogen,—and both conditions will be fulfilled.

In short, whatever the requirements demanded by the progress of the operation, this furnace places at our disposal the means of fulfilling them *instantly* and *completely*, furnishing in this respect a complete contrast to the imperfect means of control offered by the existing furnace; nor must it be forgotten, that it presents us with all these advantages, at the cost of the gaseous products of the coal, now wasted in the blast-furnace.

Having thus briefly reviewed the subject of puddling, and endeavoured to show that great improvements may be made in our present arrangements, with regard both to economy of fuel, and the production of a superior quality of iron, I will now offer a few observations on blast-furnaces.

In the figures which I employ with regard to the *theoretical* composition of the gases issuing from a blast furnace, as deduced from the composition of the fuel used, I shall principally rely upon the data on the Alfreton coal and furnaces, contained in the elaborate report on blast-furnaces, presented to the British Association in 1845, by Dr. Lyon Playfair and Professor Bunsen.

Observation has established, 1st, that the oxygen of the blast is all consumed in the immediate vicinity of the *tuyère;* 2ndly, that in hot-blast furnaces, the carbonic acid zone is of very limited extent, this gas being found entirely converted into carbonic oxide within about 3 feet from the point of the *tuyère;* 3rdly, that the coal loses all its gaseous products of distillation much above the point at which its carbon combines with the oxygen of the blast.

Each charge at the Alfreton furnaces consisted of 420 lbs. of calcined ore, 390 lbs. of coal, and 170 lbs. of limestone; and the product of iron resulting from one charge is 140 lbs. 100 parts of the coal contained—

		Parts of water capable of being heated 1° C. by the combustible.
Carbon	64·518	516,384
Carburetted Hydrogen	6·638	99,570
Carbonic oxide	1·602	3,660
Carbonic acid	1·139	
Bi-Carburetted Hydrogen	0·513	6,156
Hydrogen	0·370	13,320

Total 639,090

Contributed by the carbonaceous portion . . 516,384

Contributed by the products of distillation . 122,706

Since cast-iron contains 3·3 per cent. of carbon, there will be 1·18 abstracted from each 64·548 parts of carbon for this purpose, leaving 63·368 parts of carbon to undergo combustion.

This carbon, as has been already stated, becomes converted into carbonic oxide, by combining with the air of the blast, at a short distance above the *tuyère*, the product being—

Nitrogen . . . . : . . 295·716
Carbonic oxide . . . . 147·858

The quantity of cast-iron produced by 100 parts of coal is 35·8, containing 34·62 of pure iron ; to effect the deoxidation of which, the carbonic oxide must combine with 14·83 of oxygen, and have 25·952 of its parts changed into 40·782 of carbonic acid.

The quantity of limestone, added to 100 parts of coal, contains 18·7 parts of carbonic acid, which is evolved from the furnace with the other gaseous products, giving as a result, for the composition of the gases issuing from the mouth of the furnace—

Nitrogen. . . . . 295·716
Carbonic oxide . . . . 121·906
Carbonic acid . . . . 40·782
Ditto   do. (from lime) . . . 18·700
,   Total . . . . 477·104

Hence it appears that we shall obtain the composition of the gases from the furnace, if, to the products of the distillation of any given quantity of coal, we add the carbon of the coke formed from that coal (minus the portion entering the iron), plus the quantity of air necessary to form with this carbon, carbonic oxide, plus the oxygen of the ore (which converts a portion of this carbonic oxide into carbonic acid), and plus the carbonic acid evolved from the limestone.

From the preceding data, the composition of the gases evolved for each 100 parts of coal is found to be—

			Total of each.	In 100 pts.
	Nitrogen	295·716	295·716	60·677
	Carbonic oxide . . .	121·906	123·508	25·342
	Carbonic acid . . .	59·482	60·621	12·439
	Carbonic oxide . . .	1·602		
From dis-	Carbonic acid . . .	1·139		
tillation	Carburetted hydrogen.	6·638	6·638	1·362
of coal.	Bi-Carburetted hydrogen	0·513	0·513	·105
	Hydrogen	0·370	0·370	·075
	Total . .		487·366	100·000

Since the whole of the products of the distillation of coal escape unburnt, our task in finding the duty performed by the fuel is narrowed to an examination of the changes undergone by the carbonaceous portion.

I shall assume that carbon, in becoming $C\,O$, develops $\frac{1}{4}$ part of the heat developed by it in becoming $C\,O^2$, this being about the mean of the figures given by the three observers, Dulong, Andrews, and Grassi.

Of the total carbon in the fuel . . . . . . 64·548
    There enters the iron . . . . 1·18
    Leaves the furnace as carbonic acid . . 11·1225
    Therefore discharges its full duty . . . 12·3025
    Leaves the furnace as carbonic oxide . . 52·2455 having *apparently* discharged only $\frac{1}{4}$ of its duty.

Carbon.
Full duty of . . . . 12·3025 = 98,420
$\frac{1}{4}$ duty of . . . . 52·2455 = 139,321
                         Total . . . 237,741

If we subtract this sum from the full duty of 100 parts of coal, viz. 639,090, the remainder, 401,349, shows the duty capable of being performed by the gaseous products escaping from the furnace. Then,

as 639,090 : 237,741 : : 100 : 37·2

whence it appears that this amounts to 62·8 per cent., leaving 37·2 per cent. for the *apparent* duty realised. These figures differ so greatly from the statements of Dr.

Playfair in the report before alluded to, that I feel it necessary to advert to the subject.

The report says, " Hence follows the remarkable conclusion,
that in the furnaces of Alfreton, not less than 81·54 per cent.
of the fuel is lost in the form of combustible matter still fit
for use, and that only 18·46 per cent. of the whole fuel is
realised in carrying out the processes in the furnace."

Thus it appears that the figures given by Dr. Playfair
restrict the duty performed by the fuel to just half the
amount which my calculations have assigned to it. The
difference in the results at which we have arrived is easily
traceable to the difference in the principles which have guided
our mode of procedure, and to the difference in the ratio we
assume, that the portion of heat developed by carbon in becoming carbonic oxide, bears to the full amount it is capable
of furnishing, when it unites with 2 atoms of oxygen. To estimate the duty performed, Dr. Playfair simply takes the amount
of nitrogen, deduces the quantity of oxygen which accompanied
this into the furnace as atmospheric air, and then calculates
the heat which would be evolved by the union of this oxygen
with carbon to form carbonic oxide. It is evident that such
a method altogether ignores the duty done by the 1·18 of
carbon, which combines with the iron and descends with it
into the hearth, and also the important service towards
" carrying out the processes in the furnace " rendered by the
11·1225 of carbon, which, in the form of 25·952 of carbonic
oxide, separates 14·83 of oxygen from the ore, and becomes
converted into 40·782 of carbonic acid in so doing. The
latter process, the report observes, is attended with a thermo-
neutrality, but surely this can be no valid reason for not
reckoning it with the realised duty of the carbon. If we
revive iron by placing some iron ore in a crucible, in conjunction with charcoal, and then subjecting the vessel to the
heat of a furnace, can any one be found to contend, that the
duty performed by that portion of charcoal which combines

with the iron, and by that other portion which removes its oxygen by combining with it, is any less real, than that achieved by the charcoal of the furnace, in contributing the necessary temperature? In fact, since the *whole* of the carbon which combines with the oxygen of the ore is absolutely required for this purpose, and not a particle can be dispensed with, whilst, on the contrary, it might be a question whether some of the carbon in the furnace is not burnt in furnishing superfluous heat,—many would be disposed to accord the precedence in importance and indispensability to the services of the former—and to deny to the function it performs the title of realised duty, certainly indicates a singular inversion of ideas.

It may have been observed, that in treating on the 52·2455 parts of carbon which left the furnace as carbonic oxide, I spoke of its having *apparently* discharged only ⅓ of its duty; and however unexceptionable the mode of estimating its duty adopted, may on a superficial view appear, I am prepared to show that it inadequately represents the real services rendered by the carbon to the processes of the furnace, and that if the various parts played by the fuel are separately scrutinized, and their results analyzed with exactitude, the duty *performed* by the carbon will be found to be much more considerable. Since the whole of the atmospheric oxygen entering with the blast is converted into carbonic oxide at a comparatively short distance above the *tuyère*, affording in this form an ample supply of carbon to effect the deoxidizing process of the furnace, without the absorption of any uncombined carbon for this object; and since we know, that when we form carbonic oxide from carbonic acid by supplying the latter with another equivalent of carbon, exactly half the amount of carbon in the product must have previously been in combination with 2 atoms of oxygen, as carbonic acid, we are enabled to pronounce that the combustion of the carbon in the furnace takes place as follows :—

Deducting the 1·18 of carbon which enters the iron, precisely one-half the remainder is converted to carbonic oxide in the zone of transition which *must* exist between the carbonic acid zone (or zone of fusion) and the carbonic oxide zone, where this conversion is complete ; whilst the remaining half descends before the *tuyère*, where, by its conversion to carbonic acid, it furnishes the intense heat necessary to produce perfect fusion of the refractory minerals employed.

The fact, that *subsequently* to performing this service, the 31·684 parts of carbon, in combination with 84·49 of oxygen as carbonic acid, yields up half its oxygen to an additional portion of 31·684 parts of carbon, with a great thermometric loss, must not blind us to the amount of duty which it has previously *actually accomplished*, and which, if it left the furnace immediately afterwards, before its transformation into carbonic oxide, would be unhesitatingly admitted. Does it injure or retard the processes of the furnace, in its subsequent passage through it, that at its final exit, we curtail its claims for services previously accomplished ? So far from doing so, it increases our obligations for work performed, by separating 14·83 of oxygen from the ore, and we repay the additional service, by paring down its duty into simply yielding the measure of heat due to the formation of carbonic oxide.

Estimated according to this more accurate and precise view of its true functions, the duty performed by the fuel in a blast furnace will be as follows :

Of the total carbon in 100 parts of fuel $= 64\cdot548$
There enters the iron . . . . 1·18
Carbon (as simple carbon) burnt to $C\,O_2$ before *tuyère* . 31·684
Consequently discharges full duty . . . . 32·864
Carbon (as $C\,O$) becomes $C\,O_2$ in separating 14·83 O.,
   consequently discharges $\frac{2}{3}$ duty . . . . 11·1225

The full duty of 32·864 of carbon is 262,912 and $\frac{2}{3}$ duty

on 11·1225 is 59,820, making together 822,232 for the duty
performed by the carbon, which sum is to 689,090 the whole
duty of 100 parts of coal, as 50·4 is to 100; consequently,
the duty performed by the fuel is 50·4 per cent., or nearly
3-fold what has previously been assigned to it. That this
is the proper way of estimating the duty of the carbon is
abundantly plain and clear; and once pointed out, it will, I
think, be felt to admit of no dispute. That the duty done
by the fuel should be 50·4 per cent., and the duty capable
of being evolved from the escaping gases 62·8 per cent.,
making together the sum of 113·2 per cent., may at first
sight appear a discrepancy, but when more narrowly examined, the apparent anomaly vanishes. A given quantity of
carbon as $C O_2$, at a temperature sufficient to convert carbon
into $C O$, will, by so doing, form a compound capable of
developing by combustion 33·3 per cent. more heat than the
carbon so converted. In one case we have 100 parts of
carbon (as simple carbon), capable of developing 100 parts
of heat; in the other, 200 parts of carbon as $C O$, capable of
developing 133·3 parts of heat, according to the law that $\frac{1}{3}$
of the whole heat capable of being developed by carbon, are
developed when $C O$ becomes $C O_2$. In the case of the
blast-furnace before us for consideration, the difference will
be, the difference between the heat-developing power of
31·684 of carbon (as simple carbon) converted to $C O_2$, and
63·368 of carbon (as $C O$) converted to $C O_2 = 84,490·66$
units, and 689·090 is to 84,490·66 as 100 to 13·2 — the
amount of overplus.

Although we have found the amount of duty performed by
the fuel in blast furnaces to be much more satisfactory than
had previously been stated, amounting to 50·4 per cent.
instead of only 18·46 per cent., yet the heat capable of
being realised from the escaping products, amounting to
62·8 per cent. on all the fuel consumed, is still a loss which
it would be most desirable to curtail, either by diminishing

the *consumption*, by increasing the *duty* of the fuel, or by turning the waste products to some useful account. Their value as an article of fuel depends less upon the whole quantity of heat they are able to afford than upon the *temperature* they are capable of generating, since it is this which must determine to what objects they are applicable, and the extent of their sphere of practical utility. To ascertain the temperature capable of being attained by the combustion of these gases, we must determine their heating power from their composition, and then ascertain the specific heat of the products resulting from their combustion.*

Composition of furnace gases.		No. of degs. C. 1 part of water would be raised by the combustibles.	Result of combustion.		Totals.		Ditto reduced to 100 parts.	Specific heat.
N	295·716				652·18	N	68·9644	·27
CO₂	60·621				274·57178	CO₂	29·0344	·22
CO	123·508	282215°	{ 194·084	CO₂				
			247·016	N				
H₂ C	6·638	99570°	{ 14·9355	HO	18·925	HO	2·0012	·847
			18·2545	CO₂				
			92·9320	N				
H₂ C₂	0·513	6156°	{ 0·65957	HO				
			1·61228	CO₂				
			6·156	N				
H	0·370	13320°	{ 3·330	HO				
			10·36	N				
Totals.	487·366	401261°			945·67678		100·0000	·267

Then $\dfrac{401261}{945\cdot67678 \times \cdot267} = 1589°$ C. or 2900° Fahr. the temperature these gases are theoretically capable of yielding

* The specific heat here given for water is too high; but the error is not material to the argument.—D. K. C.

on combustion. This degree of heat is amply sufficient for raising steam, heating the blast, manufacturing gas and coke, burning bricks or lime, and a variety of other operations, but is not adequate to be employed with advantage for the manufacture of iron, as was once supposed,—an erroneous idea, which has led to the waste of much capital in abortive experiments, the success of which, a more accurate knowledge of the subject would have seen to have been *à priori* impossible.

PART III.

Section I.—FUELS: THEIR COMBUSTION AND ECONOMICAL USE.

By D. K. CLARK.

CHAPTER I.

CHEMICAL COMPOSITION OF FUELS AND FORMULAS FOR COMBUSTION.

THE fuels, or combustibles, commonly used for the generation of heat, are coal, coke, wood, charcoal, peat, and refuse tan-bark. Asphalte and petroleum, with other oils, have also been used as fuels, as well as coal-gas ; and, within the last few years, straw and megass or refuse sugar-cane have been brought into consumption for the same purpose. · The following summary of the chemical composition of fuels, exhibits the relative proportions of the combustible and other elements, in per-centages of the total weight of each fuel. They are, of course, to be taken as averages :—

AVERAGE CHEMICAL COMPOSITION OF FUELS.

FUEL.	In 100 parts by weight.					
	Carbon.	Hydrogen.	Oxygen	Nitrogen.	Sulphur.	Ash.
	per cent	per cent	per cent	per cent	per cent	per cent
Coal, desiccated . . .	80	5	8	1·20	1·25	4·55
Coke do. . . .	93·4	—	—	—	1·22	5·34
Lignite, perfect . . .	69	5	20		—	6
Asphalte.	79	9	9		·—	3
Wood, desiccated . . .	50	6	41	1	—	2
Charcoal do. . . .	79	2*	11*		—	8
Peat do. . . .	59	6	30	1·25	—	4
Peat charcoal do.	85	—	—	—	—	15
Straw, 15¾ per cent. moisture	36	5	38	·43	—	4·75
Petroleum	85	13	2	—	—	—
Petroleum oils. . . .	72·6	27·4	—	—	—	—

* These proportions are given as 2 of free hydrogen, and 11 of hydrogen, oxygen, and nitrogen.

It is convenient to formulate the relations of the constituent elements of fuels with the quantity of air entering
into chemical combination with the elements, in combustion; the quantity of the gaseous products of combustion,
the heat evolved in combustion, and the temperature of
combustion.   The following formulas and rules, for these
objects, are derived from the writer's *Manual of Rules, Tables,
and Data :* pp. 398 to 458 :—

Air consumed in the Combustion of Fuels.—Let the constituents of a fuel be expressed proportionally as per-
centages of the total weight of the fuel, by their initials, C,
H, O, S, N, respectively representing carbon, hydrogen,
oxygen, sulphur, and nitrogen.   The volume of air at 62°
Fahr., chemically consumed in the complete combustion of
1 lb. of the fuel, putting A for the volume in cubic feet of
air at 62°, is expressed by the formula,—

$$A = 1{\cdot}52 \left(C + 3 \left(H - \frac{O}{8} + {\cdot}4\ S \right) \right) \qquad (1)$$

To find the weight of the air chemically consumed,
divide the volume thus found by 18·14 ; the quotient is
the weight of the air in pounds.

NOTE.—In making ordinary approximate calculations, the
sulphur may be omitted.

Quantity of the Gaseous Products of Combustion.—Let w
be the total weight of the gaseous products of combustion,
then—

$$w = {\cdot}126\ C + {\cdot}356\ H + {\cdot}053\ S + {\cdot}01\ N \qquad (2)$$

The total volume of the burnt gases at 62° Fahr., is
given by the formula :—

$$V = 1{\cdot}52\ C + 5{\cdot}52\ H + {\cdot}565\ S + {\cdot}135\ N \qquad (3)$$

The corresponding volume of the gases at higher temperatures is given by the formula,—

$$V' = V \frac{t' + 461}{523} \qquad (4)$$

in which V is the volume at 62° Fahr., t' the new tempera-ture, and V' the new volume. That is to say, the expanded volume at any other temperature t', is formed by multiply-ing the volume at 62° by the new absolute temperature (t + 461), and dividing by 523.

Heat evolved by the Combustion of Fuels.—The total quantities of heat evolved in the combustion of one pound of the elementary combustibles with oxygen, are, according to the experiments of MM. Favre and Silbermann, as fol-lows :—

HEAT OF COMBUSTION.

Hydrogen	62,032	English units.
Carbon	14,500	„
Sulphur	4,032	„

From these data, the heating power of a fuel may be calculated, approximately, as the sum of the heating powers of its elements; excepting that, when oxygen is present in the combustible, a deduction is to be made for the equiva-lent of constituent hydrogen neutralised by it, supposed to exist in the chemical state of water in the fuel. Putting h for the total heat of combustion,—

$$h = 145 \left(C + 4 \cdot 28 \left(H - \frac{O}{8} \right) + \cdot 28\ S \right) \qquad (5)$$

The item for sulphur, ·28 S, is not considerable, and it may be omitted from calculations for ordinary purposes.

Dividing the second number of the formula by 996, the total heat of steam raised under atmospheric pressure from water at 212° Fahr., the quotient expresses the equivalent evaporative power of the fuel. Putting e for the evapora-tive power of 1 lb. of a fuel, in pounds of water from and at 212° Fahr.

$$e = \cdot 15 \left(\mathrm{C} + 4 \cdot 28 \left(\mathrm{H} - \frac{\mathrm{O}}{8} \right) + \cdot 28\ \mathrm{S} \right) \qquad (6)$$

When the total heating power, in heat-units, is known, divide it by 966, to find the equivalent evaporative power in pounds of water from and at 212°, per pound of fuel. It may be roughly approximated to by using the divisor 1000.

The Temperature of Combustion.—Though the actual temperature ever attained in practice is a doubtful subject, the nominal temperatures that would be attainable, may be calculated on the supposition that the specific heat of gases is constant for all temperatures, that the products of combustion are not diluted by surplus air, and that no decomposition or "dissociation" takes place. The principle on which it is calculable, is that the product of the weight of burnt gases per pound of fuel by the average specific heat of the gases expresses the number of units of heat absorbed in raising the temperature 1° Fahr., and the quotient obtained by dividing the total heat of combustion by this product is the total rise of temperature.

If surplus air be mixed with the products of combustion, the rise of temperature in combustion is less than when the gases are not diluted. It is calculated in the same way.

CHAPTER II.

COAL.

THE chemical history of the combustion of coal has been repeatedly and thoroughly elucidated by competent men. Of these, the earliest and most conspicuous is Mr. Charles Wyc Williams, to whom the world is much indebted for the thoroughness, impartiality, and lucidity with which he has treated the subject, and the unwearied pertinacity with which he sought to impress his views on the minds of the public. Much ridiculed he has been, no doubt; that is because, having felt strongly, he, with emphatic iteration, dwelt upon elementary truths which nobody was supposed to question. Scandalised he has been, too ; that is because, regardless of the feelings and interests of others, he denounced as empiricism and quackery that which violated his sense of " chemical propriety." Nevertheless, Mr. Williams laboured well and efficiently in the cause of coal combustion and smoke prevention.

The question of the prevention of smoke has been intimately associated with that of the combustion of coal. They are, indeed, essentially one question, for if coal be completely burned, there cannot be any smoke ; and, otherwise, if there be smoke, the coal is certainly not completely burned.

It is purposed now to add some observations on the same subject, complementary to the works of Mr. C. W. Williams and W. S. Prideaux.

Throughout all the primary and secondary conditions of the hydro-carbon compounds raised by distillation from coal, the hydrogen maintains the first claim to the oxygen present above the fuel: until it is satisfied, the carbon remains un-burned.

The following summary presents the average composition and characteristics of English, Welsh, and Scotch coals, derived from the *Report on Coals suited to the Royal Navy :—*

AVERAGE CHEMICAL COMPOSITION OF BRITISH COALS.
(From the Report on Coals suited to the Royal Navy.)

| Locality. | Constituent Elements by Weight. | | | | | | Coke left by distillation. | Water evaporated per lb. of Coal. |
	Carbon.	Hydrogen.	Oxygen.	Nitrogen.	Sulphur.	Ash.		
	per cent	per cent	per cent	per cent	per cent	per cent	per cent	lbs.
Wales . . .	83·78	4·79	4·15	0·98	1·43	4·91	72·6	9·05
Newcastle . .	82·12	5·31	5·69	1·35	1·24	3·77	60·7	8·37
Lancashire . .	77·90	5·32	9·53	1·30	1·44	4·88	60·2	7·94
Scotch . . .	78·53	5·61	9·69	1·0	1·11	4·03	54·2	7·70
Derbyshire and Yorkshire .	79·68	4·94	10·28	1·41	1·01	2·65	59·2	7·58
Total Averages	80·40	5·19	7·87	1·21	1·25	4·05	61·4	8·13

It appears, from this summary, that the composition of British coals averages about 80 per cent. of carbon, 5 per cent. of hydrogen, 8 per cent. of oxygen, 1¼ per cent. of nitrogen, 1¼ per cent. of sulphur, and 4 per cent. of ash. Also, that the coke, or fixed carbon, as distinguished from the volatilized carbon, averages a little over 60 per cent. of the weight of the raw material, leaving the difference of 80 and 60, or 20 per cent. of carbon to pass off with the hydrogen, forming hydro-carbon compounds. The disposition of the elements of 100 lbs. of average coal in combustion, would, then, be as follows :—

100 LBS. OF AVERAGE COAL IN THE FURNACE.

Volatilized hydro-carbons { Hydrogen . . .   5 lbs.
                           { Carbon . . . . 20 „
Fixed carbon, or coke . . . . 60 „
Oxygen, nitrogen, sulphur, ash . . . . . . 15 „
                                          ____
                                          100 „

CHAPTER III.

COMBUSTION OF COAL.

Much discussion has arisen on the question of the effective heating power of hydrogen raised from coal, and its value in that respect compared with carbon. One thing is clear, that in order to make the best of it, the hydrogen, once volatilized should be oxidised, as well as the carbon associated with it, in order to realise the large measure of heat generated by combustion. There existed a favourite theory, possessing at least the merit of simplicity, that the heating power of coal was measured by that of its constituent carbon. The evidence of the evaporative powers of coals, abstracted in the foregoing table, appears to support this mode of estimation, in so far as the evaporative efficiency rises generally with the percentage of constituent carbon. The percentages of constituent hydrogen vary within narrow limits, and do not afford data for marked comparison; but it may be suggested that, generally, the evaporative efficiency is less as the constituent hydrogen is greater in quantity. But, neither the variation of the hydrogen, nor those of the carbon, suffice to account for the comparatively wide differences of efficiency. On referring, nevertheless, to the next column, the constituent oxygen, it is remarkable that the efficiency of the fuel decreases regularly as the percentage of oxygen in the fuel increases. Welsh coal having about 84 per cent. of carbon, and 4 per cent. of oxygen, evaporates 9·05 lbs. of water per pound of fuel; whilst Derbyshire coal having about 80 per cent. of carbon, and 10 per cent. of oxygen, evaporates only 7·58 lbs. of water. The difference of evaporative efficiency is not sufficiently accounted for by the difference of carbon ; nor by the difference of hydrogen, the amounts of which, in fact, are practically the same. The prime cause apparently is the constituent oxygen, which is in great excess in the inferior coal; and an explanation readily occurs. It is necessary that all this oxygen be volatilized, when it absorbs a portion of heat, which is thus diverted from the business of evaporation ; and though, no no doubt, it may subsequently restore the heat thus temporarily abstracted, in combining with the hydrogen as a gas, yet, as compared with atmospheric oxygen, which, in the absence of solid oxygen, supplies its place, the solid oxygen is at a disadvantage, in so far as atmospheric oxygen is yielded at once in the half-converted and desirable condition of a gas.

It appears, then, that the evaporative efficiency of coal varies directly with the quantity of constituent carbon, and inversely with the quantity of constituent oxygen ; but that it varies not so much because there is more or less carbon, as, chiefly, because there is less or more oxygen. The percentages of constituent hydrogen, nitrogen, sulphur, and ash, are practically constant, with individual exceptions, of course ; and their united influence should be so also. Treating the question as one of evaporative efficiency, the solution of it lies between the carbon and the oxygen.[*]

Mr. C. W. Williams has expounded, fully and explicitly, the physical conditions for the complete combustion of coal in ordinary furnaces. The cardinal condition, on which he

[*] The author drew attention to the apparent inverse relation between the percentage of the constituent oxygen in coal, and the evaporative efficiency of the coal, in his work, " Recent Practice in the Locomotive Engine," published in 1858-9, page 20*. M. Perissé recognises this inverse relation in his article on " Gas Furnaces," in the " Mémoires de la Société des Ingénieurs Civils," 1874, page 803.

has forcibly insisted, is the quick and complete intermixture of the gaseous elements for combustion—the air and the combustible gases ; and, for having drilled this principle of operation into the minds of engineers, it is probable that his memory will remain ever green. The complication that usually characterises the burning of coal is both physical and chemical—physical, because an intimate mixture and a suitable proportion of the elements concerned is essential to the completeness of their conversion ; chemical, because, unfortunately for the special object of the furnace, which is to generate heat, the less important element, hydrogen, is precisely that which demands the preference, and must have its share of oxygen, before the claims of the staple element, carbon, can be really satisfied. The occasional presence of oxygen and nitrogen in the fuel, in considerable quantity, leads to further complication of the process, as they must be volatilized and driven off in the course of distillation.

That an intimate mixture of the elements for combustion s of great practical importance for completing combustion is easily proved, experimentally, by delivering one or more jets of steam of considerable presence into the fireplace above the fuel. The steam acts as a " steam-poker," to stir about and intermix the air and the gases, and so complete the intermixture wanted for effecting entire combustion. On this principle, the system of Mr. M. W. Ivison, patented in 1838, was based. A steam-pipe was led from the boiler into the interior of the furnace, where it terminated above the grate ; "streams of steam " were delivered amongst the gases rising from the fuel, in order by mixing therewith "to consume the smoke," and perfect the combustion. But obviously, unless an insufficient quantity was present with the steam, combustion could not be accomplished ; and it followed that, in many situations, Mr. Ivison's system did not answer the purpose intended.

The writer, arguing that agitation and mixture, without

air, could not fulfil the purpose any more than air in quantity without mixture, devised a system, patented in 1857 and 1861, by which the presence of a supply of air for combustion was ensured, in conjunction with jets of steam employed for completing the mixture in the furnace. A few jets of steam were, in ordinary stationary boilers, delivered from the space above the doorway over the fire, towards the bridge, and they carried with them a supply of external air, by induction, which was drawn in and delivered right into the midst of the combustible gases, with which it was instantly and completely mixed. The supply of air was delivered through circular openings, one to each jet of steam, in which it was placed concentrically, or in a sheet through the door, directed upwards by simple arrangements to meet the jets of steam. The system is thoroughly effective for the prevention of smoke.

It is nevertheless unquestionable that, regardless of the quantity of air admitted, smoke may be prevented by the simple admission of air by the doorway, leaving, for this purpose, the door a little way open. Mr. C. W. Williams,* referring to the practice of stokers on the Mersey, says that " when the furnace door is partially opened, that is, ' kept ajar,' the air enters in a restricted quantity, and in a thin film, thus presenting an extended sheet or surface for contact with the newly formed gas, and effecting its combustion. If, however, the smoke still continues to be formed, the stoker, naturally but ignorantly expecting a good result from the same cause, opens the door wider. . . . . By the first operation, keeping the door slightly ajar, the effect is useful, both in respect to steam and smoke. When, however, the door is opened wider, the air, instead of entering as a thin sheet or film as before, rushes in in a body, with the force and effect of an onward current. The vigorous current

* "Letter on the Operation of the Smoke Nuisance Act," 1856, page 7.

necessarily counteracts or defeats the required mixing or diffusive action between the gases and the air; the latter passing rapidly towards the flues in an unbroken mass, and producing the cooling effect already mentioned."

Thus, even the apostle himself unbends, not hesitating to agree that a sufficient degree of diffusion may result when the air is admitted in the form of a thin sheet at the doorway, though not necessarily subdivided into jets. M. E. Bède * states that M. Combes, after having tried an arrangement similar to that of Mr. C. W. Williams for smoke prevention, arrived at the same result by simply delivering the air horizontally beyond the bridge, from two conduits in the sides of the furnace. Through sight-openings it could be observed that when, after firing, these openings were closed, dense black smoke was discharged, and that the smoke vanished when they were opened, and was replaced by brilliant flame. But, unfortunately, though smoke was effectually prevented, there was no saving of fuel.

Sir William Fairbairn † made many experimental observations on the action of Mr. C. W. Williams's system, the results of which were embodied in a Report to the British Association. He announced, as the general result of these experiments, that an economy of 4 per cent. of fuel was effected by the use of Mr. Williams's system, compared with the ordinary system of furnaces and stoking. This amount of economy was considerably less than that which was effected under Mr. Williams's direction. Sir William Fairbairn deduced from his experiments that the area of permanent opening for air above the fuel should not exceed 1 square inch per square foot of fire-grate for double flue-boilers, and 1¼ square inches for cylindrical boilers [under-fired, probably].

* " De l'Economie du Combustible," 1863, page 100.
† " Useful Information for Engineers," 1856, page 49.

Mr. Thomas Wicksteed * testified to the remarkable economy that may be derived by detaining, and actively and repeatedly mixing, the products of the furnace, and the air, in the flues. To effect this object, the floor or flame-bed beneath the boiler, when under-fired, or the lower portion of the tubular flue of internally-fired boilers, was formed as a succession of " semi-elliptical chambers," as Mr. Baker called them—being, in fact, a floor constructed like a series of curling sea-waves. The points of the ridges thus formed, entangled and diverted the inflamed currents, rolling them over and accelerating their intermixture. A supply of air was admitted by two successive openings at the back of the bridge. Burning small Newcastle coal of inferior quality, in two comparative experiments, made with plain Cornish boilers, and with the same boilers adopted with Mr. Baker's system, Mr. Wicksteed found that when the quantity of water evaporated per hour was about the same in the two cases, the evaporation per pound of coal amounted to 6·92 lbs. and 7·70 lbs. of water, respectively, from the initial temperature, 90° to 95° Fahr. ; or, reckoned from 212°, the water per pound of coal was respectively 8·22 lbs. and 9·27 lbs., showing an economy of 12¼ per cent. in favour of the wave-system of flue. The system was not tested as a smoke-preventer, but the quantity of visible smoke was materially diminished.

Mr. William Gorman,† in 1859, demonstrated the advantage of separate and alternate supplies of air above and below the grate in ordinary furnaces, "consuming the gas and the coke of coal at different intervals of time." After a fresh charge of coal, the ashpit was nearly closed, and the

---

* Report of Mr. Thomas Wicksteed on the Steam Boiler Furnace of Mr. Henry F. Baker ; dated June 21, 1848.

† See Mr. Gorman's paper " On the Combustion of Coal," in the " Transactions of the Institution of Engineers in Scotland," 1858-59, page 78.

M

air-apertures in the doorway were opened for the admission
of air above the fuel, to consume the gases which were raised
from the fresh fuel by the heat of the incandescent coke.
When the gases were burned off, the air-openings at the doorway were closed, and the ashpit-valve was opened for air to
pass through the grate and burn off the coke. From the
results of comparative experiments made with a small boiler ;
in which the coal was burned with the fire-doors constantly

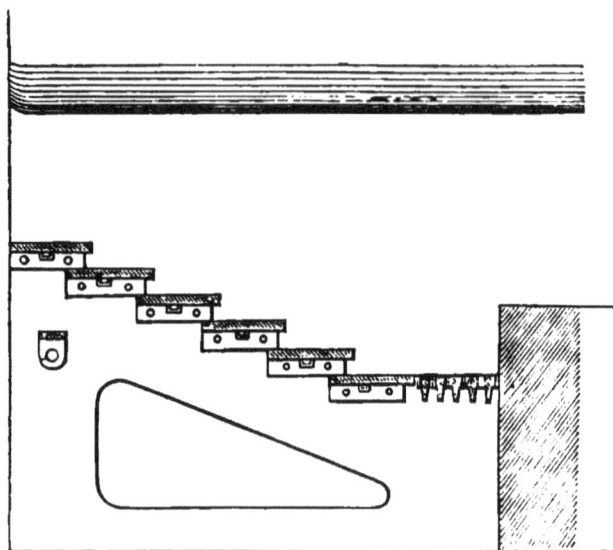

Fig. 117.—Step-Grate.

closed, and then burned on the alternate system of air-supply,
it appeared that in the second case, an evaporative duty
85 per cent. more was rendered than in the first case.
These are instructive results, pointing to the true method
of management in the combustion of fuel.

The step-grate was designed for the purpose of burning
small coal, slack, and decrepitating fuels, like some lignites
and dry coals. It consists of a number of cast-iron plates,
Fig. 117, arranged like steps in descending order, and supple-

mented by an ordinary flat grate. The coal is charged upon the upper plates, and is pushed down successively upon the lower plates. These grates have long been in use, principally in Germany. The small particles of fuel are not liable to fall through the step-grate, as they would do through ordinary grates; whilst the air-passages may bo made of any width. The conditions are favourable for the prevention of smoke. The greatest producers of smoke,—the coals of Mons and of Denain, have been burned with facility on this grate, without any smoke except at the time of charging fresh fuel. When these coals were mixed with one-fifth part of the dry coals of Charleroi, no smoke was produced even at the time of firing. It is, at the same time, to be remarked that the step-grate does not lend itself readily to the economical employment of dry coals, having a short flame, since by the disposition of the grate, the fuel is too far from the boiler.

For a 30-horse power boiler, M. Marsilly, who had much experience of the step-grate, employed the following dimensions :—

Length of the plates	3 feet $3\frac{1}{2}$ inches
Width „ „	8 inches
Advance of each plate, on the plate immediately above it	2 „
Thickness of the plates	$1\frac{3}{16}$ „
Vertical distance apart	$1\frac{3}{16}$ „
Distance of the first or highest plate from the boiler	$12\frac{1}{2}$ „
Distance of the lowest plate . . .	2 feet
Thickness of the bars of the horizontal or flat grate	$1\frac{3}{16}$ inches
Width of air-space between the bars .	$\frac{1}{4}$ „
Number of bars	5 „

From these data, the width of the flat grate is $7\frac{1}{2}$ inches; the width of the horizontal projection of the inclined grate is 3 feet 2 inches, and the total horizontal width of grate is

M 2

3 feet 9 inches.   Multiplied by 3 feet 3¼ inches, the length of the bars, the area of the grate amounts to 12¼ square feet. The sum of the widths of the air-spaces in the grate-surface is equal to 10 inches, or about one-fourth of the total width of grate.   With the step-grate now described, the production of steam was at the rate of 6·10 lbs. per pound of coal:—a result which was considered to be satisfactory.

CHAPTER IV.

EVAPORATIVE PERFORMANCE OF COAL IN A MARINE BOILER AT NEWCASTLE-ON-TYNE.

THE principal portions of a report embracing the results of an instructive course of experiments made in 1857, are given in the Appendix, p. 174. Those experiments, which were conducted by Messrs. Longridge, Armstrong, and Richardson, were made to test the evaporative power of the steam-coal of the Hartley district of Northumberland. The experimental boiler was of the marine type, 10 feet 8 inches long, 7 feet 6 inches wide, and 10 feet high ; with 2 internal furnaces, 8 feet by 3 feet 3 inches high, and 135 flue-tubes above the furnaces, in 9 rows of 15 each, 8 inches in diameter inside, 5¼ feet long. The dead-plates were 16 inches long, and 21 inches below the crown of the furnace. As the result of many preliminary trials, two standard lengths of fire-grates were fixed upon—4 feet 9 inches, and 3 feet 2¼ inches, with a fall of ¼ inch to a foot ; and the fire-bars were cast ½ inch thick, with air-spaces from ⅝ to ¾ inch wide. The fire-doors were made with slits ¼ inch wide and 14 inches long, for the admission of air. The chimney was 2 feet 6 inches in diameter. A water heater was applied at the base of the chimney, in the thoroughfare ; it contained 76 vertical tubes, 4 inches in diameter, surrounded by the feed-water.

Total area of fire-grates, 4 feet 9 inches long, 28½ square feet.
       ,,   ,,   3 ,, 2¼ ,,   ,, 19¼   ,,
Heating surface of boiler (outside), 749 square feet.
       ,,   water heater . . 320   ,,

Ratio of larger grate-area to heating surface of boiler, 1 to 26·28
„   smaller   „   „   „   1 to 38·91

Two systems of firing were adopted, as " standards of practice : "—First, ordinary or spreading firing, in which the fuel was charged over the grate, and the whole of the supply of air was admitted through the grate, Second, coking-firing, in which the fuel was charged, 1 cwt. at a time, upon the dead plate, and subsequently pushed on to the grate, making room for the next charge ; and air was admitted by the door-way as well as by the grate. Four systems of furnace were tried, of which Mr. C. W. Williams's was adjudged by the experimentalists to have rendered the best performance. According to this system, air was admitted above the fire at the front of the furnace, by means of cast-iron casings, having apertures on the outside, with slides, and perforated through the inner face, next the fire, with numerous ⅜ inch and ¼ inch holes, having total area of 80 square inches, or 5·33 square inches per square foot of grate. Alternate firing was adopted by Mr. Williams. The general results of the experiments are given in the following table.

NEWCASTLE COALS (OF THE HARTLEY DISTRICT OF NORTHUMBERLAND)—RESULTS OF EVAPORATIVE PERFORMANCE IN AN EXPERIMENTAL MARINE BOILER AT NEWCASTLE-ON-TYNE. 1857.

(Compiled from the Report of Messrs. Longridge, Armstrong, and Richardson, to the Steam Collieries Association of Newcastle-on-Tyne.)

Numerical Order.	Plan of Furnace.	Area of Fire-grate.	Coal consumed per hour.	Coal per sq. foot of grate per hour.	Water consumed from 60° per hour.	Water per sq. foot of grate per hour.	Water evaporated as from 212° per pound of coal.	Remarks on the Prevention of Smoke, &c.
		sq. ft.	cwt.	lbs.	cu. ft.	cu. ft.	lbs.	
1	Standard grate, ordinary management	28·5	5·38	21·15	74·80	2·62	8·04	{Air admitted entirely through the grate. Much smoke, often very dense.
2	Standard grate, best management	„	4·88	19·00	79·12	2·93	11·13	{Air admitted through both the grate and the door. No smoke.
3	Standard grate, ordinary management	19·25	3·61	21·00	56·01	2·91	10·00	{Air through the grate alone; used 100 cubic feet per pound of coal. Temperature in uptake, 448°. Much smoke.
4	Standard grate, best management	„	3·00	17·25	57·78	2·995	12·53	{Air through the grate and the door; used 70 cubic feet through the grate and 88 cubic feet through the doors, per pound of coal. Temperature in uptake, 480°. No smoke.
5	C. W. Williams's plan	22	3·33	17·27	61·59	2·80	11·70	Prevention of smoke, practically perfect.
6	„ „ „	„	5·30	26·98	88·96	4·04	10·80	Do.
7	„ „ „	18	4·40	27·36	76·92	4·31	11·37	Do.
8	„ „ „	15·5	5·18	37·40	85·30	5·51	10·63	{Prevention of smoke, practically perfect. Temperature at base of chimney above 600°.
1		2	3	4	5	6	7	8

NOTES TO TABLE.—1. When the temperature was 600° in the uptake of the boiler, it was reduced by from 40° to 50° after having passed through the water-heater.

2. In another case, working with Williams' apparatus, when no air was admitted through the door, and with much smoke, the temperature in the uptake was 600°. With one aperture in the door opened, it was raised to 625°; with two apertures, 683°; with three, 686°; with five it fell to 620.

3. The quantities in column 7 have been recalculated.—D. K. C.

The experimentalists reported that Mr. Williams's plan gave the best results, and they concluded: "1st. That, by an easy method of firing, combined with a due admission of air in front of the furnace, and a proper arrangement of fire-grate, the emission of smoke may be effectually prevented in ordinary marine multitubular boilers whilst using the steam-coals of the Hartley district of Northumberland. 2nd. That the prevention of smoke *increases* the economic value of the fuel and the evaporative power of the boiler. 3rd. That the coals from the Hartley district have an evaporative power fully equal to that of the best Welsh steam-coals, and that, practically, as regards steam navigation, they are decidedly superior."

These gentlemen made a trial of Aberaman Welsh coal, and they found that its practical evaporative power, when it was hand-picked, and the small coal rejected, was at the rate of 12·35 lbs. of water per pound of coal, evaporated from 212°; this may be compared with the best result from Hartley's coal, large and small together, in the table, which was 12·53 lbs. water from 212° per pound of coal, or with another result of experiment with Hartley coal, not given in the table, showing 12·91 lbs. water per pound of coal. As a check on these results, they ascertained the total heat of combustion of the two coals here compared, by means of an apparatus constructed by Mr. Wright, of Westminster, so contrived that a portion of coal is burned under water, and the products of combustion actually passed through the water, which absorbs the whole heat of combustion. The following are the comparative values:—

	Water practically Evaporated per Pound of Coal.	Total Heat of Combustion in Evaporative Efficiency.
Welsh coal, hand-picked .	. 12·35 lbs. .	. 14·30 lbs.
Hartley coal, large and small	12·91 ,,	. 14·63 ,,

The experimentalists also point out the " elasticity of

action " of the Hartley coals : they burned them at rates varying from 9 to $37\frac{1}{2}$ lbs. per square foot of grate per hour, without difficulty, and without smoke. The Welsh coal, burned at the rate of $34\frac{1}{2}$ lbs. per foot per hour, melted the fire-bars, it is said, after an hour and a half's work.

CHAPTER V.

EVAPORATIVE PERFORMANCE OF COAL IN LANCASHIRE AND GALLOWAY BOILERS AT WIGAN.

AN important and extensive series of experiments on the evaporative power of the coals of South Lancashire and Cheshire, were conducted at Wigan in 1866—68, by Mr. Lavington E. Fletcher. The results of these trials have been abstracted by the writer in his *Manu al of Rules, Tables, and Data :* from which the following notice is derived.

The coal selected for trial was Hindley Yard coal, from Trafford Pit, which ranks with the best coals of the district. Three stationary boilers were selected : 1st, an ordinary double-flue Lancashire boiler, 7 feet in diameter, and 28 feet long ; the flue-tubes were 2 feet 7½ inches in diameter inside, of ⅜ inch plate. 2nd, another Lancashire boiler of the same dimensions, in which the tubes were of $\frac{16}{1}$ inch steel plate. 3rd, a Galloway, or water-tube boiler, 26 feet long, and 6 feet 6 inches in diameter ; with two furnace-tubes 2 feet 7¾ inches in diameter, opening to an oval flue 5 feet wide by 2 feet 6½ inches high, containing 24 vertical conical water-tubes. The first and second boilers were new and specially constructed for the trials ; the third boiler was a second-hand one. These three boilers were set side by side, on side walls and with two dampers. The flame passed through the flue-tubes, back under the boiler, then along the sides to the chimney. The chimney was 105 feet high, above the floor ; octagonal, 6 feet 10 inches wide at the base, and 5 wide at the top, where the sectional area was 21 square feet.

Total grate-area in each boiler :—6 feet long ; 31·5 square feet.
     "   "   4   "   21·0   "

	Lancashire.	Galloway.
Heating surface, in flue-tubes	464·34 square feet.	431·12 square feet.
In external flues	303·08 "	288·24 "
Total surface	767·42 "	719·36 "
Ratio of grate-area, 6 feet long, to heating surface	1 to 24·4 "	1 to 22·8 "
Ratio of grate-area, 4 feet long, to heating surface	1 to 36·5 "	1 to 34·3 "
Circuit, or length of heating surface, traversed by the draught from the centre of the grate. . . .	80 feet.	74 feet.
Total distance from centre of grate to base of chimney	117 "	101 "
Height of chimney above level of floor.	100 feet.	
Height of chimney above level of grates.	96 feet 9 inches.	

A Green's fuel-economizer was placed in the main flue ; it had 12 rows of 4½ inch cast-iron pipes, 8 feet 9 inches long, placed vertically—84 tubes in all—having a collective heating surface of 850 square feet, exclusive of the connecting pipes at top and bottom. The feed-water was passed through the economizer on its way to the boiler, and absorbed a portion of the waste heat.

The fire-grates were tried at two lengths, 6 feet and 4 feet. The shorter grate gave the more economical result, but it generated steam less rapidly. Three modes of firing were tried—spreading, coking, and alternate firing. With round coal, on the whole, the greatest duty was obtained by coking firing, with the least smoke. With slack, alternate-side firing had the advantage.

Fires of different thicknesses were tried: 6 inches,

9 inches, and 12 inches. It was found that 9 inches was better than 6 inches, and 12 inches better than 9 inches. Air admitted at the bridge gave a slightly better result than by the door; and the admission of air in small quantity on the coking system, prevented smoke. The doors were double, slotted on the outside, and pierced with holes on the inner side. The maximum area of opening was 31½ square inches for each door, being at the rate of 2 square inches per square foot for the 6-feet grates, and 3 square inches for the 4-feet grates. The amount of opening was regulated by a slide.

The standard fire adopted for trial was 12 inches thick, of round coal, treated on the coking system, with a little air admitted above the grate, for a minute or so after charging.

The water was evaporated under atmospheric pressure.

The quantity of refuse from the Hindley Yard coal, averaged, in the trials with the marine boiler, to be afterwards described, 2·8 per cent. of clinker, 2·8 per cent. of ash, and 0·8 per cent. of soot, in all, 6·4 per cent. Making allowance for the difference of soot, the total refuse may be taken, in the trials of the stationary boilers, at 6 per cent.

General Deductions.—The advantage of the 4-feet grate over the 6-feet grate, was manifested by comparative trials with round coal 12 inches thick, and slack 9 inches thick. With the 4-feet grate, the evaporative efficiency, taking averages, was 9 per cent. greater than with the 6-feet grate; though the rapidity of evaporation was 15 per cent. less, at the same time that 19½ per cent. more coal was burned per square foot per hour.

When equal quantities of coal were burned per hour, the fires being 12 inches thick, 8 per cent. more efficiency and 12 per cent. greater rapidity of evaporation were obtained from the shorter grate. Thus :—

Coking Firing.

Length of grate . . .	6 feet.	4 feet.
State of damper . . .	two-thirds closed.	fully open.
Coal per hour . . .	4·0 cwt.	4·14 cwt.
Coal per square foot of grate per hour	14 lbs.	23 lbs.
Water at 100° evaporated per hour	65 cubic feet.	72·6 cubic feet.
Water at 212° per pound of coal	10·10 lbs.	10·91 lbs.
Smoke per hour :—		
Very light. . . .	4·3 minutes.	4·1 minutes.
Brown	0·4 ,,	0·3 ,,
Black.	0·0 ,,	0·0 ,,

To compare the performances with coking and spreading firing, having 12-inch fires for round coal, and 9-inch fires with slack:—Whilst, with round coal, the rapidity of evaporation was the same with both modes of firing, the efficiency was from 3 to 4 per cent. greater with coking. With slack, on the contrary, the spreading fire evaporated a fourth more water per hour than the coking fire, though with 4¼ per cent. less efficiency.

With thicknesses of coking fire, 6 inches, 9 inches, and 12 inches, for round coal; and 6 inches and 9 inches for slack; the results were in all respects decidedly in favour of the thicker fires than the thinner fires. Comparing the thinnest and the thickest fires, from 5½ to 20 per cent. more water was evaporated per hour by the thickest fires, and from 11 to 18 per cent. more per pound of fuel.

The effect of the admission of air above the grate, continuously or intermittently, for the prevention of smoke, as compared with that of its non-admission, was ascertained with round coal, and with slack. The averaged results showed that by admitting the air above, the evaporative efficiency was increased 7 per cent.; but that the rapidity of evaporation was diminished 3½ per cent.

Comparing the admission of air above the fuel at the door, and at the bridge through a perforated cast-iron plate; it

was found that the admission at the bridge made a better
performance, by about 2½ per cent. than at the door.

To try the effect of increasing the supply of air above the
fuel, the door-frame was perforated to give an additional
square inch of air-way per foot of grate, making up 3 square
inches ; an allowance of 1 square inch was also provided at
the bridge.   Round coal was burned on the coking system,
12 inches thick, on 6-feet grates, with a constant admission
of air above the fuel.   When the supply by the door was
increased from 2 inches to 3 inches per square foot of grate,
the evaporative efficiency fell off 8½ per cent., and the
rapidity 3 per cent.   When an extra inch was supplied at
the bridge, making up 4 square inches per foot of grate, the
evaporative efficiency only fell off 0·65 per cent., and the
rapidity 1½ per cent.   The effect of this evidence is, that
the bridge is the better place for the admission of air, and
that if the air be admitted by the bridge alone, the area of
supply may be beneficially raised to 4 square inches per
square foot of grate.

Comparing the effect of the admission of air in a body,
undivided, with that of its admission in streams, on a 6-feet
grate, with coking fires 12 inches thick of round coal, there
was 6½ per cent. of loss of efficiency by the admission in a
body, though the smoke was equally well prevented.

Mr. Fletcher concludes that the greatest rapidity of
evaporation was obtained when the passages for the admission of air above the fuel were constantly closed ; that the
next degree of rapidity was obtained when they were open
only for a short time after charging, and the lowest when
they were kept open continuously.   He also concludes that,
whilst, in realizing the highest power of a free-burning and
gaseous coal, smoke is prevented ; yet, in realizing the
highest power of the boiler, smoke is made.

In burning slack, smoke was prevented as successfully as
in burning round coal, though its evaporative efficiency was

from 1 to 1¼ lb. of water per pound of fuel less than with round coal.

To work out the problem of firing slack without smoke, and without loss of rapidity of evaporation, trials were made at the boilers of 16 mills, when the slack was fired on the alternate-side system. No alterations were made in the furnaces in preparation for these trials; in many instances, the fire-doors had no air-passages through them. The grates were from 3 feet 7 inches to 7 feet long; they averaged 6 feet in length.

Number of boilers fired . . . . . .   65 boilers.
Slack burned per boiler per week of 60 hours .   17·35 tons.
Slack per square foot of grate per hour . . -   19·25 lbs.
Smoke per hour:—
   Very light . . . . . .   11·5 minutes.
   Brown . . . .   2·3 „
   Black . . . .   0·3 „
                   14·1

In 12 instances, no black smoke whatever was made. It is said that the steam was as well kept up, and the speed of the engines as well maintained, as before the trials were made.

The performance of the double-flue boilers amounted practically to the same as that of the water-tube boiler.

CHAPTER VI

COAL-BURNING IN LOCOMOTIVES.

MR. MICHAEL REYNOLDS, in his excellent work on "Locomotive Engine Driving," thoroughly and authoritatively discusses the management of the fire in a locomotive engine, so as to keep up steam, to prevent waste, and to economise the fuel. "Sometimes," he says, "Welsh coal, probably from being too much wetted, hangs together in the firebox, and prevents the engine from steaming evenly, causing the fire to burn hollow, and draw air. An engine fire will sometimes run for miles without any variation of pressure in the boiler of ½ lb. per square inch, either way, unless the feed happens to be put on. The first thing required to be done with such a stupid fire is to get a shovelful of small coal, and scatter them pell-mell over the top of the fire, but chiefly along the sides and front of the box; the effect will be that some of the small coals put on in this way will fall into the very hole or holes through which the engine is drawing air. Tobacco-smokers sometimes do this kind of thing to get their pipes to burn. They slightly move the tobacco with their finger on the top, and that knocks a morsel of weed into the air-hole, and the pipe afterwards burns charmingly. When this plan is of no effect—and this can be soon ascertained by watching if the needle of the pressure-gauge begins to rise—the dart should be thrust into the centre of the fire, and the fire gently raised so as to open it if it close, or to close it together when it has burned hollow. Provided

that a driver can see his way with such a fire, it is, in point
of economy, best to leave it undisturbed, for, sooner or later,
the action of the blast and the vibration of the road will
bring it round ; but, of course, on fast and important
trains, action is required to be taken at once, and either
of the two mentioned remedies will seldom fail to move the
needle.

"The coal-fire of haycock shape, eminently associated
with failures through want of steam, is made by shovelling
the coals into the middle of the firebox, a practice about as
far behind the times, comparatively speaking, as the use of
the flint and the tinder-box would be in the year 1878. The
characteristics of such a fire are—uncertainty as regards
making steam, and certainty as regards destruction to fire-
boxes and tubes. It generally draws air at the walls of the
box, and, in consequence, the fire-irons are always in the
fire, knocking it about, and wasting the fuel. As such fires
are formed on the centre of the grate, they weigh down the
fire-bars in the middle, and may even cause them to drop
off their bearers or supports. But there are greater evil
consequences even than these : the cold air being admitted
into the firebox up the sides, instead of in the middle, comes
into direct contact with the heated plates and stays, doing
them a deal of damage by causing intermittent expansion
and contraction.

"That the fire in a locomotive firebox should maintain
steam under all circumstances of load and weather, should
consume its own smoke, should burn up every particle of
good matter in the coal, and, in fine, should be worked to the
highest point of economy, it requires to be made in the
beginning, and maintained, to a form almost resembling the
inside of a tea-saucer, shallow and concave, where the
thinnest part of the fire is in the centre. A fire of this
form makes steam when other fires do not, being built on
a principle that never yet misled either the driver or the

fireman.   It has brought a man a good name many a time.

"How to fire?   This is a very important question.

"The first shovelful of coal should find a billet in the left-hand front corner; the second shovelful in the right-hand front corner; the third shovelful in the right-hand back corner; the fourth shovelful in the left-hand back corner; the fifth shovelful under the brick arch close to the tube-plate; the sixth, and last, under the fire-door.   To land this one properly, the shovel must enter into the fire-box, and should be turned over sharp to prevent the coals falling into the centre of the grate or the fire.

"It will at once be seen that this fire is made close against the walls of the firebox, and in actual contact with the heating surface; also that the principal mass of the coals lies over the bearers which carry the fire-bars.   The centre of this kind of fire is self-feeding, for, by the action of the blast and the shaking of the engine, the lumps in the corners are caused to roll or fall towards the centre.   On this system, the centre is the thinnest part of the fire, quite open and free from dirt; the dirt falls down by the sides of the copper plates, and assists in preventing the cold air from touching the plates.   With a fire of this description, the air or oxygen can only get into the firebox and into the neighbourhood of the tubes through the centre—through fire— and, mingling with the flame, it becomes instantly heated to a very high temperature before entering the tubes, which are thereby assisted in maintaining an even pressure in the boiler.

"Coals of the same description have been delivered to two different drivers, having engines of the same class, working on the same day, running the trains over the same ground, with equal average loads, and the result has been, that while one driver could do anything with the coals, the other man was 'afraid' of them.   The former put his coals against the walls of the firebox, and the latter put them in the centre of the grate.

"The secret of first-rate firing is to fire frequently, a little at a time." *

* "Locomotive Engine Driving," page 78.

CHAPTER VII.

COKE.

COKE, as has already been stated, is the solid residuum of coal from which the volatilizable portions have been removed by heat—a process which is illustrated in the action of ordinary furnaces, in which the gasified elements of coal are first burned off, then the fixed or residuary coke.

Quantity of Coke yielded by Coal.—The quantity of coke produced from coal, excluding anthracite, is found, by laboratory analysis, to be as follows :—

COKE (*Excluding Anthracites*).

English coals	50 to 72 per cent.
American coals	64 to 86 ,,
French coals	53 to 76 ,,
Indian coals	52 to 84 ,,

The percentage of coke obtained from coal, excluding anthracite, is thus seen to vary from 50 to 86 per cent., and it varies as much in quality. Anthracite cokes scarcely deserve the name of coke; they are without cohesion and pulverulent, or powdery. The best coke is produced from coals of bituminous quality: it is clear, crystalline, and porous, and is formed in columnar masses. It has a steel-grey colour, possesses a metallic lustre, with a metallic ring when struck, and is so hard as to be capable of cutting glass.

Coke comprises, besides the fixed carbon of coal, the ash, or incombustible element of coal; and, therefore, though a given coal may yield a large percentage of coke, the coke may be of inferior quality, and may do less duty than a smaller yield of coke from another coal which contains a less quantity of ash. For instance, Australian coal gives 68·27 per cent. of coke, containing 8·38 per cent. of ash ; whilst the Nagpore coal yields 76 per cent. of coke, containing 18·73 per cent. of ash. But though the yield of coke from the Nagpore is the greater, yet its gross efficiency must be the less, since the Australian contains 60 per cent. of fixed carbon, and the Nagpore only 57 per cent. of the total weight of coal, after deducting the percentage of ash.

The quality of coke obviously depends in a great measure on the proportion of the constituent hydrogen and oxygen of the coal from which it is made, which regulate the degree of fusibility of the coal when exposed to heat. Taking for example the particulars of the coke produced from French coal, and arranging the average for each kind of coal in the order of the quantity of hydrogen in excess, the nature of the coke produced, as described by M. Péclet, was as follows :—

Averages.	Hydrogen.	Oxygen and Nitrogen.	Hydrogen in excess.	Nature of the Coke.
	per cent.	per cent.	per cent.	
Anthracite	2·67	2·85	2·43	pulverulent
Dry coals, long flame .	5·23	16·01	3·09	in fragments
Bituminous coals, long flame . .	5·35	8·63	4·15	porous
Bituminous hard coals	4·88	4·38	4·27	,,
Bituminous coking coals	5·08	5·65	4·30	very porous

Showing a series of five coals, with an ascending series of hydrogen in excess, from 2·43 to 4·30 per cent. The nature of the cokes advances correspondingly from pulverulent, or powdery, to very porous or excessively fused and raised. The first is, in fact, a failure as a coke, and the second, with 3·09 per cent. of hydrogen in excess, barely coheres, being in fragments; the third and fourth, with about 4·20 per cent. of hydrogen in excess, produce a porous and cohesive coke; and the fifth, an excessively porous coke, bright, but comparatively light for metallurgical operations.

From this it appears that coal that has less than 3 per cent. of hydrogen in excess is unfit for coke-making; and that, for the manufacture of good coke, coal containing at least 4 per cent. of free hydrogen is required. The hydrogen, being in combination with carbon, in various proportions to form tar and oil, softens the fixed carbon, and forms a pasty mass, which is raised like bread by the expansion of the confined gases and vapours seeking to escape.

Coke of good quality weighs from 40 lbs. to 50 lbs. per cubic foot, solid, and about 30 lbs. per cubic foot, heaped. The average volume of 1 ton is 75 cubic feet. In composition, coke varies within the following limits :—

			Average of 19 Cokes.
Carbon	. . .	85 to 97½ per cent.	. . . 93·5 per cent.
Sulphur	. . .	¾ to 2 „	. . . 1·2 „
Ash	. . .	1½ to 14½ „	. . . 5·3 „
			100·0

Coke is capable of absorbing from 15 to 20 per cent. of its weight of water from the atmosphere. Exposed to the atmosphere for a length of time, it commonly holds from 5 to 10 per cent. of moisture.

The best experience of the combustion of coke has been derived from the practice of locomotives. A rapid draught is required for effecting the complete combustion of coke, preventing the reaction which is likely to take place when currents of carbonic acid traverse ignited coke, and become converted into carbonic oxide. The writer, in 1852, showed by a process of mechanical analysis,* that the combustion of coke in the firebox of the ordinary coke-burning locomotive was practically complete. The total heat of combustion of 1 lb. of good sound coke was found ordinarily to be disposed of as follows, when the temperature in the smoke-box did not exceed 600° Fahr. :—

78 per cent. in the formation of steam
16½   „   by the heat of the burnt gases in the smoke-box.
 5½   „   draw-back by ash and waste.
───
100

This appropriation of the performance of one pound of coke is based on the chemical fact, that the maximum evaporative power of absolutely pure coke,—entirely carbon,—is expressed by a trifle more than 12 lbs. of cold water supplied at 60° Fahr., evaporated into high-pressure steam, by 1 lb. of such coke. Of this ultimate performance, 78 per cent. represents the evaporation of 9½ lbs. of water, and 16½ per cent. represents the heat carried off by the products of combustion, which, if economised, would evaporate additionally 2 lbs. of water.

These conclusions were subsequently corroborated by the results of a chemical analysis in 1853, of the products of combustion of coke in the engines of the Paris and Lyons Railway, by MM. Ebelmen and Sauvage. They experimented with passenger and with goods engines ; and they found that the proportion of carbonic acid contained in the gases collected from the tubes at the smoke-box was greater than was found in the gases from ordinary boiler-furnaces, whilst the proportion of free oxygen due to surplus air in the gases, was less in the locomotive. In the passenger engines and mixed-traffic engines it was found that the proportion of carbonic acid varied from 12 to 18½ per cent.

* "Railway Machinery," page 122.

of the total volume of the gases, without any trace of
carbonic oxide, proving that the carbon was entirely converted into carbonic acid.   In the goods-engines with deep
charges of coke, a greater proportion of carbonic oxide was
produced.   When the fire was 40 inches deep, there was
7½ per cent. of oxide, by volume, representing an equal
volume of carbonic acid displaced.   The total volume of
carbonic acid in the gases of completely burned coke.
averages 20½ per cent. of the total volume, and it appears
that in this instance, a third of the carbon was discharged
as carbonic oxide.   But a depth of 40 inches is excessive,
as a matter of ordinary practice ; and to exemplify the
influence of draught upon the state of the combustion, it
may be added that, when steam was shut off, the production
of carbonic oxide rose as high as 12 per cent. of the entire
volume.

CHAPTER VIII.

LIGNITE, ASPHALTE, AND WOOD.

LIGNITE, or as it is occasionally called, brown coal, though it is often found of a black colour, belongs to a more recent formation,—the tertiary,—than coal. It is, in fact, an imperfect coal. Brown lignite is sometimes of a woody texture, sometimes earthy. Black lignite is either of a woody texture, or it is homogeneous, with a resinous fracture. Some lignites, more fully developed, are of a schistose character, with pyrites in their composition. The coke produced from various lignites is either pulverulent, like that of anthracite, or it retains the forms of its original fibres. Lignite is less dense than coal.

Asphalte, like lignite, has a large proportion of hydrogen, but it has less of oxygen and nitrogen: and, having 8¼ per cent. of free hydrogen, it yields a firmer coke. The average composition of perfect lignite and of asphalte, may be taken in round numbers, as follows :—

	Lignite.	Asphalte.
Carbon	69 per cent.	79 per cent.
Hydrogen	5 „	9 „
Oxygen and nitrogen . .	20 „	9 „
Ash	6 „	3 „
	100	100

The lignites are distinguished from coal by the large proportion of oxygen in their composition,—from 13 to 29 per cent.,—which goes far to neutralise the hydrogen.

N

The woods of resinous trees are nearly identical in chemical composition, which may be taken as averaging thus :—

PERFECTLY DRY WOOD.

Carbon	50 per cent.
Hydrogen	6 ,,
Oxygen	41 ,,
Nitrogen	1 ,,
Ash	2 ,,
	100

showing that there is only 56 per cent. of combustible matter, that there is a large quantity of oxygen, nearly sufficient to neutralise the whole of the hydrogen, and that there is only 2 per cent. of ash. The composition of ordinary firewood, including hygrometric water, is as follows :—

Hygrometric water	25 per cent.
Carbon	37·5 ,,
Hydrogen	4·5 ,,
Oxygen	30·75 ,,
Nitrogen	0·75 ,,
Ash	1·5 ,,
	100·00

English oak weighs 58 lbs. per cubic foot, and yellow pine 41 lb. per cubic foot. A cord of pine wood,—that is, of pine wood cut up and piled,—in the United States, measures 4 feet by 4 feet, by 8 feet, and has a volume of 128 cubic feet. Its weight, in ordinary condition, averages 2,700 lbs. equivalent to 21 lbs. per cubic foot.

It has been ascertained in America, that 1 ton of Cumberland coal, best quality, is equal to 2·12 cords, or 2·55 tons of pine wood. From this it would follow that 1 lb. of coal is equivalent to 2·55 lbs. of pine: or, that pine has in practice only two-fifths of the evaporative power of coal, equal to about 2¼ lbs. of water evaporated per pound of pine. According to the results of other experiments in locomotives, 1 lb. of coal is equivalent to 3 lbs. of pine wood. This indicates an evaporative power of only 2 lbs. of water per pound of fuel. Mr. Haswell states that from 2¼ lbs. to 2¼ lbs. of pine are equal to 1 lb. of the best coal; and, allowing 6 lbs. of water to be evaporated per pound of coal, —in 1850, at the time of the observation,—the water evaporated per pound of pine was 2½ lbs. Professor W. R. Johnson found, in 1844, that 1 lb. of dry pine would, by careful management, evaporate 4·69 lbs. of water.

The results of recent experiments made in 1869—1874, with unseasoned pine wood and with kiln-dried wood, will be afterwards noticed, as given by Mr. William Anderson, in the Chapter on *Peat*. They show that in a stationary double flue-boiler, wood which had been cut one year, and was then damp, evaporated 3¼ lbs. of water from and at 212° Fahr., per pound of wood; and that the desiccated wood evaporated 5 lbs.

CHAPTER IX.

PEAT.

PEAT is the organic matter, or vegetable soil of bogs, swamps, and marshes,—decayed mosses or sphagnums, sedges, coarse grass, &c.—in beds varying from 1 or 2 feet to 20, 30, or 40 feet deep. The peat near the surface, less advanced in decomposition, is light, spongy, and fibrous, of a yellow or light reddish-brown colour; lower down, it is more compact, of a dark-brown colour; and in the lowest strata, it is of a blackish brown, or almost a black colour, having a pitchy or unctuous feel, the fibrous texture nearly or altogether obliterated.

Peat, in its natural condition, generally contains from 75 to 80 per cent. of its entire weight, of water. The constituent water occasionally amounts to 85 per cent. or even to 90 per cent.; in this case, the peat is of the consistency of mire. It shrinks very much in drying; and its specific gravity, when dry, varies from ·22 or ·34 to 1·06; the surface peat being the lightest, and the lowest peat the densest. If peat be masticated, macerated, or milled, whilst it is wet, so that the fibre is broken, crushed or cut, the contraction in drying is much increased by the treatment; and the peat becomes denser, and is better consolidated than when it is dried as cut from the bog. Peat so prepared is known as *condensed peat;* and the degree of condensation varies according to the natural·heaviness of the peat. Peat from the lowest beds, so treated, is condensed only to a small extent; but peat from the middle and the upper beds, becomes con-

densed, when dry, to from two to three times its natural
density. So effectively is peat consolidated and condensed
by the simple process of destroying the fibres whilst wet,
that no merely mechanical force of compression is equal in
efficiency to mastication. Mr. A. McDonnell gives the composition of average " good air-dried " peat and "poor air-
dried " peat, analysed by Dr. Reynolds, as in the annexed
table. An analysis by Dr. Cameron of dense peat from
Galway is added.

COMPOSITION OF IRISH PEATS.

Description.	Moisture.	Carbon.	Hydrogen.	Oxygen.	Nitrogen.	Sulphur.	Ash.	Coke.
	per cent	per cent	per cent	per cent	per cent	per cent	per cent	per cent
Good air-dried	24·2	45·3	4·6	24·1		—	1·8	—
Poor air-dried	29·4	42·1	3·1	21·0		—	4·4	—
Dense, from Galway }	29·3	42·0	5·1	17·5	1·7	·6	3·8	31·3
Averages .	27·8	43·1	4·3	21·4		·2	3·3	—

Ordinary air-dried peat contains from 20 to 30 per cent. of
its gross weight, of moisture. If dried in air in the most
effective manner, it contains at least 15 per cent. of moisture ;
and even when dried in a stove, it seldom holds less than
7 or 8 per cent.

The weight of a solid cubic foot of air-dried peat varies
from 15 lbs. to 66 lbs. according to the original formation of
the peat. Condensed peat weighs from 60 lbs. to 80 lbs. per
cubic foot solid. In heaps, the weight per cubic foot of air-
dried peat is, of course, much less ; it varies from 6 lbs. to
22½ lbs. per cubic foot. From this, it follows that a ton of
the lightest air-dried peat may occupy a space of 370 cubic
feet ; a ton of the densest air-dried peat occupies 100 cubic
feet of space ; whilst a ton of condensed peat only occupies
a space of from 40 to 50 cubic feet.

British peat and foreign peat are very much like Irish peat in composition; the principal variation takes place in the proportion of ash.

The average evaporative performance of dry peat in steamboilers is one half of that of good coal, weight for weight. But the proportion varies either way from the average. Mr. William Anderson, in an instructive Note,* gives an account of comparative experiments conducted by M. Keerayef, at the Abouchoff Steel Works, near St. Petersburg, on the evaporative performances of coal, wood, and peat, in double-flue multi-tubular cylindrical boilers. The peat was of a compact quality, was moulded by hand into 4-inch balls, and airdried till the moisture did not exceed 14 per cent. The wood consisted of a mixture of red pine and white pine in billets. These results, together with those of some experiments made by Mr. Anderson in a boiler of the same class, at Erith Iron Works, are given below. The fire-grate area amounted to 30 square feet, and the heating surface to 696 square feet :—

RESULTS OF COMPARATIVE EXPERIMENTS ON THE EVAPORATIVE POWERS OF COAL, WOOD, AND PEAT.
(*Mr. Anderson's Experiments.*)

Locality and Description of Fuel.	Fuel consumed per hour.	Water evaporated per hour, at 100·4° F.	Water evaporated at 212° F., per lb. of fuel.
	lbs.	cubic feet.	lbs.
Erith Iron Works, 1869-70 :—			
Good Newcastle coal . . .	335	48	9·79
Good Welsh coal	351	50	10·09
Abouchoff Steel Works, 1870-74			
Superior coal	450	49·5	7·57
Inferior coal	515	51	6·76
Wood cut 1 year; still damp .	796	38·6	3·25
Wood dried artificially . . .	538	40·4	5·0
Peat	—	—	4·26

* "Notes of a visit paid to some Peat Works in the neighbourhood of St. Petersburg, in May, 1875," in the *Proceedings of the Institution of Civil Engineers*, vol. 41, 1874-75, page 202.

Here is to be noted the superior efficiency of desiccated wood compared with damp wood, already noticed. The peat, containing 14 per cent. of moisture was more efficient for evaporation than the undried wood ; and its performance, 4·26 lbs. of water per pound of peat, is precisely one half of the average performance, 8·55 lbs. per pound of fuel, of the four coals in the table :—a ratio which is corroborative of the commonly accepted value of peat compared with that of coal.

CHAPTER X.

TAN, STRAW, AND COTTON-STALKS.

TAN, or oak-bark, after having been used in the process of tanning, is burned as fuel. The spent tan consists of the fibrous portion of the bark. According to M. Péclet, five parts of oak-bark produce four parts of dry tan, and the heating power of perfectly dry tan, containing 15 per cent. of ash, is 6,100 English units, whilst that of tan in an ordinary state of dryness, containing 80 per cent. of water, is only 4,284 English units. The weight of water evaporated from and at 212° Fahr. by one pound of tan, equivalent to these heating powers, is—

> For perfectly dry tan . . . . . . 6·31 lbs.
> For tan containing 30 per cent. of moisture . 4·44 „

These results are in accord with the results of experiments made by Professor R. H. Thurston on the evaporative power of spent tan, fresh from the leaches, containing from 55 to 59 per cent. of moisture as compared with air-dried tan, weighing 42½ lbs. per cubic foot. By the combustion of the wet tan, from 3½ lbs. to 4¼ lbs. of water was evaporated per pound of the tan; or, allowing for the excess of moisture in the tan, from 4·41 lbs. to 5·68 lbs. of water per pound of air-dried tan.

Mr. John Head* states the results of many experiments

* See "A Few Notes on the Portable Steam Engine," by Mr. John Head, 1877; page 42.

with straw and cotton-stalks as fuel in portable-engine boilers; from which it appears that from $2\frac{1}{8}$ lbs. to $2\frac{1}{4}$ lbs. of water is evaporated per pound of straw containing 16 per cent. of moisture, and from $2\frac{3}{4}$ lbs. to 3 lbs. of water per pound of cotton-stalks or brushwood.

CHAPTER XI.

LIQUID FUEL—PETROLEUM.

PETROLEUM, though as a liquid fuel last in order, is certainly not the least powerful fuel. It stands, on the contrary, first in heating power. The average composition is as follows:—

Carbon	85 per cent.
Hydrogen	13 ,,
Oxygen	2 ,,
	100

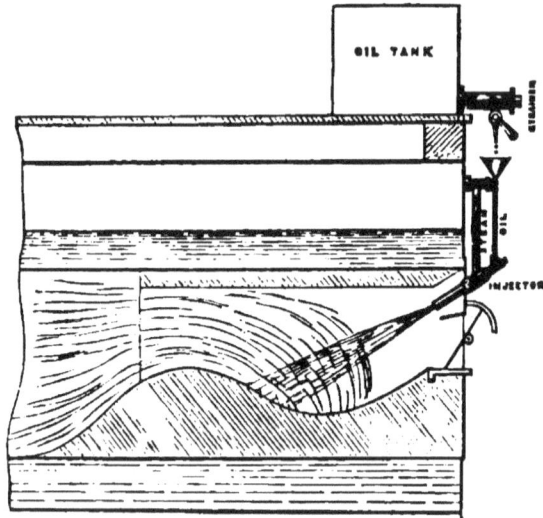

Fig. 118.—Liquid Fuel for Steam Boiler.

In a valuable paper on liquid fuels,* Mr. Harrison Aydon gives the results of many experiments made with petroleum

* On "Liquid Fuels," in the *Proceedings of the Institution of Civil Engineers*, 1877-78, vol. lii., page 177.

as a fuel. Burnt under steam boilers, on the system of
Messrs. Wise, Field, and Aydon,—a mixture of petroleum,
air, and superheated steam,—the fuel has been proved to be
capable of evaporating, under ordinary circumstances, 20 lbs.
of water from and at 212° Fahr., per pound of fuel. In
one instance, it appeared that 25·2 lbs. of water was evaporated at 85 lbs. pressure, per pound of oil at 50° Fahr.: equivalent, when reduced for 212° Fahr., to 28·9 lbs. of water per
pound of fuel. The arrangement of the furnace with which
this performance was accomplished is shown in Fig. 118.
The liquid fuel is injected over the door into the furnace,
with steam, either plain or superheated, so as to convert the
oil into vapour, and at the same time to mix with it just a
sufficient proportion of air to insure complete combustion.
The boiler was of the Cornish type, 80 feet long by 7 feet in
diameter; the flue, 3 feet in diameter, was made up with
firebrick to form the furnace, as shown.

CHAPTER XII.

TOTAL HEAT OF COMBUSTION OF FUELS.

THE annexed table * shows, in a small compass, the total heat of combustion of one pound of combustibles, and their equivalent evaporative powers, with the weight of oxygen and the quantity of air chemically consumed.

TOTAL HEAT EVOLVED BY COMBUSTIBLES, AND THEIR EQUIVALENT EVAPORATIVE POWER, WITH THE WEIGHT OF OXYGEN AND QUANTITY OF AIR CHEMICALLY CONSUMED.

Combustible.	Weight of oxygen consumed per lb. of combustible.	Quantity of air consumed per lb. of combustible.		Total heat of combustion of 1 lb. of combustible.	Equivalent evaporative power of 1 lb. of combustible, under one atmosphere, at 212° F.
1 lb. weight.	lb.	lb.	cubic ft. at 60° F.	units.	lb. of water from and at 212°
Hydrogen	8·0	34·8	457	62,032	64·2
Carbon, making carbonic oxide .	1·33	5·8	76	4,452	4·61
Carbon, making carbonic acid .	2·66	11·6	152	14,500	15·0
Carbonic oxide . .	0·57	2·48	33	4,325	4·48
Light carburetted hydrogen . .	4·0	17·4	229	23,513	24·34
Bi-carburetted hydrogen, or olefiant gas . . .	3·43	15·0	196	21,343	22·09
Sulphur	1·00	4·35	57	4,032	4·17
Coal of average composition . .	2·46	10·7	140	14,133	14·62
Coke, desiccated . .	2·50	10·9	143	13,550	14·02
Wood ,, . .	1·40	6·1	80	7,792	8·07
Peat ,, . .	1·75	7·6	100	9,951	10·30
Lignite	2·03	8·85	116	11,678	12·10
Asphalte	2 73	11·87	156	16,655	17·24
Straw, 15¾ per cent. moisture . . .	·98	4·26	56	5,196	5·56
Petroleum	4·12	17·93	235	27,531	28·50

* Abstracted from *A Manual of Rules, Tables, and Data*, page 405.

CHAPTER XIII.

GAS-FURNACES:—FUNCTION AND OPERATION OF GAS-FURNACES.

GAS-FURNACES have been employed on the Continent, particularly in Styria, for upwards of 35 years. It was early apprehended that, whilst the principles of the combustion of fuel in an ordinary grate were explicit enough, the work of combustion and of generation of heat was, for the most part, but roughly completed in reality, and that the defects of the open-grate system of combustion comprised not only a conversion of the elements in a greater or less degree imperfect, but also a maximum temperature frequently much lower than that which was chemically attainable. The inferiority of the temperature which is made available in this manner, is, for many applications of heat, of much more serious import than the actual loss of heat by quantity; and it has long been recognised that the only system of heating which carries with it a complete remedy for the shortcomings of the ordinary fire, and the imperfections of various fuels, is the system of heating by gases generated from the fuel, by a process resembling distillation, which, being collected and conducted to the region where the heat of combustion is to be utilised, are mixed with air, ignited, inflamed, and consumed on the spot. These gaseous substances arriving without the accompaniment of ash or cinder, at the place for action, do not alter or affect the surfaces of the bodies destined to receive the heat; and the heat is discharged, by

radiation as well as by conduction, just where it is required.
It appears that the idea of transforming solid combustibles into
gaseous combustibles, as well as of making the first practical
application of the idea, is due to M. Ebelmen, who, in
January, 1842, read a paper at the Academy of Sciences, in
which he explained the results of experiments which he had
made at the forges at Audincourt during the preceding year.
He drew attention to the availability of the débris of charcoal, and of other combustibles of little value, by subjecting
them to the action of a blast of hot air, by which they were
converted into carbonic oxide, and of employing this gas
as a combustible for the operations of iron manufacture.
He clearly indicated the elementary principles of the process
for the manufacture of gas-fuel :—the use of air, either forced
or not forced ; either hot or cold, for producing the gases or
for burning them ; the use of steam ; the division of the
air and of the gas into thin sheets, for the purpose of producing a perfect mixture, and the recovery of the heat
previously absorbed by the air.

Mr. William Gorman,* in 1859, highly appreciating the
superiority of the system of combustion by preliminary conversion of the fuel into gas, foresaw the advantage of the
system in the facilities afforded by it for placing the combustion where it would be most effective. " This power,"
he says, " of transferring the greater part of the combustion
of the coal from the grate to the body of the furnace, together with complete combustion of the coal gas, promises
to be of great use in the manufacture and working of
iron."

But it was twenty years after Ebelmen had first drawn
attention to the advantages of converting solid combustibles
into gas-fuels—when Dr. C. William Siemens had accomplished the remarkable practical results of his system of

* "On the Combustion of Coal," in the *Transactions of the Institution of Engineers in Scotland*, vol. ii., 1858-59; page 79.

regenerative furnace—that the gas-furnace came into general practical use in the manufacturing industries. "*Gazogènes*," or gas-generators, or gas-producers, are now generally employed, especially on the Continent, in metallurgic operations, in the manufacture of glass, and many other industries. In the greater number of industries, gas-furnaces exclusively are employed.

There are two modes in which gaseous fuel may be generated. The first consists in a distillation of the combustibles —that is to say, heating them in retorts or close vessels, without the intervention of air. This treatment is only applicable to such fuels as contain hydro-carbonaceous compounds susceptible of being volatilised by heat—such as yield gases better adapted for lighting than for heating; and it may be passed over without further comment. The second method differs radically from the first, since every portion of combustible, whether volatilisable or not, is converted into gaseous matter, with the exception of the ash. The portion of the fuel next to the grate, deprived of its volatile elements, and in immediate contact with air, is converted into carbonic acid. Passing upwards through the fuel, this gas, taking up an additional equivalent of carbon, is converted into carbonic oxide; whilst from the upper portions of the fuel, the volatile hydro-carbons and other gases are driven off by distillation. The result is a gaseous mixture, consisting mainly of carbonic oxide, hydro-carbon, and nitrogen, which, by a due administration of air at the required point, is consumed, and results in the usual products of combustion—carbonic acid, steam, and nitrogen,— which are passed off into the chimney.

Dr. Siemens concisely describes the functions of the gazogene :—"In the lower portion the fuel is burned, and this may be called the zone of combustion ; higher up, the carbonic acid takes up a further equivalent of carbon, becoming carbonic oxide, and this may be called the zone of

carbonization; whilst, at the uppermost layer of the producer, hydro-carbons are produced in what may be called
the zone of distillation."

The functions of gas-furnaces are, then, to pistil and vola-
tilise the fuel into carbonic oxide and hydro-carbon gases, in
the gazogene; to conduct the gaseous mixture into a combustion-chamber,—the spot where it is required to develop
the heat; and then to mingle with it the proper proportion
of atmospheric air required for effecting its complete combustion. The combustible and the air are in the same
physical condition—gaseous; so that they may be intimately mixed in suitable proportions, and with but a slight
excess of air.   Herein is an important source of economy of
fuel in the gas-furnace; with the regularity and constancy
of the supply of combustible, there is great facility for
measuring with exactitude the necessary quantity of air, and
for regulating its admission with precision.   An ordinary
grate, immediately after having been fired with fresh fuel,
should—notwithstanding Mr. C. W. Williams—be supplied
with twice as much air as it needed before having been stoked.
When the needful supply is wanting, the hydro-carbons
which are immediately generated are carried off unburned,
taking with them the heat expended in volatilising them.
But since, in the producer of the gas-furnace, the layer of
fuel is of constant thickness, the rate at which the gas is
generated and delivered is regulated by the damper once for
all; and exactly the same conditions are uniformly present for
the supply of air to be mingled with the gases.   Much labour
is saved, as the fuel need only be supplied at intervals of
from eight to twelve hours; and inexperienced labourers
can without difficulty be trained to become good firemen.
Very considerable indeed is the money-saving, apart from
the economy of fuel, to be effected by the employment of
gas-furnaces, simply by the facility with which small coal
and fuel of inferior quality may be utilised.

Gazogenes are employed under various forms, suited to the nature of the fuels and of the duty to be performed. The normal form is typified by the generator employed by Dr. Siemens, described further on. There is a grate at the lower part, nearly horizontal, upon which a bed of fuel lies, of considerable thickness; the thickness may vary from 2 feet to 4 feet in depth, according to the kind of fuel used. The upper part of the gazogene is generally at the level of the ground, made with one or several boxes or simple openings through which the fuel is charged. There are also holes through which the fuel may be picked or loosened, when necessary, and arches that are formed by bituminous fuels broken down. These holes are also useful to enable the stoker to inform himself of the state of the fire, and to judge of the proper time for introducing more fuel. The depth of ordinary gazogenes varies from 8 feet to 10 feet. A fire-proof damper is adapted to the producer to regulate the supply of gas, and to close at any moment the communication with the furnace.

CHAPTER XIV.

APPLICATION OF GAS-FURNACES FOR THE MANUFAC-TURE OF GAS.*

The employment of gas-furnaces at the gas works at Montreuil has been successfully introduced there to the designs of MM. Muller and Eichelbrenner, the results of many experiments made with the assistance of M. Fichet. The consumptions of these furnaces may be compared with those of the ordinary grate-furnaces, in terms of the quantity of coke consumed as fuel for the distillation of a ton of coal in the retorts, and assuming that the coal loses 30 per cent. of its weight, leaving consequently 70 per cent. of coke. In the old furnaces, with ordinary grates, containing eight retorts grouped in batteries of sixteen furnaces, or more, working altogether, the quantity of coke consumed amounted to 24¼ per cent. of the weight of the coal distilled. In single furnaces, holding five, six, or seven retorts, the quantity of coke consumed amounted to from 30 to 35 per cent. of the weight of the coal distilled; and, in many instances, when the furnaces were out of order, to 50 per cent. of the weight of coal distilled. In the gas-furnaces applied by MM. Muller and Eichelbrenner, heating five, six, and seven retorts, in single groups, or in pairs of groups side by side, the quantity of coke consumed as fuel amounted to 17½ per cent. of the weight of coal distilled.

* See " Etudes sur la Combustion," by M. Fichet, in the " Mémoires de la Société des Ingénieurs Civils," 1874, page 670.

Thus, whilst in the ordinary furnaces, from 35 to 50 per cent. of the quantity of coke manufactured was consumed ; in the gas-furnaces, the consumption did not exceed 25 per cent. The reason of the difference is, in some degree, to be traced to the difference of the management of the furnaces. The ordinary furnaces were filled up with coke as far as possible, for the sake of making long intervals between the firings. Incomplete combustion resulted from the accumulation of fuel, and carbonic oxide in large quantity was carried off by the chimney. The fires, on the contrary, were occasionally neglected, and got low ; admitted a large surplus of air, and so lowered the temperature. In either case, the distillation within the retorts was inferior. It became better, when the fire was regularly maintained, though demanding, for this purpose, additional care.

With the gas-furnace, such inconveniences are avoided. The thickness of the bed of fuel traversed by the air being constant, it suffices, once for all, to regulate the valves of the furnace to insure the regularity of the production of gas and of the supply of air, so as to avoid an excessive supply of air, and to maintain a uniform supply of heat. As the delivery-orifices for the gas are distributed throughout the length of the furnace, and are adjustable at will, a regular and uniform temperature is produced in all parts of the furnace ; to attain which is a difficulty with ordinary furnaces, in which the heat is unavoidably more powerful at one part than at another. The furnace-doors need never be opened, and no currents of cold air can be admitted ; nor is there any chance of local accumulations of ash. As the fuel is charged only once in eight or twelve hours, the economy of labour is considerable, whilst the labour is simple. By reference to the illustrations, Figs. 119 to 121, it may be seen how these results have been obtained. Fig. 119 is a cross section, and Fig. 120 is a vertical longitudinal section of the furnace, taken through one of the retorts. Fig. 121 is a vertical section of the gas-producer, the position of which is shown in Fig. 120, behind the two furnaces which are supplied by it, and between them. The producer, Fig. 121, consists of a hopper A as high as the furnace, having dimensions proportioned to the quantity of gas required in twenty-four hours. A step-grate, B, placed

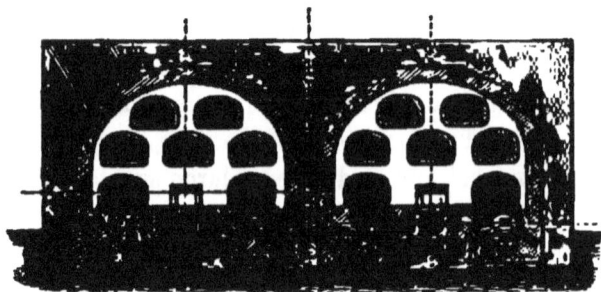

Fig. 119.—Gas Furnaces for the Manufacture of Gas.

at the lower part, prevents the fuel from falling out, and serves also for the admission of air. The hopper is completely filled with coke. This coke is in a state of combustion throughout the mass comprised between the grate and the

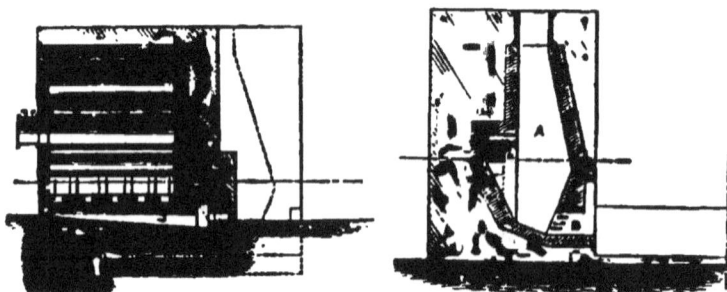

Figs. 120 and 121.

orifices h, by which the gas is conducted into the furnace; and the thickness of the bed of combustible so traversed is such that the oxygen of the air, after having been transformed into carbonic acid, is converted into carbonic oxide, which, escaping by the openings h, is delivered into the conduit S, placed in the axis of the furnace, at the lower part. The flow of the gas into the conduit is regulated by a damper. Through a number of openings in the upper part of the canal S, the gas is discharged into the furnace, and so forms a line of flame with the heated air, which is delivered from a passage *a*, on each side of the canal, in directions inclined to and upon each stream of gas. The rising flame is baffled by a horizontal slab erected above the openings from the canal S. The air for combustion is admitted into the passages *a*, controlled by dampers. It circulates in the thickness of the mason work or in metal pipes heated in the flues, prior to its being delivered into the furnace.

The hopper is built of brick, with a lining of firebricks. It is closed at the upper end by a cast-iron plate luted with clay ; or the plate is formed with a flange which lies in a groove filled with powder. The work of the stoker is to fill the hopper twice or three times in the twenty-four hours, and to clear the grate once a day. The proportions of the air and the gas are precisely adjusted by means of dampers ; and with the damper at the chimney, the temperature of the furnace is adjusted. When these three dampers are once regulated, they are not touched again so long as the fires are alight, which may continue for a year or more. In consequence of the regularity of action, the durability of the retorts is improved, and the wasting of fuel is prevented.

CHAPTER XV.

GAS-FURNACES FOR STEAM-BOILERS.

MANY varieties of gas-furnace have been employed for heating steam-boilers ; and many have been forgotten. At the tobacco factory of Gros-Caillou, the application of gas-furnaces to steam-boilers has been the subject of much practical investigation.

M. Fichet, when he applied gas-furnaces to steam-boilers, under arrangements similar to those which had given so satisfactory results when employed for the manufacture of coal-gas, naturally expected similar success. But he was disappointed. By the rapid cooling of the flame in contact with the surface of the boilers, he was led to the adoption of producers very differently arranged, for the service of steam-boilers. When he used dry combustibles, and admitted a quantity of air very little in excess of the quantity chemically consumed in combustion, the flame was extinguished when it came into contact with the boiler, and the products of combustion proved, on being analysed, to consist of a mixture of free oxygen and carbonic oxide, with nitrogen and carbonic acid.

The combustible gases supplied by rich coals, yielded smoke in addition ; whereas smoke had never been produced in the preceding applications. M. Fichet, therefore, entered upon a series of experiments with the object of studying the mode of the formation of smoke, with different arrangements of producers and furnaces, guided by the analyses of the products of combustion more or less complete, and the observations of temperature made with the calorimeter.   The result of his experimental observations led him to the principle on which complete combustion was to be attained, without any excess of air mingled with the gaseous products of combustion.   The enunciation of the principle is but an echo of the first principle laid down by Mr. C. W. Williams. Still, it is gratifying to learn that M. Fichet should have distinctively arrived at the same conclusion by an independent course of experiment.   The principle consists in intimately mixing, within an enclosure consisting of substances refractory at a high temperature, the combustible gases and the air, each of them having been divided into thin threads; in permitting the combustion to be completed within this enclosure or combustion-chamber, and in preventing the contact of any but completely converted gases with the surfaces of the boiler.

The application of the gas-furnace constructed on these principles, to an ordinary French boiler, at the iron works of M. Muller, Ivry, is illustrated by Fig. 122.   The heaters are 24 inches in diameter, and the boiler is 43½ inches.   The heating surface amounts to 560 square feet, exclusive of that of the feed-water heating apparatus; the producer, of a form different from that which has before been described, is placed in front of the boiler and below the level of the floor.   The fuel is supplied through the box entrance at the top in quantities of 200 lbs. at a time, every hour; and the cover is closed with a sand-joint.   The box is so constructed that the charge of fuel may be delivered without the chance of any reflux of gases through the openings when the pressure in the producer happens to be greater than that of the atmosphere ; or, on the contrary, of air being drawn in, when the internal pressure is less. It is, in short, the charging-box ordinarily adapted to producers.   The fuel falls, when the valve is opened, into a hopper where the temperature is not considerable, and where, for want of air, combustion cannot take place. Thus the fuel is dried and is gradually heated as it descends, until it arrives in a hotter region. When it has passed into the vault, it comes into contact with the hot gases in the lower part of the producer : it begins slowly to distil at the surface, at the same time falling gradually, until it arrives within reach of the air, when it is converted into coke,

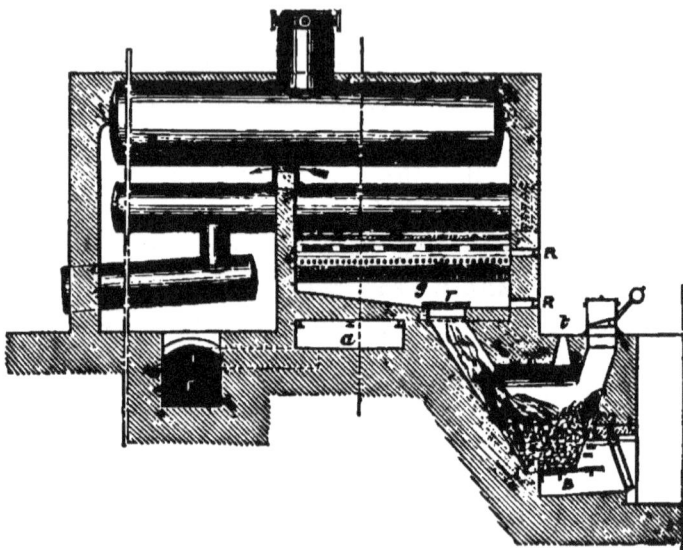

Fig. 122.—Gas-furnace for Steam Boiler.

under the pressure of the superincumbent load. During the descent and the progressive distillation, the small coal, which has been charged above, becomes agglomerated and yields a dense coke which does not go to pieces during combustion, and is nevertheless sufficiently porous to admit of the circulation of air.

To obviate loss of heat by radiation through the grate, a swing-door is hung at the entrance for air, formed double, and perforated with air-holes through which air passes, and by its circulation prevents the doors from becoming over-heated.

As the fuel descends below the crown of the vault, it spreads outwards and downwards according to the angle of repose, the sides of the hopper being suitably inclined to facilitate the natural action of the fuel. The gases, as they are produced, ascend and are directed through an inclined opening into the chamber *g* under the boiler, the roof of which is constructed with flat pieces in fireclay, formed with grooves or interspaces, by which the gases ascend into the combustion-chamber *i*. The flow of the gases from the producer is regulated by the damper *r*. When the production of steam is to be suspended for some time, this damper is completely closed; so also is the damper for air. The gazogene may thus continue alight for several days; and it may be restored to its usual state of activity in a few hours, when the dampers are reopened. The air for combustion is supplied by a pipe *a*, which passes through the exit-flue F to the chimney, and is heated in its course. It is delivered into a chamber *a* below the gas-chamber *g*, at each side of which it rises, when it passes to the combustion-chamber. The air receives additional heat in skirting the gas-chamber, and thus acquires a degree of ascensional force by which it is delivered with velocity through the orifices where ignition takes place. It may, therefore, be divided into thin streams, or jets, which meet the streams of gas arriving in another direction. Resulting eddies take place, which facilitate the mixture of the elements, by which complete combustion is accomplished. The fuel is charged into the hopper every hour. The stoker finds with satisfaction that the less the fire is touched, the better it goes.

The boiler which was fitted with the gas-furnace just described, lay alongside another boiler of equal dimensions, heated with an ordinary grate-furnace. The comparative performances of the two boilers were tested, using the same

o

quality of coal and the same quality of water. The coal
was the bituminous coal of the north of France; the regular
evaporative performance of the boiler with the ordinary grate,
amounted very uniformly to about 6 lbs. of water per pound
of coal—slightly less than that. The gas-furnace boiler was
found, by long and careful trial, to evaporate from 8·60 lbs.
to 9·20 lbs. per pound of coal—say, an average of 8·90 lbs.
This result shows an augmentation amounting to 48 per
cent. of evaporative efficiency; otherwise, an economy of

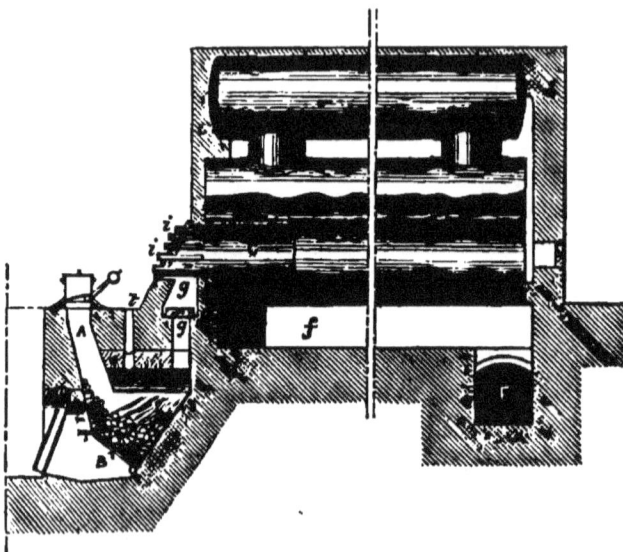

, Fig. 123.—Gas-furnace for an Internally Fired Boiler.

32 per cent. for equal quantities of water. The coal was con-
·sumed in the gas-furnace at the rate of 84 lbs. per hour: the
heating surface of the boiler amounted to 559 square feet.
The temperature of the products of combustion in the flues,
after having passed the feedwater-heaters, varied from 400°
to 600° Fahr.

In applying the gas-furnace to internally fired boilers, it
was foreseen by M. Fichet, after his experience with the

ordinary French boiler, that special precautions must be taken to provide against the premature cooling of the gases, and the extinction of the flame. The arrangement which he employed is shown in Fig. 123. The gazogene is placed, as for the previous boiler, in front and under the level of the ground. The gas is delivered into a passage g, provided with a damper, whence it passes into the firebrick combustion-chamber C, which is constructed within the flue-tube at one end. The outer end of the combustion-chamber is provided with an inclined door, lined with firebrick, and pierced with a number of holes, into which numerous iron tubes are fixed, open at both ends, through which the air for combustion is admitted into the chamber C. The tubes act as nozzles, through which jets of air are blown into the body of combustible gas, setting up very active combustion. The temperature for combustion is maintained by the fire-brick lining, so that the combustion is completed before the flame can touch the surface of the metal. It is sometimes found of advantage to raise a perforated firebrick wall or diaphragm at the inner end of the firebrick chamber, sub-stituting, at the same time, two long vertical sheets of air through the door, for the multitubular jets. The products of combustion, after circulating round the boiler, pass off by the flues f and F, to the chimney.

M. Fichet states, as the result of numerous observations, that in the gases supplied by the gazogene, there are fre-quently not any traces of carbonic acid, sometimes ½ per cent., rarely 1 per cent. In the gaseous products of com-bustion, there is not a trace of free carbonic oxide, and often there is not a trace of free oxygen. But for the strongly bituminous coals, it is necessary to admit an excess of oxygen, of from 1 to 1½ per cent., to ensure the complete absence of smoke. This proportion of free oxygen repre-sents an excess of air amounting to from 4 to 6½ per cent.

CHAPTER XVI.

DECOMPOSITION OF FUEL IN GAZOGENES.

Coals.—It has been stated that the air arriving in contact with the incandescent fuel next the grate, produces at first carbonic acid, represented symbolically as C O² ; and that in passing through the bed of fuel, more or less coked, the greater part of it is converted into carbonic oxide, C O. The thickness of the bed of coal should be sufficient for effecting this transformation, at least nearly completely ; it should be greater where the passage through is the easier, since the surfaces of contact are less. But however favourable the conditions may be, the gas always contains a fraction of carbonic acid ; for the conversion from the acid into carbonic oxide is never completely effected. In the laboratory, it is true, the transformation of the whole of the carbonic acid may be effected ; but that is only accomplished after a very prolonged action of the gaseous element upon the carbon, which acts only at the surface.

The combustible gases contain not only carbonic acid and carbonic oxide, but also hydrogen and hydro-carbons, for, in the decomposition of coals by the application of heat, these gases are always disengaged. Nitrogen, of course, is present, from the air, and also from the coals ; and oxygen also, when the thickness of the bed of fuel is insufficient, indicating the passage of free air.

Coke, Charcoal, Peat.—M. Felix Leblanc gives the following analysis of the combustible gases produced from coke, in the Siemens process.

SIEMENS GENERATOR. COKE.

Carbonic oxide . . . . . 26·0 per cent.
Carbonic acid . . . . . 4·5 „
Nitrogen . . . . . . 67·5 „
Oxygen . . . . . . 0·5 „
Hydrogen . .. . not observed.

M. Ebelmen's analyses of charcoal gases, taken from charcoal gazogenes, are as follows: The first is an average analysis of samples taken from a close generator, supplied with a blast of dry air. The second is an average analysis taken from the same furnace supplied with air and water-vapour.

EBELMEN'S GAZOGENE. CHARCOAL.

	Dry air.	Air and steam.
Carbonic oxide . . .	33·3	27·2
Carbonic acid . . .	0·5	5·5
Nitrogen	63·4	53·3
Oxygen	—	—
Hydrogen	2·8	14·0
By volume . . .	100·0	100·0

It is to be remarked that the first of these analyses, with dry air, closely approximates to the exact chemical proportions for the entire conversion of carbon into carbonic oxide, without any carbonic acid, which are :—

$34\frac{1}{2}$ per cent. of carbonic oxide
$65\frac{1}{2}$ „   nitrogen
—————
100

For peat, the production in the gazogene is imperfect; there is a large proportion of carbonic acid :—

GAZOGENE.—PEAT.

	Ponsard Gazogene.		Ebelmen's Analysis.	
	1	2	3	4
Carbonic oxide . .	21	23	22·63	21·04
Carbonic acid . . .	11	9	7·32	10·79
Nitrogen	—	—	64·13	58·81
Hydrogen, &c. . .	—	—	5·92	9·36

In No. 1 of these analyses, the peat held 50 per cent. of water. In No. 2 the peat held 28 per cent. of water. The peat employed for Nos. 1 and 2 was analysed after having been desiccated with the following result :—

Volatilised matter	58·5
Solid matter (including 8 per cent. of ash) .	41·5
	100·0

For Nos. 3 and 4, the peat was air-dried, and contained 18 per cent. of water.

CHAPTER XVII.

IRON FURNACES.—ORDINARY IRON FURNACES.

THE old puddling furnace, illustrated by Fig. 124, consists of a rectangular structure of iron plates, nearly 6 feet high, 6 feet wide, and 12 feet in length, lined throughout with firebrick. The fireplace or grate is at one end, about 3 feet square, and is separated from the hearth of the furnace by a brick bridge. The hearth is six or seven feet long, and 3½ feet wide at its widest part. It rests on a cast-iron bottom plate, which is nearly on a level with the fire-grate, and is covered with oxide of iron, or "fettling." The hearth is arched over, so that the heat may be reverberated or radiated from the arch upon the metal. The

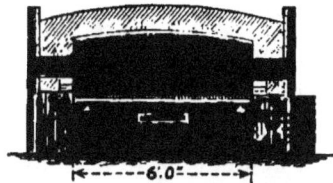

Fig. 124.—Old Puddling Furnace.

farthest end of the furnace is contracted to 18 inches in width, forming the flue leading to the chimney, which is from 35 to 40 feet in height, and is fitted with a damper at the top. The furnace is arched over with brickwork, and, to prevent the side from being thrust out by the expansion of the heated bricks, the side plates are tied together with a number of iron bolts, to receive and resist the thrust of the arch. A working doorway is made in one side of the body of the furnace, the bottom of which stands at eight or ten inches above the level of the hearth. The door is moved vertically by a balanced lever.

Great improvement has been made in the construction and the efficiency of furnaces both for puddling and for heating, or re-heating, iron. In relation to the economical production and application of heat, they may be ranked in three classes: the improved furnaces, substantially of the old type, comprising a fire-grate, a hearth, and a chimney; the gas-furnace, in which the fuel is converted into combustible gases prior to its being burned for the development of heat in the furnace; and the gas-furnace, in which the heat remaining in the departing products of combustion, is in a greater or less degree utilised for superheating the combustible gases and heating the air, before they are brought into combustion.

The quantity of coal consumed in ordinary single puddling furnaces, in the north of England, averages about 24 cwt. per ton of puddled iron produced from the furnace; pig iron being treated in charges of 4½ cwt. or 5 cwt. at a time. In the old single furnaces at Round Oak Iron Works, and at other places in Staffordshire, 30 cwt. of screened slack is consumed per ton of puddled iron; the weight of the charges of pig iron being 4½ cwt. The production of iron amounts to about 2½ tons in 24 hours. In double furnaces, the quantity of fuel consumed is less by 5 cwt. per ton of iron produced, and it is well ascertained that the greater the charge, the less fuel is required per ton of iron produced. The average waste of Staffordshire pig iron in the furnace amounts to 7½ per cent. At the Royal Gun Factory, Woolwich, the consumption of unscreened Gawber Hall coal, from Yorkshire, is in the ordinary puddling furnaces from 23½ cwt. to 15 cwt. for charges of pig of from 5 cwt. to 15 cwt. Here, there is the usual reduction of fuel in proportion as the charges are increased. Of the gross weight of the charge of pig iron at Woolwich, 95 per cent. is produced as puddled iron—showing a loss of 5 per cent.

In heating, or re-heating, furnaces, for raising wrought iron to a welding heat, less fuel is consumed, as may be easily understood, than is necessary for puddling an equal weight of iron. Ordinary re-heating furnaces in South Wales, employed for heating rail-piles, consume 8 cwt. of coal per ton of iron. At Woolwich, in ordinary furnaces, working day-shift only, 9½ cwt. of coal is consumed per ton of iron, and when worked day and night continuously, 8 cwt. of coal is consumed.

To form an estimate of the quantity of heat generated, and the proportion in which it is distributed :—The total heat of combustion of 1 lb. of coal of average composition is 14,133 English units ; the weight of the gaseous products of combustion is 11·94 lbs. ; and the specific heat of these gases taken together is ·246. The quantity of heat required to raise the temperature of the whole of these gases 1° Fahr. is $(11·94 \times ·246 =)$ 2·935 units; and the temperature of combustion, supposing the initial temperature to be 62° Fahr., is $(14133 \div 2·935 = 4815) + 62° = 4877°$ Fahr. But, in practice, this temperature is not reached, as an allowance is to be made for the inevitable surplus of free air which accompanies the products of combustion, and it may be taken at one-half of the quantity of air that is chemically consumed for the complete combustion of the coal. The weight of the surplus air is, then, 5·35 lbs., to be added to the weight of the burnt gases, 11·94 lbs., making altogether 17·29 lbs. per pound of coal consumed. The mean specific heat of this mixture is ·243, and the quantity of heat required to raise its temperature 1° Fahr. is $(17·29 \times ·243 =)$ 4·207 units. Dividing the total heat of combustion of 1 lb. of coal by this product, the quotient $(14133 \div 4·207 =)$ 3359° Fahr. is the temperature of combustion in the furnace, above 62° Fahr.; or, in all, 3421° Fahr. Here is an example of the influence of the admixture of free air in lowering the maximum temperature attainable ; for, instead

o 3

of the possible maximum, 4877° Fahr., the actual maximum may not exceed 3421° Fahr.

To determine now, in the first place, the quantity of heat absorbed by the metal under treatment, in the furnaces:— In the puddling furnace, the cold pig-iron is raised in temperature to the melting-point, which is, say, 2,000° Fahr., and melted. The quantity of heat expended in both these operations, taken together, was determined by M. Clement; * it was equivalent to 504 English units per pound of pig-iron, at the temperature attained in a blast-furnace. The heat absorbed in melting one ton of pig-iron therefore amounts to $2240 \times 504 = 1,128,960$ English units; and the quantity of coal which generates this quantity of heat in complete combustion, taking the quantity of heat generated per pound of coal as 14,133 units, is therefore $(1128960 \div 14133 =) 79\cdot9$ lbs.—say 80 lbs. That is to say, the net quantity of heat absorbed by one ton of the melted iron is that which is equivalent to the heat of combustion of 80 lbs. of coal.

But this is not all: after the iron has been melted, it is puddled, and ultimately becomes wrought iron; the constituent heat of which at high temperatures is greater than that of cast iron. Taking the temperature at 2900° Fahr., something over the temperature of welding heat, with the specific heat, ·185, the heat contained in one pound of puddled iron in the furnace is 536·5 units; and in one ton the total quantity of heat contained is (2240 lbs. $\times$ 536·5=) 1,201,760 English units, for the generation of which the quantity of coal required is (1,201,760 $\div$ 14133 =) 85 lbs. Evidently, this is the maximum estimate that can be made of the heat of the coal, actually utilised, since there is a considerable quantity of heat generated by the combustion of the carbon driven off from the pig-iron; which, whatever be its function, is not now taken into account, though it would

* See "A Manual of Rules, Tables, and Data," 1877, page 497.

lead to a reduction, to some extent, of the quantity of coal required.

Adopting, then, the quantity, 85 lbs., of coal as the net quantity utilised per ton of puddled iron,—not making any allowance for small loss of weight, in the conversion of the iron; and taking the total coal consumed as 18 cwt. or 2016 lbs. per ton, in ordinary double furnaces, the efficiency of the furnace, measured in terms of the fuel thus utilised, amounts to $\left(\dfrac{85 \times 100}{2016} = \right)$ 4·21 per cent., nearly 4¼ per cent.

In the heating furnace, although the process consists simply in heating wrought iron up to the welding point, it may be taken that the heat absorbed by the metal is the same as that absorbed in the puddling furnace, represented by the combustion of 85 lbs. of coal of average composition. The total quantity of coal consumed is 8 cwt. or 896 lbs. per ton of metal; and the efficiency of the furnace amounts to $\dfrac{85 \times 100}{896} = 9\cdot49$ per cent.—say 9½ per cent.

The heat which is carried off by the burnt gases must necessarily form a large proportion of the total heat which is generated. As the melting-point of wrought iron is about 2900° Fahr., it may be assumed that the temperature of the passing gases, as they escape from the puddling furnace, is at least as much as that. If the interior of the chimney be examined through an aperture, situated, say, half-way up, it may be seen that bright red heat, or even white heat, is usually maintained. From the datum, 2900° Fahr., the quantity of heat carried off for each pound of coal consumed is ascertained by multiplying it by 4·207, the number of units of heat in the escaping gases per 1° Fahr., and is equal to (2900 × 4·207 =) 12,200 units per pound of coal. Thus (2016 lbs. × 12200 ÷ 14133 =) 1740 lbs. is the quantity of coal of which the total heat of combustion is equal to the quantity of heat escaping from the puddling furnace, and it amounts to (1740 × 100 ÷ 2016 =) 86 per cent. of the total coal consumed.

In the heating furnace, it is certain that the temperature of the escaping gases is lower than that in the puddling furnace. According to observations to be afterwards noticed, it may be inferred that there is a difference of probably 500° Fahr.; from which it follows that the temperature may be taken at (2900° − 500° =) 2400° Fahr. The number of units of heat in the escaping gases is equal to (2400 × 4·207 =) 10097 units per pound of coal. The coal consumed per ton of iron heated being 8 cwt., or 896 lbs., in the heating furnace, the quantity of coal of which the total heat of combustion is equal to the quantity of heat escaping in the gases, is equal to (896 lbs. × 10097 ÷ 14133 =) 640 lbs., amounting to (640 × 100 ÷ 896 =) 71 per cent. of the total coal consumed.

CHAPTER XVIII.

UTILISING THE WASTE HEAT OF ORDINARY IRON FURNACES BY GENERATING STEAM.

THE large proportions of heat which pass off in the spent gases of iron furnaces, are partially utilised in heating boilers for the generation of steam. The employment of the gases as a means of raising steam is sound in principle, since a higher temperature than 500° or 600° Fahr. is not required for inducing the maximum intensity of draught in a chimney. It has been estimated that, for the production of the steam consumed in the manufacture of wrought iron by the independent application of heat, with economical stoking, 4 cwt. of coal would be required per ton of finished iron produced. Considering that at least 25 cwt. of coal is consumed for puddling and re-heating per ton of finished iron, an appropriation of the heat of 4 cwt., or about one-sixth of the total coal consumed, does not appear to be excessive. On the contrary, it is probable that the proportion of the total heat generated, utilised in generating steam, is greater than what is represented by 4 cwt. of coal. The probability of such an assumption is supported by the results of very careful observations made at large ironworks in the north of France, cited by M. Ponsard. One pound of the coal burned on the grates of re-heating furnaces yielded 2½ lbs. of steam from the horizontal boilers heated by the waste gases. This was the average of many observations. In another series of observations, 1 lb. of coal yielded 2·69 lbs. of steam.

Again, it is generally admitted, in a metallurgical district referred to by M. Ponsard, that 25 horse-power is yielded by welding furnaces consuming from 1,200 lbs. to 1,800 lbs. of coal in 24 hours ; and, assuming that 40 lbs. of steam be consumed per horse-power per hour, there would be a total consumption of steam equal to $(40 \times 25 \times 24 =)$ 24,000 lbs. for 24 hours, which would be at the rate of $(\frac{24000}{1200} =)$ 2 lbs. of steam per pound of coal consumed. In another district, where vertical steam-boilers are employed, it is said that 3 lbs. of steam are produced per pound of coal.

Taking the first of the three data first given, $2\frac{1}{4}$ lbs. of steam yielded per pound of coal consumed in the furnaces, as a fair average, it may be compared with the average evaporative performance of 1 lb. of coal burned under a steam-boiler, say, $7\frac{1}{4}$ lbs. of water. The comparison shows that the effective economy by the utilisation of the waste heat in raising steam amounts to $\dfrac{2\frac{1}{4}}{7\frac{1}{4}} = \frac{1}{3}$ of the total fuel burned in the iron furnaces.

The distribution of the fuel representing the heat of combustion, according to the foregoing determinations, is tabulated as follows :—

Distribution.	Puddling furnace.		Heating furnace.	
	cwts.	per cent.	cwts.	per cent.
Directly utilised, absorbed by the iron	·75	4¼	·75	9½
Retrieved by generating steam	6·00	33	2·67	33
Total fuel utilised . .	6·75	37¼	3·42	42½
Lost by the chimney . . .	9·54	53	3·04	38
Lost by radiation and conduction	1·71	9¾	1·54	19½
Total fuel lost	11·25	62¾	4·58	57½
Total fuel consumed	18·00	100	8·00	100

From this table it appears that, taken generally, ordinary double iron-furnaces for puddling and heating with large charges, utilise, in heating the iron and generating steam, about 40 per cent. of the total heat of combustion of coal, and that 60 per cent. is wasted. Here is a wide margin for progress in efficiency; and much has already been done in reducing the consumption of fuel.

CHAPTER XIX.

ATTEMPTS were made, as early as in 1850 or 1851, to economise the waste heat of furnaces by heating the air for combustion, and it is obviously the best and most direct method
of retrieving the inevitably departing heat. By thus raising
the temperature of the air supplied for combustion, the temperature of combustion is likewise raised. It may be assumed
that the heat of combustion is increased by the quantity of
heat imparted to the entering air; and that, approximately,
the effective work of the heat of combustion is increased in
the same ratio. Mr. Prideaux, page 204, has shown reason
for supposing that the effective work is augmented in a
greater proportion than the simple augmentation of temperature and heat. But it is well to assume an equal ratio only,
for the losses by conduction and radiation are likely to be
augmented with the temperature.

Suppose, for example, that the temperature of the incoming air is increased by 500° Fahr. To find the quantity
of heat absorbed in raising the temperature, the specific
heat of air at constant pressure is ·2377: such that 1 lb. of
air raised 1° Fahr. in temperature, absorbs ·2377 unit of
heat, or nearly a quarter of a unit. The quantity of heat
absorbed by 1 lb. of air for 500° rise of temperature is therefore (·2377 × 500 =) 119 units. The quantity of air

chemically consumed for the complete combustion of 1 lb. of coal of average composition, is 10·7 lbs.; and, when the surplus air admitted to the furnace amounts to one-half of the air that is chemically consumed, the total quantity used is (10·7 × 1½ =) 15·85 lbs., say 16 lbs., per pound of fuel. The quantity of heat absorbed by the air per pound of fuel, therefore, is equal to (119 × 16 =) 1,904 units for a rise of 500° Fahr. of temperature; and this is the additional heat of combustion supplied. Now the heat of combustion actually generated in the combustion of 1 lb. of coal is 14,183 units ; so here is a clear addition of 1,904 units, or $\left(\dfrac{1,904 \times 100}{14,183} = \right)$ 13·4 per cent., or about one-seventh, the signification of which is that 1 cwt. out of 8 cwt. of coal may thus be saved.

If, similarly, the temperature of the incoming air be raised 1,000° Fahr., the heat so retrieved would amount to (·2377 × 1,000 =) 238 units per pound of air, or (238° ×16 =) 3,808 units per pound of fuel, or 27 per cent. of the total heat of combustion, which would be equivalent to a saving of one out of every 5 cwt. of coal.

Applying these proportions of saving to the normal consumptions for puddling and heating furnaces, they could thus be reduced from 18 cwt. and 8 cwt. respectively, to 15¾ cwt. and 7 cwt., for a rise of 500° Fahr. of temperature of the incoming air by the absorption of waste heat; and to 14½ cwt. and 6½ cwt. for a rise of temperature of the air, of 1,000° Fahr.

But there is a distinct and important means of economy in the substitution of fuel in the gaseous state for fuel in the solid state. The economy may proceed from two sources :— by the facility for adjusting the quantity of air admitted to the chemical requirements, with a very small surplus ; and by the facility for effecting the combustion in the locality where the heat is required to be absorbed, instead of the

locality of a grate, which is more or less distant from the
locality for the useful absorption of the heat.

With regard to the first source of economy—the minimis-
ing of the quantity of surplus air necessary for effecting the
complete combustion of the fuel,—coal,—there is reason for
believing that, under proper regulation, the needful surplus
may not exceed 10 per cent. of the quantity chemically consumed. The temperature of the products of combustion
must be relatively higher, and it may be calculated in terms
of the specific heat of the mixture. There are 11·94 lbs. of
burned gases from 1 lb. of coal, having the specific heat ·246.
The surplus air is 10 per cent. of 10·7 lbs., the weight of air
chemically consumed, or 1·07 lbs., of which the specific heat
is ·2377. From these data, the specific of the mixture is thus
obtained :—

1 lb. of coal.	Weight.		Sp. heat.		Units.
Burnt gases . . .	11·94 lbs.	×	·246	=	2·935 per 1° Fahr.
Air	1·07 „	×	·2377	=	·254 „
	13·01	×	·245	=	3·189 „

The quantity of heat to raise the temperature of the mixed
gaseous products, is 3·189 units, and the temperature of
combustion amounts to (14,133 ÷ 3·189 =) 4,432° Fahr.,
above, say, 62°. Now, the temperature of combustion of the
gaseous products, which contained 50 per cent. excess of
air, was found to be 8,359° Fahr., above, say, 62° Fahr. ;
which is now exceeded by the higher temperature in the
ratio of 3,359° to 4,432°, or 1 to 1·32 ; that is to say, the
gain of temperature is 32 per cent., or one-third more. If
the useful performance be in the same proportion, the same
quantity of puddled iron would be produced by a reduced
consumption of $\left(\dfrac{1}{1·32} =\right)$ ¾ths of the coal used with an excess of air of 50 per cent., making a saving of 25 per cent.

Applying this reduction to the normal consumptions for

puddling and for heating furnaces, 18 cwt. and 8 cwt., these would be reduced to 13$\frac{1}{4}$ cwt. and 6 cwt. respectively. There is, in addition, an economy, already pointed out, arising from the preliminary heating of the air by the waste heat ; but the economy from this source will be less than it was before calculated, since the quantity of air to be heated per pound of coal is now less. The quantity of air per pound of coal, if there be only 10 per cent. in excess, is (10·7 lbs. + 1·07 lbs. =)11·77 lbs., or, say, 12 lbs., and as 119 units of heat are absorbed per pound of air in raising the temperature 500° Fahr., the total heat so applied amounts to (119 × 500 =) 1,428 units per pound of coal, or to $\left(\dfrac{1,428 \times 100}{14,133} = \right)$ 10 per cent. of the total heat of combustion of 1 lb. of coal. This is equivalent to a saving of 1 cwt. out of 11 cwt. of coal, or 9 per cent.

But, if the air be raised 1000° Fahr. in temperature, the heat absorbed by the air would amount to 2856 units, or 20 per cent. of the total heat of combustion of 1 lb. of coal : equivalent to a saving of 1 cwt. out of every 6 cwt. of coal, or 17 per cent.

Combining the two economies thus estimated, by a process of compound reduction, the ultimate reduced consumptions are given :—

	Reduced consumption of coal. per cent.	Reduced consumption of coal. per cent.
Surplus air reduced to 10 per cent. .	75	75
Entering air heated	500° F. 91	1000° F. 83
Ultimate reduced fuel consumed . .	68	62

Showing that, by the combined influence of the reducing of the surplus air, and the heating of it by the waste heat, in the given proportions, the consumption may be reduced to about two-thirds, more or less. Accordingly, the normal consumptions of coal, 18 cwt. and 8 cwt., for puddling and for heating furnaces respectively, may thus be reduced to about 12 cwt. and 5 cwt. : estimates the reasonableness of which has been amply demonstrated by experience.

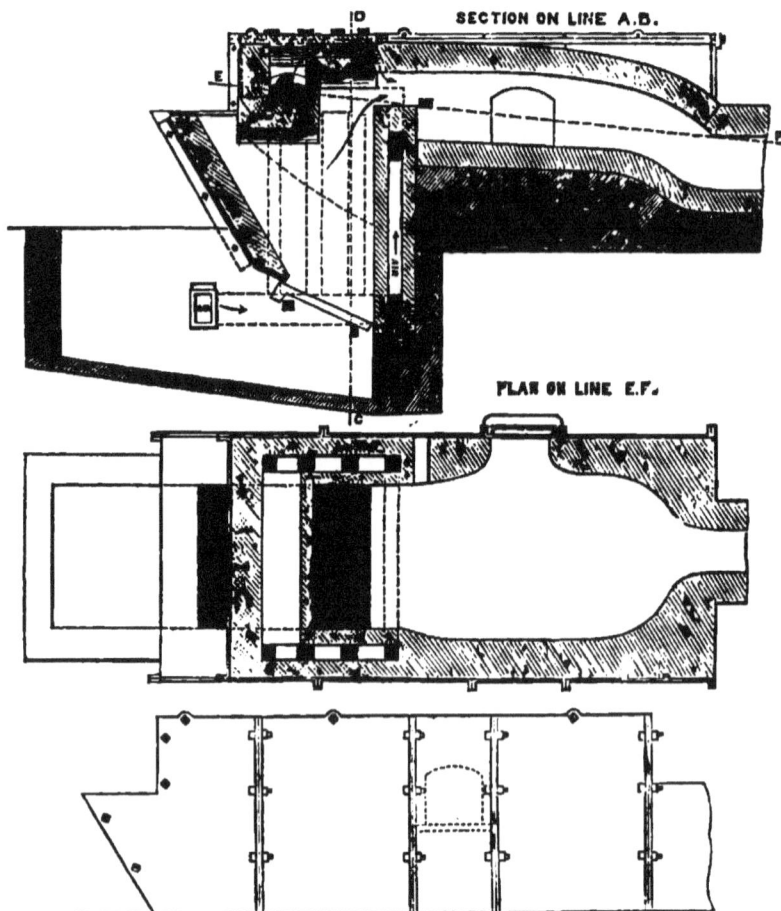

Figs. 125.—The Boetius Heating Furnace.

Mr. J. F. Boetius patented, in August, 1865, a system of gas-furnace for melting glass, heating iron, and other purposes for which a high temperature is required. The furnace, in one of its forms, is illustrated by Figs. 125, 126. Here the

ordinary gas-furnace, with sloping hearth and grate, is employed. The side walls and the roof of the generator are
surrounded by air-passages, through which the air used for
combustion is passed. The air
enters by an opening in each
side of the ash-pit, and ascends
through subdivided passages,
meeting in the passage over
the crown of the generator,
whence it is delivered at the
bridge in an inclined direction,
so as to impinge upon and mix
with the combustible gases,
as they enter the furnace. The
effect of the evidence on the
performance of the Boetius
furnace, when employed for
glass melting, is clear on the
fact of economy of fuel in
quantity, and of the facilities
afforded for the employment of
fuels of inferior quality. The

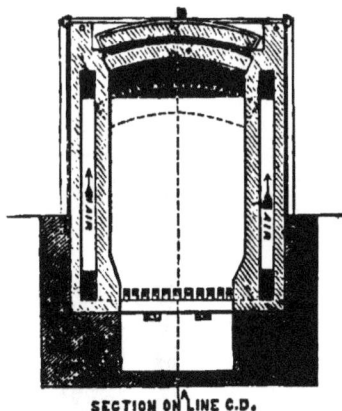

SECTION ON LINE C.D.

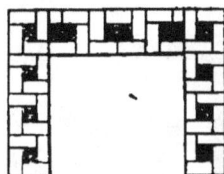

Fig. 126.—The Boetius Heating
Furnace.—Air-heating Passages.

saving of fuel, according to the evidence, ranged from
15 to 30 per cent., by the substitution of the gas-furnace
for the old fire-grate furnace. This proportion of economy
is quite in harmony with the general evidence of the superior
efficacy and economy of gas-fuel compared with the solid
coal used directly as fuel :—an economy which is due,
first, to the facility for intimately mixing the combustible
gases and air, and thus reducing to a very low margin
the surplus of air required for the accomplishment of
complete combustion; secondly, to the facility for generating the heat in the place where it is to be absorbed,—in
contact, practically, with the object to be heated; and
thirdly, to the preliminary heating of the air supplied for

Figs. 127.—The Bicheroux Puddling Furnace.

combustion. It does not appear to what degree the air is heated; but a comparatively small rise of temperature has been found to make a considerable difference for the better, in the action of the furnace.

Mr. Bicheroux, in May, 1872, patented a modification (Figs. 127) of the Boetius gas-furnace, in which the grate of the generator is made wider and shallower than in the Boetius furnace; there is a less amount of heating surface provided for heating the air; and, notably, an intermediate mixing chamber is introduced between the generator and the furnace or heating chamber. This chamber operates beneficially in promoting the intermixture of the combustible gases, prior to their meeting with the supply of air for combustion, at the entrance to the heating chamber. M. J. de Macar * gives the comparative results of performance of the old puddling furnace, the Boetius furnace, and the Bicheroux furnace, at the iron-works of MM. Piedboeuf and Bisénius, at Dusseldorf.

OLD FURNACE.

8 charges of 4½ cwt. . . . . 36 cwt.
Consumption of coal per ton . . . . 20 cwt.
Waste, 12 to 13 per cent.

BOETIUS FURNACE.

8 charges of 6½ cwt. in 12 hours . . 52 cwt.
Consumption of coal per ton. . . . . . . 13¾ cwt.
Waste, 8 to 10 per cent.

BICHEROUX FURNACE.

8 charges of 8 cwt. in 12 hours . . . 64 cwt.
Consumption of coal, per ton . . . . . . 12 cwt.
Waste, 9 to 10 per cent.

The "Newport furnace" was first erected and developed at the Newport Rolling Mills, at Middlesborough, about the year 1870. It is, for the most part, an ordinary furnace with an ordinary grate. The peculiar principle of the furnace

* *Revue Universelle des Mines.* Tome I, 1877; p. 205.

consists in the employment of jets of steam to induce a blast
of air for the supply of the furnace, and the heating of
the combined current on its passage to the furnace by the
heat of the gases in the chimney. The blast is delivered
into a closed ash-pit, at the temperature 550° Fahr., according to the results of experiments made by Mr. Jeremiah
Head.* He communicated, in 1872, the relative performances of the old puddling furnaces, and the Newport furnace,
at the Newport Works, which averaged as follows:—

	Old Furnace.	Newport Furnace.
Coal consumed per ton of puddled bar, for long periods	24½ cwt.	12·8 cwt.
Pig iron per ton of puddled bar . . .	20·62 „	20·7 „

These results show a saving of nearly one half of the coal,
and the saving is attributable to the united economy derived from the use of heated air, and to the reduction of the
quantity of air necessary approximately to that which is
chemically consumed in effecting complete combustion. Mr.
Head measured the quantity of air admitted: the measurement appeared to show that the exact supply chemically
necessary had been admitted, and no more. Indeed, he
had some difficulty in measuring up or proving the whole
of the necessary quantity. It may be admitted that the
steam which was injected with the air took some part in
the combustion;—being decomposed, in the first instance,
into its elements, hydrogen and oxygen, of which the
oxygen, in the second instance, took up an equivalent of
carbon, forming carbonic oxide, to be ultimately converted
into carbonic acid.

Mr. Head observed that the temperatures of the burnt
gases in the chimneys of an ordinary furnace and a Newport furnace, working under precisely similar circumstances,
were respectively 2,033° Fahr. and 1,577° Fahr., showing a

* See Mr. Head's paper "On the Newport Puddling Furnace," in
The Journal of the Iron and Steel Institute, vol. i., 1872, page 220.

reduction of 456° Fahr. in the Newport furnace : corresponding fairly with the raising of the temperature of the ingoing air 500° Fahr., especially if the respective specific heats of the burnt gases and the air be taken into account.

Again, the waste heat of each furnace was partly utilised in generating steam in a boiler. By the gases from the ordinary furnace, 20·4 cubic feet of water per hour was evaporated into steam of 50 lbs. pressure ; by the gases from the Newport boiler, only 10·1 cubic feet of water per hour, at 180° Fahr., was evaporated into 50 lbs. steam— evidence that the air-heating stove had been doing its work.

At Blaenavon, where the Newport furnace had been at work since 1870, the consumption of coal-slack amounted to 14¼ cwt. per ton of puddled bar.   Messrs. Jones Brothers report, as the result of a month's comparative trials, that, in their Newport furnaces, 16¼ cwt. of coal was consumed per ton of puddled iron; whilst, in the ordinary furnaces, 24 cwt. was consumed.   The grate-bars, it is stated, lasted one-third longer time in the Newport furnace.

In the Casson-Dormoy puddling furnace, which was introduced in 1872—73, it was attempted to perfect the main features of the old reverberatory puddling furnace of Cort, by intensifying and concentrating its action, and reducing the working cost.   In this furnace, which was double, shown in longitudinal section, Fig. 128, the fire-grate was long and narrow, reaching alongside the bed ; thus the locality in which the heat was generated was brought up as close as was practicable to the bed.   The bed consisted of a cast-iron basin or dish, resting on a number of iron friction-balls laid in a cast-iron pan containing water.   The basin easily adjusted itself on the balls for expansion and contraction ; and it might also, when required, be turned about to change its position in order to ensure equable tear and wear.   Beyond the bed, there was a supplementary hearth, on which the pigs were deposited and heated nearly up to melting-point,

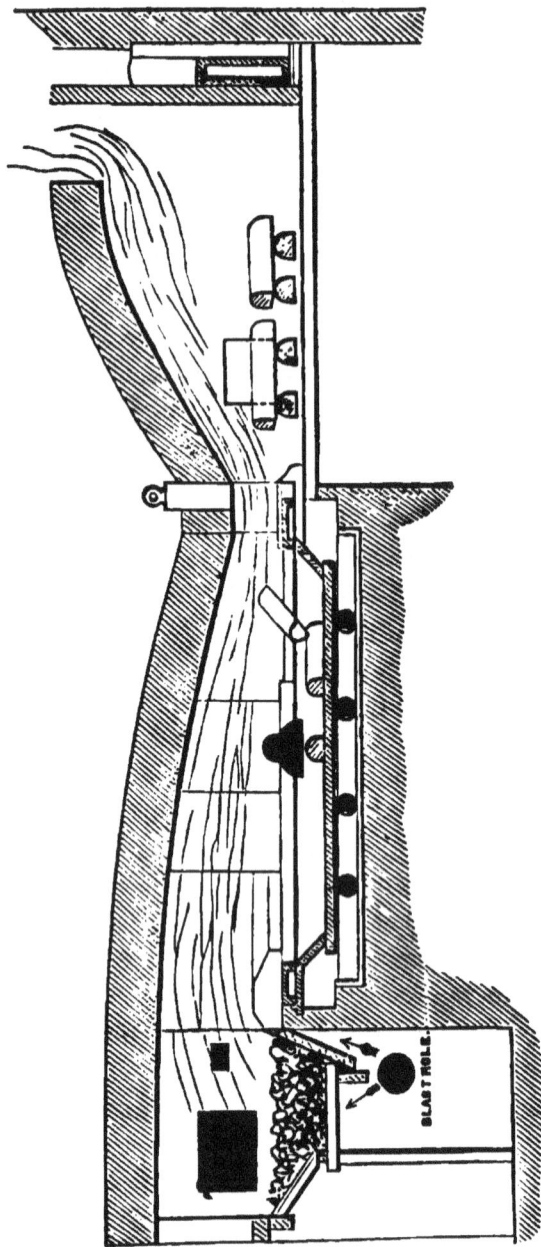

Fig. 128.—The Casson-Dormoy Puddling Furnace.

by the heat of the spent gases passing over them on the way to the chimney.

The grate was 5 feet 10 inches long by 1 foot 10 inches wide. It was fitted with a cast-iron plate of the full width of the furnace, which sloped upwards to the bridge at an angle of 30°. The back part of the grate also sloped upwards, and it consisted of cast-iron fire-bars laid close together. The bottom or lower grate was horizontal, and consisted of ordinary round or square bars. The fireplace, thus, had the form of a trough laid alongside the bridge. A blast of air was admitted through each side into an air-tight ash-pit, below the grate. The sloping back was cooled by the air-blast, and thus the formation of clinkers was prevented. Pig iron was charged in loads of 10 cwt. at a time. By the employment of a blast, under control, either an oxidizing or a reducing flame could be produced at will; whilst, with the peculiar form of grate, screened-slack might be used instead of forge-lumps. The consumption of coal amounted to 18 cwt. per ton of iron produced; and 6 tons of iron were produced in 24 hours. This performance was certainly an improvement on that of the primitive furnaces with small charges; but it was due, for the most part, to the adoption of comparatively heavy charges. In other furnaces of the ordinary construction,—those at Woolwich, for instance,—as high a degree of efficiency of the fuel had been obtained: with fuel of a better quality, perhaps, than that used in the Casson-Dormoy furnace.*

After three or four years of work, this furnace was improved by substituting for the grate a gas-producer, the invention of Mr. Smith-Casson, which was started at Round Oak Works in July, 1876.

The producer is supplied by a blast of air from below, as shown in the figure. The air is previously heated to

* See a detailed account of the Casson-Dormoy furnace in the *Journal of the Iron and Steel Institute*, 1876, page 109.

some extent before it enters the ash-pit, by being led over
the surface of the nearest stack.   The coal used is screened
slack; it is filled into a hopper, and gradually worked down
into the producer by a revolving ratchet-wheel.   Air for producing combustion is conducted above, below, and over the
sides of the gazogene, and is partially heated when it arrives
at the entrance to the furnace, where it meets the combustible
gases.   Briefly, the main features of the furnace are the
application of gas and hot blast combined to the puddling of
iron.   It does not clearly appear what the consumption of
fuel amounts to; but it is stated, on apparently good
authority, that one ton of puddled iron is produced by the
consumption of 12 cwt. of rough Staffordshire slack.   In
Mr. Casson's furnace, there is instance of the superior
efficiency of the gas-furnace worked with heated air, compared with ordinary grate-furnaces.

The evidence in favour of the employment of heated air,
as well as of heated fuel, gaseous and solid, for efficiency
and rapidity of action, is incontestable, in view of the remarkably good performance of Mr. John Price's retort-
furnace, in which the air and the fuel are both heated by
the waste-heat from the furnace, prior to their entering into
combustion.   Several of Price's furnaces have been at work
at the Royal Gun Factory, Woolwich, since 1874, when the
first of them was erected.   The double-furnace is shown
sectionally in Figs. 129, 130.   The hearth, A, is nearly
circular in plan, and measures 7 feet 9 inches long, by 7 feet
8 inches wide between the doorways, with a maximum height
of 2 feet $9\frac{1}{2}$ inches above the cast-iron bed-plates, reduced
to 2 feet clear above the cinder-bottom.   The fireplace, B,
contains a grate of the ordinary kind, 6 feet wide across
the furnace, and 3 feet long, having an area of 18 square
feet.   It is separated from the hearth by a bridge which is
15 inches wide, and 12 inches above the level of the grate.
The combustion-chamber is 2 feet 4 inches high, above the

grate. At the far end of the hearth, the gases escape into the dandy, C, whence they return, over the hearth and the combustion-chamber. Here they are delivered into the casing by which the fuel-retort F is surrounded; from which they pass downwards into the sunk chamber G, enclosing the cast-iron reservoir H; and thence by the flue I, to the chimney K. The air-blast for the furnace is supplied by a Lloyds' fan, under a pressure of 8 inches of water.

The retort F is, for the lower part, constructed of fire-brick; the upper part is of cast iron, weighing 15 or 16 cwt. It stands 11 feet 6 inches high above the sole, and is 2 feet 10 inches in diameter at the lower end, narrowing upwards to a diameter of 20 inches at the top. It is fitted with a hopper L, at the top, into which the coal is introduced, and from which it is charged at intervals into the retort, by the action of the segmental damper or valve M. The fuel is accumulated on the sole of the retort, where it is subjected to a coking process by exposure to the heat from the furnace. When the gases are driven off, the coke is pushed forward into the combustion-chamber, where the combustion is completed. The stoking of the fuel is effected by the doorway N, which is opened for the purpose, and is at other times closed air-tight. The air-blast is conveyed in a pipe which is carried within the brickwork completely round the smoke chamber G, and delivers the air into the central reservoir H. From this reservoir the air is discharged by a blast-pipe into the close chamber or ash-pit O, under the grate. The pipes lie freely in a comparatively open space within the brickwork: the space being in free communication with the chamber G, by means of a number of openings, which admit of the circulation of the hot gases around the air-pipe.

The action of the furnace is as follows:—A fire is lighted on the grate, and burns in the usual manner, until the

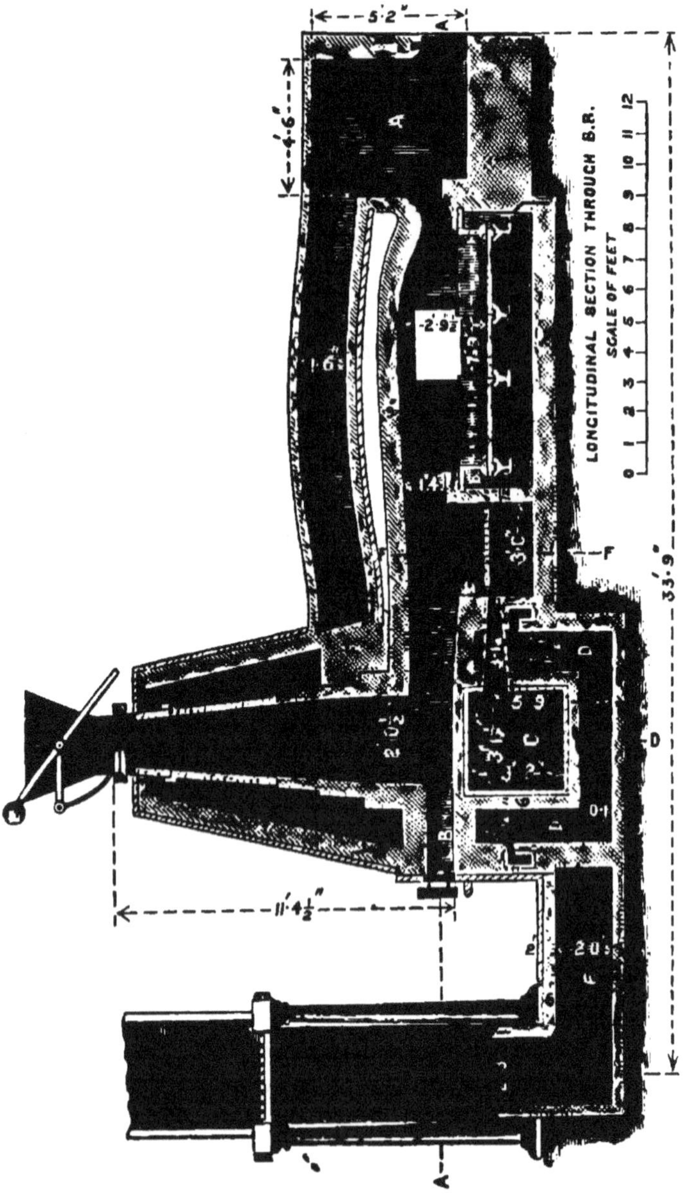

Fig. 129.—Price's Retort Furnace.

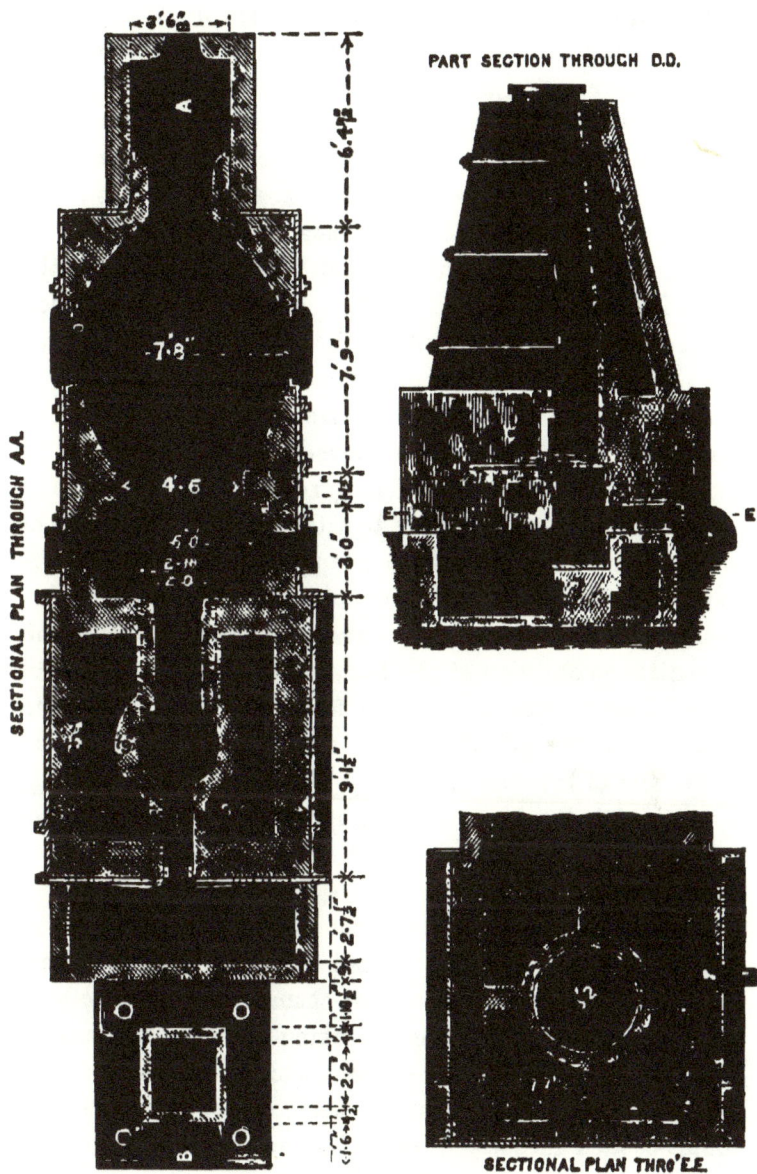

PART SECTION THROUGH D.D.

SECTIONAL PLAN THROUGH AA.

SECTIONAL PLAN THRO' E.E.

Figs. 130.—Price's Retort Furnace.

furnace becomes well-heated. The retort is then filled up with fuel, after which stoking takes place from the retort towards the grate. The burnt gases which circulate around the retort maintain it at a dull red heat; and the fuel within it is gradually carbonised, as it descends, in a manner similar to the action of a common gas-retort, until, when it reaches the bottom, it is deprived of nearly the whole of its gaseous constituents, and a fuel nearly in the condition of coke remains, to be pushed forward to the grate. As the retort is made quite air-tight, the combustible gases which are generated are driven off by the lower end of the retort into the combustion-chamber, where they meet the free air arriving through the grate, and are burned with it, at the same time that the coke is burned on the grate. The temperature of the combustible gases arriving in the combustion-chamber is from 800° to 1,000° Fahr. The coke, of course, arrives at the same high temperature, whilst the air-blast is delivered at a temperature of about 500° Fahr. These conditions are very favourable for the complete and immediate combustion of the fuel, and for the production of a very high temperature in the furnace. They involve, to some extent, the principle of the regenerator : the utilising of the heat of the spent gases, by which, it appears, the coal is carbonised, the air is heated, and the pigs receive a preliminary scorching in the dandy before being passed on to the hearth.

The Price furnace, in its earliest form, had no dandy, and the burnt gases were discharged into the chimney direct from the upper end of the retort-chamber. Mr. J. Lothian Bell * states that this furnace, working in 1875, was capable of the following performances. For a single-bedded puddling furnace, working 12½ tons of pig-iron per week; and a double-bedded furnace, working 25 tons per week; both of

* See his paper on "Price's Patent Retort Furnace," in the *Journal of the Iron and Steel Institute*, 1875, page 455.

them working by the draught of the chimney only, and with cold air direct from the atmosphere :—

CONSUMPTION PER TON OF PUDDLED IRON AND SCRAP IRON BALLS PRODUCED (COLD AIR).

	Single-bed. Cwts.	Double-bed. Cwts.
Pig and scrap iron	20·70	20·97
Fettling	·46	·21
Coal	14·02	10·71

The coal used was unscreened Gawber Hall, a Yorkshire coal of good quality. The same kind of coal was used in all the trials.

For the next trials, a fan-blast was used, and the air propelled through heated pipes in the manner already described, by which its temperature was raised to 300° Fahr. The furnace was double-bedded, and with the heated blast it reduced 26½ tons of pig in ten shifts, being 1½ tons more than was treated with cold air.

CONSUMPTION PER TON OF PUDDLED IRON AND SCRAP BALLS PRODUCED (HEATED BLAST OF AIR).

	Double-bed. Cwts.
Pig and scrap iron	21·06
Fettling	·38
Coal	9·44

Showing a reduction of 1¼ cwt. of coal, 11½ per cent., by using the heated blast in place of cold air.

Mr. Price stated, in discussion, that the respective quantities of coal consumed per ton of iron produced in the furnace, as it was then constructed, were as follows : —

	Charge of Pig. Cwt.	Coal per ton in ordinary furnace. Cwt.	Coal per ton in Retort furnace. Cwt.	Gain. per cent
Single furnace . .	5	23½	13½	42½
Double furnace . .	10	18	9½	47
Do. do. . .	15	15	7½	50

The ordinary reheating furnaces consumed 9½ cwts. of coal per ton of iron, working day-shift only; or 8 cwt. per ton, working day and night. When they were adapted with the retort system, the consumptions were as follows :—

		Coal per ton.
Cinder bottom.	Day shift only, including lighting up . .	5·25 cwt.
Do. do.	Day and night work	4·25 ,,
Sand bottom.	Day shift only, including lighting up . .	4·50 ,,
Do. do.	Day and night work	3·75 ,,

showing a gain of one-half.

A pressure of blast 3½ lbs. per square inch was tried by Mr. Whitham, of Leeds, but the heat was excessive :—the bricks melted, and the furnace was burned through in the course of twenty-four hours. With a pressure of ½ lb. per square inch, equivalent to 13½ inches of water, the damage was obviated.

The temperature in the flues of a reheating retort-furnace, during several heats, ranged from 1,100° to 1,400° Fahr., the average temperature being 1,260° Fahr.

In the flues of puddling retort-furnaces, the temperature ranged higher; but it was not accurately observed, although it was estimated that it ranged from 1,600° to 1,800° Fahr.

The results of the performance of the double-bed retort-puddling furnace at Woolwich, averaged for a long period of time, are as follows :—

	Per ton of iron produced Cwt.
Pig and scrap iron charged	18·68
Fettling	4·12
Coal	9·40

Mr. Whitham reports that, in a retort-puddling furnace, taking charges of 15 cwt., he has produced a ton of iron with a consumption of 7½ cwt. of coal. This was done in Yorkshire, where the iron is weaker and more easily puddled than that used at Woolwich.

That the temperature of the retort-furnace is high, has been proved by the melting of 40 lbs. of wrought-iron in each of three crucibles at once, in the course of three hours. From analyses which were made of the combustible gases, it would appear, as Mr. Lothian Bell observes, that the nature of the combustion may be controlled so that a flame, more or less reducing in its character, may be maintained. In one instance, two large piles of wrought-iron were placed in a heating furnace, and projected above the level of the bridge, where they were exposed to cutting action by the flame. To obviate waste from this cause, the quantity of the blast was moderated to about half the usual volume. Under these conditions, the gases contained a large proportion of carbonic oxide, and a proportion of hydrogen, thus :—

COMPOSITION OF THE BURNT GASES, AT HALF-BLAST.

	By volume.	By weight.
Carbonic oxide	13·07	13·29
Carbonic acid	7·76	12·49
Hydrogen	7·35	·53
Nitrogen	70·82	73·59
	100·00	100·00

The temperature of the blast was 500° Fahr., and that of the burnt gases when they arrived at the entrance to the chimney, amounted to 1,500° Fahr., where the flue was red-hot.

In the experiment which followed, the full blast of air, heated to 550° Fahr., was supplied, the object of the experiment being to ascertain the maximum intensity of heat that was available by this furnace.

COMPOSITION OF THE BURNT GASES, AT FULL BLAST.

	By volume.	By weight.
Carbonic acid	15·9	22·8
Oxygen	2·2	2·3
Nitrogen	81·9	74·9
	100·0	100·0

Here it is clearly indicated that not only was the carbon effectually burned, but, by the presence of oxygen, that there was an excess of air. The proportion of this excess may be estimated from the 2·3 per cent. of oxygen, which corresponded to 7·7 per cent. of nitrogen. This per-centage of nitrogen amounted to fully one-tenth of the whole per-centage of nitrogen in the gases; and it showed that the surplus air amounted to 10 per cent. of the air that was required for chemical consumption. The temperature in the furnace sufficed for the melting of 26 lbs. of wrought iron in 2¼ hours. The temperature of the escaping gases only reached 900° Fahr.: the flue was not visibly red-hot.

The cost of a double puddling furnace for receiving charges of 15 cwt. is about £400: about twice as much as hat of an ordinary furnace of equal capacity.

A puddling and heating furnace, patented in September, 1877, by Messrs. Caddick and Mabery, presents a very simple and effective combination for supplying a heated blast of air for combustion with the fuel. They provide what they call a generator of combustible gases, A, Figs. 131; which is, in fact, an ordinary fireplace, in which the solid fuel is burned on a grate. It is constructed of firebrick, enclosed in a casing of plate iron. This is surrounded by a second iron casing enclosing an air-space, into which air is blown from the pipe B. The air circulating round the generator is heated in the air-spaces, and is discharged partly into the enclosed ashpit C, and partly into the combustion-chamber D, above the fire, through several openings made for the purpose. The air that enters the ashpit receives an augmentation of heat from the hot ashes and cinders.

It appears that, in a double-bed furnace,* with averaged charges of 11 cwt. each, 19 cwt. of "stamps," or puddled

---

* See an article in *The Engineer*, September 21, 1877; page 210.

bars, were produced per ton of pig iron and scraps charged, with a consumption of coal, 12¼ cwt. per ton charged, and 13¼ cwt. per ton of bar.

In the old single puddling furnace, 18¼ cwt. of stamps

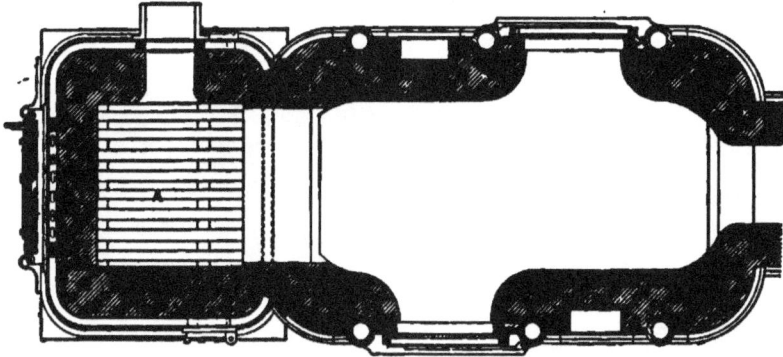

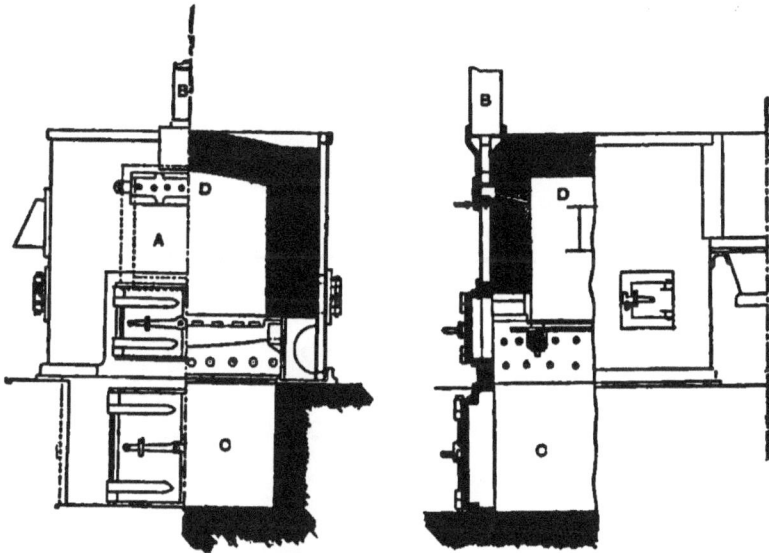

Figs. 131.—Caddick and Mabery's Furnace.

were produced per ton charged, for a consumption of fuel of 23 cwt. per ton of stamps. From these results, it is seen

that a saving of 44 per cent. in fuel was effected.   The
saving in fettling amounts to 50 per cent.

Mr. T. R. Crampton's unique system of heating iron
furnaces by the combustion of powdered fuel—coal-dust—
will be described further on, in the chapter on Powdered
Fuel.

CHAPTER XX.

BLAST-FURNACES.

It is outside the scope of this little work to follow out the intricacies of combustion in blast-furnaces, or to trace the progress of economy of fuel in this connection. Suffice it to subjoin an abstract of the detailed estimates formed by Mr. J. Lothian Bell of the appropriation of the heat of Durham coke in the Cleveland blast-furnaces, reported in the *Journal of the Iron and Steel Institute*, 1872—1875.*

Durham coke, it is assumed, consists of 92·5 per cent. of carbon, 2·5 per cent. of water, and 5 per cent. of ash and sulphur. To produce 1 ton of pig-iron, there are required 11 cwt. of limestone, and 49 cwt. of calcined iron-stone; the iron-stone consists of 18·6 cwt. of iron, 9 cwt. of oxygen, and 21·4 cwt. of earths. There is formed 7·26 cwt. of slag, of which 1·1 cwt. is formed with the ash of the coke, and 6·16 cwt. with the limestone. There are 21·4 cwt. of earths from the iron-stone, less ·74 cwt. of bases taken up by the pig-iron and dissipated in fume; say, 20·66 cwt. Total of slag and earths, 27·92 cwt.

Mr. Bell assumes that 30·4 per cent. of the carbon of the fuel which escapes in a gaseous form is carbonic acid; and that, therefore, only 51·27 per cent. of the heating power of the fuel is developed and the remaining 48·73 per cent.

* The abstract is derived from "A Manual of Rules, Tables, and Data," 1877, page 498.

leaves the tunnel-head undeveloped. He adopts, as a unit of heat, the heat required to raise the temperature of 112 lbs. of water 1° Centigrade.

DISTRIBUTION OF THE HEAT GENERATED IN THE BLAST-FURNACE FOR THE PRODUCTION OF 1 TON OF PIG-IRON.

		Units.	Per cent.
Evaporation of water in coke, and chemical action, in smelting		48,354	54·1
Fusion of pig-iron		6,600	7·4
Fusion of slag		15,356	17·2
Expansion of blast		3,700	4·1
Appropriated for the direct work of the furnace		74,010	82·8
Loss by radiation through the walls . .	3,600		4·0
Carried away by tuyere-water . . .	1,800		2·0
Sensible heat of gaseous products . .	10,000		11·2
Waste		15,400	17·2
Total heat generated in the furnace . .		89,410	100·0

The undeveloped heat of the fuel amounts proportionally to 89,410 × $\frac{84,980}{89,410}$ = 84,980 units. Add to this, the sensible heat of the gaseous products, 10,000 units, and the sum, 94,980 units, is disposed of as follows :—

DISTRIBUTION OF THE WASTE AND UNDEVELOPED HEAT OF THE FUEL REQUIRED FOR THE PRODUCTION OF 1 TON OF PIG-IRON.

		Units.	Per cent.
Generation of steam for blast-engine and various pumps connected with the work . . .		28,080	29·6
Heating the blast to 905° F.		11,920	12·5
Appropriated for direct work		40,000	42·1
Loss by radiation from the gas tubes . .	3,320		3·5
Loss of heat escaping by the chimneys .	21,660		22·8
(temperature, 770° F., from boilers)			
(„ 640° F., from stoves)			
Radiation at boilers and stoves, 25 per cent.	16,240		17·1
Waste		41,220	43·4
Loss of gases from blast-furnaces, in charging, 5 per cent.		4,740	5·0
Sundry		9,020	9·5
Total waste and undeveloped heat .		94,980	100·0

For the performance of the duty according to these analyses, Mr. Bell states that 19·08 cwt. of carbon, or 20·62 cwt. of coke, is required, per ton of iron produced from ore yielding 41 per cent. of iron. In a furnace having 18,000 cubic feet of capacity, 80 feet high, 1 ton of No. 3 pig-iron was produced with 21½ cwts. of ordinary Durham coke, from Cleveland iron-stone.

In recent years, by raising the temperature of the blast to 485° C., or 905° F., the consumption of coke, with a furnace 48 feet high, was reduced to 28 cwt. per ton of iron. With a cold blast, more than 60 cwt. would probably have been required.

It is stated, that at Barrow works, where the Siemens-Cowper regenerative stove is employed for heating the blast to 1,100° F., the quantity of coke consumed is 20·08 cwt. per ton of iron.

CHAPTER XXI.

THE SIEMENS REGENERATIVE GAS-FURNACE.

In the system of heating known as the regenerative gas-furnace of Messrs. C. W. and F. Siemens, ordinary fuels after having been converted into combustible gases in a gazogene, the principle of which has already been described, are cooled down in a cooling tube through which they are passed, in order to precipitate by condensation the aqueous vapour in mixture with them, whilst at the same time a proportion of tarry matters is likewise precipitated. The combustible current, thus purged, is next passed through and heated by a mass of hot firebricks. The air also is passed through, and heated by a mass of hot firebricks. The currents of combustible gases and air, after having thus been raised to a high temperature, are conducted to the furnace, and brought into contact and mixture; combustion ensues, and intense heat is generated in the furnace. The products of combustion are then led off through other masses of firebrick, to which they communicate their surplus heat, previously to their passing away by the chimney. Professor Faraday described the Siemens regenerative furnace in the following terms:—" The gaseous fuel is obtained by the mutual action of coal, air, and water, at a moderate red heat. A brick chamber, perhaps 6 feet by 12 feet, and about 10 feet high, has one of its end walls converted into a fire-grate; that is, about half-way down it is a solid plate, and for the rest of the distance consists of strong horizontal plate-bars where air enters, the whole being at an inclination such as that which the side of a heap of coals would naturally take. Coals are poured through openings above upon this combination of wall and grate, and, being fired at the under surface, they burn at the place where the air enters; but, as the layer of coal is from 2 to 3 feet thick, various operations go on in those parts of the fuel which cannot burn for want of air. Thus the upper and cooler part of the coal produces a large body of hydro-carbon; the cinders, or coke, which are not volatilised, approach, in descending, towards the grate; that part wh'ch is nearest the grate, burns with the entering air into carbonic acid, and the heat evolved ignites the mass above it; the carbonic acid, passing slowly through the ignited carbon, becomes converted into carbonic oxide, and mingles in the upper part of the chamber (a gas-producer) with the former hydro-carbons. The water, which is purposely introduced at the bottom of the arrangement, is first vaporised by the heat, and then decomposed by the ignited fuel, and rearranged as hydrogen and carbonic oxide, and only the ashes of the coal are removed as solid matter from the chamber at the bottom of the firebars.

"These mixed gases form the gaseous fuel. The nitrogen which entered with the air at the grate is mingled with them, constituting about one-third of the whole volume. The gas rises up a large vertical tube for 12 or 15 feet, after which it proceeds horizontally for any required distance, and then descends to the heat-regenerator, through which it passes before it enters the furnaces. A regenerator is a chamber packed with firebricks, separated so as to allow of the free passage of air or gas between them. There are four placed under a furnace: the gas ascends through one of these chambers, whilst air ascends through the neighbouring chamber, and both are conducted through passage outlets at one end of the furnace, where, mingling, they burn, produc-

ing the heat due to their chemical action. Passing onwards
to the other end of the furnace, they (that is, the combined
gases) find precisely similar outlets, down which they pass,
and traversing the two remaining regenerators from above
downwards, heat them intensely, especially the upper part,
and so travel on in their cooled state to the shaft or chimney.
Now the passages between the four regenerators and the gas
and air are supplied with valves and deflecting plates, which
are like four-way cocks in their action ; so that, by the use
of a lever these regenerators and airways, which were carrying off the expended fuel, can in a moment be used for conducting air and gas into the furnace, and those which just
before had served to carry air and gas into the furnace, now
take the burned fuel away to the stack. It is to be observed
that the intensely heated flame which leaves the furnace for the
stack, always proceeds downwards through the regenerators, so
that the upper part of them is most intensely ignited, keeping
back, as it does, the intense heat; and so effectual are they
in their action, that the gases which enter the stack to be
cast into the air are not usually above 300° Fahr., of temperature. On the other hand, the entering gas and air always
pass upwards through the regenerators, so that they attain a
temperature equal to a white heat, before they meet in the
furnace, and there add to the carried heat that is due to their
mutual chemical action. It is considered that when the
furnace is in full order, the heat carried forward to be evolved
by the chemical action of combustion is about 4,000° Fahr.,
whilst that carried back by the regenerator is about 3,000°
Fahr., making an intensity of power which, unless moderated
on purpose, would fuse furnace and all exposed to its action.

" Thus the regenerators are alternately heated and cooled
by the outgoing and entering gas and air; and the time for
alternation is from half-an-hour to an hour, as observation
may indicate. The motive power on the gas is of two kinds :
a slight excess of pressure within is kept up from the gas

producer to the bottom of the regenerator, to prevent air entering and mingling with the fuel before it is burned; but from the furnace, downwards through the regenerators, the advance of the heated medium is governed mainly by the draught in the tall stack or chimney.

"Great facility is afforded in the management of these furnaces. If, whilst glass is in the course of manufacture, an intense heat is required, an abundant supply of gas and air is given. When the glass is made, and the combustion is to be reduced to working temperature, the quantity of fuel and air is reduced. If the combustion in the furnace is required to be gradual from end to end, the inlets of air and gas are placed more or less apart, the one from the other. The gas is lighter than the air; and if a rapid evolution of heat is required, as in a short puddling furnace, the mouth of the gas inlet is placed below that of the air inlet. If the reverse is required, as in the long tube-welding furnace, the contrary arrangement is used. Sometimes, as in the enameller's furnace, which is a long muffle, it is requisite that the heat be greater at the door end of the muffle and furnace, because the goods being put in and taken out at the same end, those which enter last are withdrawn first, and remain, of course, for a shorter time in the heat at that end; and, though the fuel and air enter first at one end and then at the other alternately, still the necessary difference of temperature is preserved by the adjustment of the apertures at those ends.

"Not merely can the supply of gas and air to the furnace be governed by valves in the passages, but the very manufacture of the gas-fuel itself can be diminished, or even stopped, by cutting off the supply of air to the grate of the gas-producer; and this is important inasmuch as there is no gasometer to receive and preserve the aëriform fuel, for it proceeds at once to the furnaces.

"Some of the furnaces have their contents open to the

fuel and combustion, as in the puddling and metal-melting arrangements; others are enclosed, as in the muffle-furnaces, and flint-glass furnaces.

" The economy in the fuel is esteemed practically as one-half, even when the same kind of coal is used, either directly for the furnace or for the gas-producer; but, as in the latter case the most worthless kind can be employed, such as slack, &c., which can be converted into a clean gaseous fuel at a distance from the place of the furnace, so, many advantages seem to present themselves in this part of the arrangements."

The essential principle of the Siemens furnace is, then, to intercept the heat of the products of combustion escaping from a furnace, and cause it to heat the furnace anew. The transformation of the fuel into a combustible gas was a necessary adjunct of the employment of the departing heat, consequent on the unsuccessful efforts made by heating the air only, and directing it to the fuel in the solid state, by natural draught cnly. When, subsequently, the fuel was converted into gas, it was naturally considered advantageous to raise the temperature of the gas, as well as that of the air, before these two elements were brought into mixture for combustion. In this way, the temperature was elevated without difficulty. Before the introduction of the regenerative furnace, the heat of the departing gases was but partially economised; besides the heat lost in the production of steam, in calcination, &c., heat was lost in many furnaces, sometimes in enormous quantities, as in glass-furnaces, when chambers of large capacity were to be heated to a high and equable temperature at all points. The flame from the glass furnaces carried off the greater part of the heat of combustion into the chimney—heat which could not be utilised in generation of steam, for which there was not any demand. In such cases, the economy effected by the Siemens regenerator has been most remarkable, and has been influential in rapidly promoting the adoption of the means of heating to high temperatures, according to that system.

Formerly, the air and the gas were heated to temperatures of from 400° to 800° Fahr., which indicated but moderate accessions of heat when compared with the actual temperature of combustion. In the Siemens furnace, on the contrary, the elements may be heated up to from 1,800° to 3,600° Fahr., and more, before their entering into chemical combination ; thus not only economising heat, but producing higher temperatures than before. Suppose, for instance, a block of steel is to be melted on the old system, for which the melting point is, say, 4,000° Fahr., coal of the best quality would be requisite, the dimensions of the furnace must be restricted, the exact proportion of air to be admitted must be precisely adjusted, and every precaution must be practised in order to produce a temperature of at least 4,000° Fahr. But the conditions and the results are entirely changed, when, by the modern process, the steel may be plunged into an atmosphere of 5,500° Fahr.

The elements of the Siemens system will now be described more in detail, with general reference to the illustrations, Figs. 132, 133, 134.

Gazogene.—The gas-generator or producer or gazogene, is a brick-chamber 8 feet 2 inches deep, formed as a vault, with an arched roof, two vertical side walls, from 5 feet to 6½ feet apart, a vertical back wall, and a front wall B forming an inclined plane at an angle of from 45° to 60°. The front wall is supported by iron plates ; and it is continued towards the bottom by a step-grate at the same inclination, about 2 feet 8 inches deep, joining to an ordinary horizontal grate, c. This grate stands at a height of 16 inches above the floor of the structure. The lower side of the step-grate is from 4 to 6 inches clear above the horizontal grate, to leave room for clearing out the cinder and ash. The lowest bar of the step-grate suffers most, and is either made hollow and filled with

water to keep it cool, or is made thicker than the others.
The bars should be of wrought iron, and their number should
be limited to three or four.   The bars of the horizontal grate

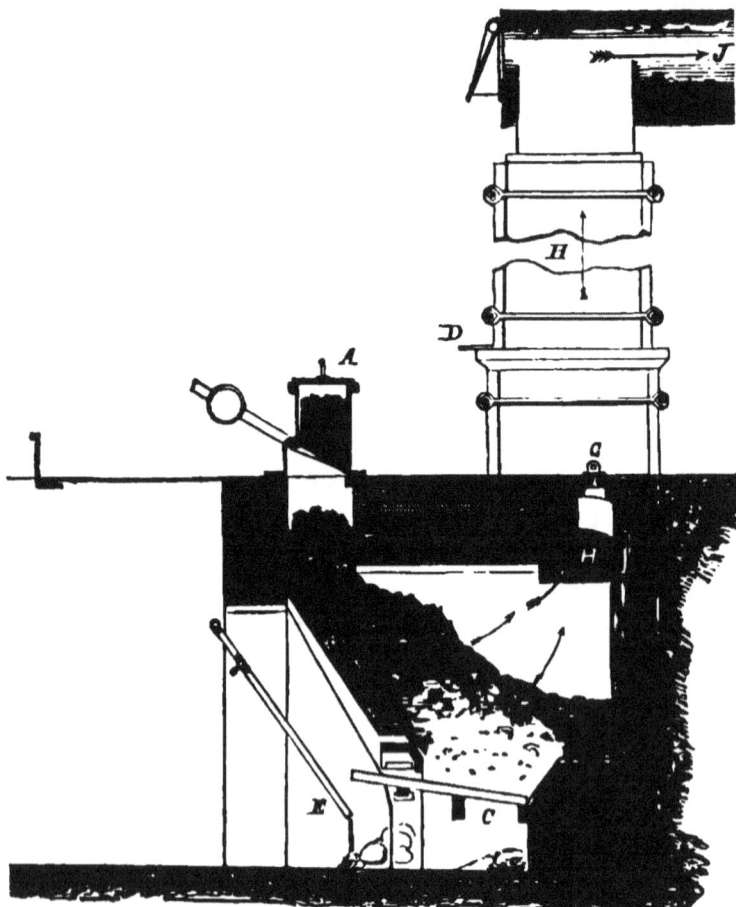

Fig. 132.—The Siemens Regenerative Furnace.—Vertical Section through
the Gas-generator.

may be few in number, when the coal lies on a bed of clinker.
The roof of the gazogene is formed with orifices, through
which the fuel may be guided, and agglomerations may be
broken down by pickers.   Two of the openings are made

larger and are fitted with hoppers, A, through which the fuel
is supplied.   Other openings, as G, are made through the
roof, through which the fire may be inspected, and the fuel
may be pushed down.   To prevent access of air directly to
the gases above the fuel, the back wall is stepped forward

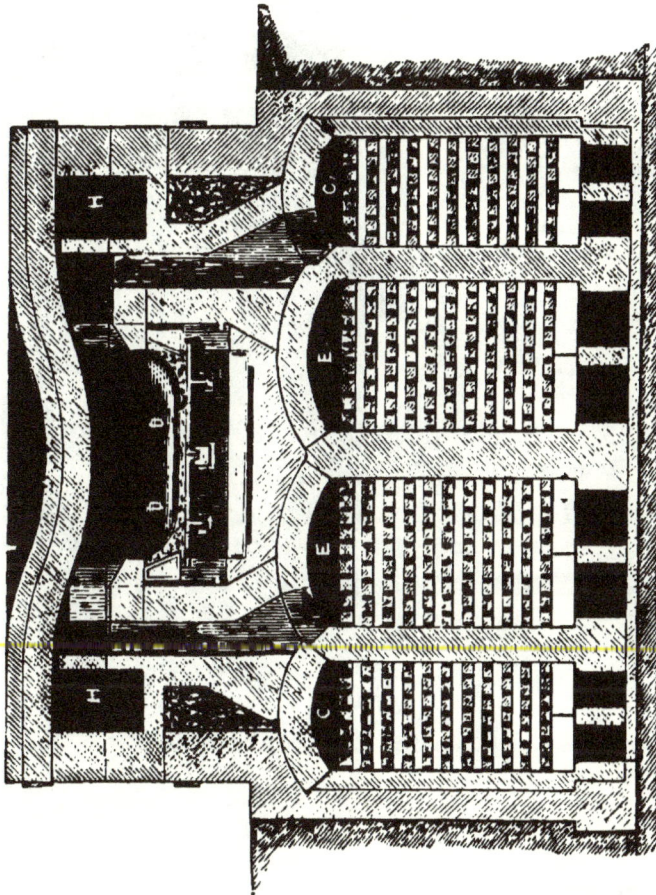

Fig. 133.—The Siemens Regenerative Furnace.—Transverse Vertical Section through the Heating-chamber and the Regenerators.

to the extent of 10 or 12 inches, at a level a little above the
horizontal grate. The fuel rests upon the bench thus formed,
and it effectually debars the entrance of free air next the
wall. By this simple precaution, the quality of the gas is
considerably improved.

Q

Charging the Gazogene.—The depth of the fuel and the inclination of the grate vary, as before stated, with the nature of the fuel. The larger the pieces of the fuel, the

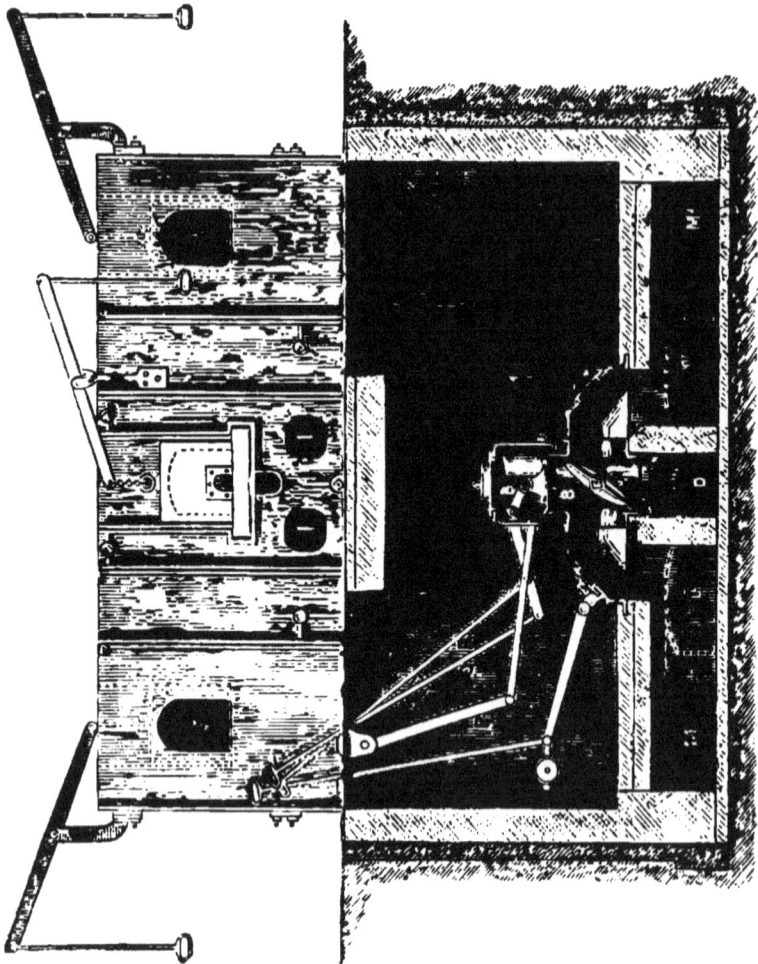

Fig. 134.—The Siemens Regenerative Furnace.—Vertical Section showing the Reversing Valves.

less is the tendency to descend, and the greater is the inclination required; which amounts to saying that the inclination should be equal to that of the natural slope of the fuel. The thickness, which may vary from two to three feet, should increase with the size of the coal, as the greater is the tendency to form interstices by the agglutination of the parts.

The quantity of coal that may be burned per day of twenty-four hours in a gazogene depends upon its dimensions and the nature of the fuel. For a generator of the given dimensions, the quantity varies from 3,000 lbs. to 4,000 lbs. per day. It is generally preferable, especially with rich coals, to moderate the formation of gas; for thus the combustion of gas is better regulated, and the gas is more easily cooled. But the generator must be maintained at a sufficiently high temperature. Coke used as fuel should be disposed in beds of from 3 feet to 3½ feet thick, and it may be burned in more considerable quantities.

Coal, from the time it is charged until it is entirely consumed, remains in the converter for a period of from 36 to 60 hours according to the capacity of the converter and the rate of combustion. The grate is usually cleaned once a day; and, when there are several generators, they are cleaned successively, at regular intervals. The cinders are sifted and returned to the generators with the ordinary coal.

Whatever the quality of coal used, a great quantity of ash in the generator is always a great obstacle to its proper action, and it is strongly recommended to wash dirty coal. The washing of coal is easily and cheaply done.

Heat is utilised by employing water, which is supplied in limited quantities by a pipe, E. The vapour, coming into contact with the carbon of the fuel, is resolved into its elements, oxygen and hydrogen, of which the former takes up an equivalent of carbon, and forms carbonic oxide, free from nitrogen. The carbonic oxide and the hydrogen are afterwards burned in the furnace. The water lies in a pool below the grate, into which the hot clinkers fall, and there generate steam. The quantity of vapour that may thus be absorbed is necessarily limited, since the vapour exercises a refrigerating action in the generator: it depends upon the heat which is disposable, from the combustion of the carbon by the air. The mass of coal in the generator is in best condition when, seen from the sight-holes, it has the clear dark-red appearance of a coke-fire, without either flame or smoke, the pieces of coke being less hot than the interspaces between them, by which the gas rises. A cherry-red should not be exceeded. When the colour becomes bright-red, it indicates that the combustion is too active, and that a considerable proportion of the carbonic acid escapes being converted into carbonic oxide. Two evils result from such excessive activity: the gas is impoverished, and it is wastefully raised in temperature. Air, besides, is drawn in through the interspaces, and burns the gases above the fuel. The gas should not be so hot as to inflame when it is discharged into the air. Dr. Siemens states that the temperature in the chamber of the generator does not exceed 750° Fahr.

From the results of various observations it has been found that, to convert 1 ton of coal in 24 hours, an area of opening of grate equal to from 4½ to 4¾ square feet is required. For an area of 4½ square feet the rate of consumption reduced to the ordinary measure would amount to 500 lbs. per square foot for 24 hours, or to 20·83 lbs. per square foot per hour. Greater rapidity of combustion, even twice as much, has been practised; but it is not good practice. When the generator becomes too hot, it is reduced to the proper temperature by the admission of steam under the grate. The grates should always be well cleaned, and the coal next the grates should burn with a bright red colour.

There is another reason for moderating the rapidity of conversion, especially of rich coals. The coal, when subjected to too high a temperature, runs together and cakes, forming vaults and incurring irregular descents with void spaces. The very bituminous coal, besides, has an aptitude for producing liquids,—tar, by which the passages are obstructed, and soot, which lines the pipes to inches of thickness, and is deposited chiefly in bends and at valves, wherever there is change of direction or of section. These deposits, of course, demand frequent cleaning out.

By the expansiveness of the hot gases in the gazogene, a slight excess-pressure is set up which is useful for preventing ingress of air into the passages on the way from the gazogene to the regenerators. If the gazogenes can be placed on a lower level than the furnaces, the gas, in virtue of its lightness and ascensional power, would produce by its upward movement a sufficient pressure. Dr. Siemens states that, for this object, a difference of level of 10 feet would be sufficient. But, in the absence of difference of level, the desired pressure is produced by the employment of a cooling tube. The hot gas is delivered from the generator into a vertical flue, H, from 13 to 15 feet high, built of brick, to preserve the heat; thence it passes into the horizontal "elevated cooling-tube," J, of sheet-iron, where it is cooled, and its density is augmented, and whence it descends by a tube which completes the syphon. The height required for the cooling-tube to yield the pressure, which would be equivalent to a difference of level of 10 feet between the generator and the furnace, may be calculated.

Thus, a certain amount of the energy of sensible heat is transformed into that of pressure:—that there may not be any leakage of air into the gas-flue, and that the gas may be delivered with a slight outward pressure at the furnace.

The temperature of the gas on leaving the generator has been found to be 600 C., or about 1,100° Fahr.; and, in the cooling tube, the gas is cooled to 104° Fahr. The required height amounts to about 14 feet, and the pressure is measured by $\frac{1}{10}$ inch of water.

By the cooling of the gas, the steam which necessarily passes over with it is, for the most part, condensed. For furnaces dealing with iron or with steel, when moist fuels are consumed, it is absolutely necessary to completely cool the gases, in order to separate by condensation the aqueous vapour, which, when permitted to enter the furnace, oxidises the metal. For the reheating furnace, at the iron-works of Munkfors, in Sweden, the gaseous products of green saw-dust, holding 45 per cent. of hygrometric water, hold, after condensation, no more than 2 out of the 33 per cent. of vapour existing in the gases before condensation.

The gases from the several generators are collected in a horizontal conduit, whence they pass into a single vertical brick shaft, on their way into the cooling tube. For a furnace consuming 3,300 lbs. of coal per day of 24 hours, the cooling tube should present a surface of at least 65 square feet; and, of course, the greater the number of generators, the more extended must be the cooling tube. The descending tube is most usually made of sheet iron, and delivers the cooled gases into a conduit from 28 to 40 inches square, which may be of a length of from 30 to 300 feet, by which they are conducted to the regenerators. The greater part of the tar is collected in a reservoir placed at the lower end of the descending tube.

A regulating valve is placed in the conduit, to adjust the supply of gas, or to shut it off entirely. It is of cast-iron, varying from 16 inches to 24 inches in diameter, and may be lifted to give from 6 inches to 12 inches of opening. The gas next reaches the reversible valve,—a flat valve, turning on a horizontal axis, and commanding three ways—two to the right and the left, alternately opened for the passage of the gas to the regenerators by turning the valve through an arc of 90°. The third way is the exit-flue to the chimney, by which the spent gases, deprived of their surplus heat, are delivered into the chimney. When the generators are being lighted up, the valve is placed vertical, and the gases go direct to the chimney. The valve is of cast-iron, usually

rectangular ; but sometimes elliptical, when it is placed in
a three-way cast-iron pipe.

The gas arrives at the lower part of one of the regenerators, which are placed under the furnace in which combustion is to take place. As the gas ascends through the
regenerator, it becomes heated by contact with the bricks,
until it is delivered into the furnace ; the draught of the
chimney should be adjusted so as not to be felt until the
products of combustion pass from the furnace, to make their
descent into and through the regenerators.

The air for effecting combustion is admitted by a mush-
room-valve of the same dimensions as that employed for the
admission of the gases. It is usually lifted only half as
much as the gas-valve.   Having passed the regulating-valve,
the air arrives at a three-way valve, adapted like the other
valve for gas, to direct the air towards one or the other of
the compartments of the generator where it is to be heated.
Having passed through this, it is delivered into the furnace.

Regenerators.—The regenerators consist of four compartments, c, c′, e, e′, Fig. 183.   Each compartment is a
rectangular chamber, usually from 16 to 32 inches wide,
8 feet to 12 feet long, and from 8 feet to 10 feet high.
Bricks are placed in tiers in this chamber—10, 15, or
30 tiers, according to their dimensions,—arranged to over-
hang and to afford a thoroughfare between and amongst
them.   The conduits from which the air and the gases
are delivered into the compartment are 18 or 20 inches
high, and they run under for the whole length of the
compartment.   The conduits are over-arched by bricks,
leaving large openings or interspaces through which the air
or the gas ascend into the compartment.   The bricks are
supported upon this arch.   The best arrangement of bricks
is that which admits of the lodgment of a great weight of
bricks in a given space, maintaining a clear way by placing
them in ranges : making sufficient passage-way for the gases

to pass on slowly, and offering a suitable area of surface. Dr. Siemens has found that a superficial area of 6 square feet is required for taking up the heat from one pound of coal consumed per hour, which would be equivalent to 560 square feet for each ton of coals consumed in 24 hours, or to 280 square feet per ton per 24 hours, for each of two compartments. The volume of space required for the four compartments amounts to 4½ cubic yards per ton of coal per 24 hours ; of which one-half, or 2¼ cubic yards, is taken as empty, and the other half as filled. Firebrick weighs 1·353 tons per cubic yard solid ; and (1·353 × 2¼) 3 tons for four compartments, or 15 cwt. for one compartment. This proportion is greater than what may be deduced in terms simply of the specific heat of firebrick, which is 0·21, and that of the gases, which is 3·44 ; but, necessarily, the lower ranges of brick are much less heated than the upper ranges. Their maximum temperature is not usually more than about 212 Fahr.; and, obviously, a greater body of brick is needed to take up the heat of the departing gases than if the whole of the brick were alike heated to the highest temperature in the regenerator.

The firebricks manufactured with the Dinas clay, of the Vale of Neath, are the best for the purpose of the regenerator. This clay consists almost entirely of silica: the composition is as follows :—

Silica	98·31
Alumina	·72
Protoxide of Iron	·18
Lime	·22
Potash and Soda	·14
Water in Combination	·35
	99·92

The powdered rock is mixed with one per cent. of lime, and just enough water to cause the mixture to cohere under pressure. The lime acts as a flux to cement together the particles of quartz.

The compartments of the regenerator are placed together in two groups, most commonly underground. They are surrounded by a thick wall, which may be built of two thicknesses with an interspace filled with sand. The compartments are separated from each other by walls which are carefully built, to prevent any direct communication between the gas and the air. The groups are connected with special reversing valves B, B, Fig. 134, so arranged that the gas-regenerator of the admission pair is connected with the gas-producer, and the air-regenerator with the atmosphere, whilst the products of combustion pass through the exit-regenerators to the chimney. The heating-chamber D, Fig. 133, is placed above the regenerators, and there are two sets of ports leading from it to the two pairs of regenerators.

The currents of the gas and the air meet for combustion at their exits from the regenerator. They may be inclined toward each other, and may meet in a mixing or combustion-chamber before entering the furnace; and if, at the same time, the air be admitted in sufficient quantity, a complete combustion is obtained, with a short flame, and a great degree of heat at the entrance. For long furnaces, like glass furnaces, where the flame is required to be prolonged, the currents of gas and air are delivered in parallel directions, so that the mixture takes place gradually, and the combustion is not completed until the elements have advanced to some distance from the origin. And again, if the quantity of air be sufficiently restricted, combustion and flame may be prolonged over the whole extent of the hearth. The gas and the air may also be delivered at different levels. If the gas be the lower current, it rises by its superior levity, and burns rapidly with the air. Thus, the length of the flame may be raised from 2 feet or 2 feet 6 inches, for melting crucible steel, to 30 feet as in large glass furnaces. When the mixing chamber is suppressed, the air is delivered by the upper opening and spreads like a sheet under the roof; and the gas, arriving from below, gradually ascends, mixes, and inflames; whilst the material on the hearth is protected against oxidation by air in contact. The flames descend through the regenerators for about a fourth of their depth, and the temperature in the upper region is of course equal to that of the flames. The products of combustion should never, if possible, leave the regenerator at a temperature higher than 410° Fahr. When the capacity of the regenerator is sufficient, the temperature of the departing gases need not exceed 212° Fahr. By a damper in the flue to the chimney, the force of the draught is controlled.

The chimney is constructed with a diameter of from 28 inches to 40 inches, and a height of from 30 feet to 65 feet. It is sufficient for the service of several furnaces; and it is put in requisition for supplying an active draught for lighting the furnaces, and getting up the temperatures: for this purpose a temporary fire is lighted at the base of the chimney. The sectional area is proportioned to give from 30 to 45 square inches per ton of coal consumed in 24 hours.

GENERATION AND DISTRIBUTION OF THE HEAT OF THE
REGENERATIVE GAS-FURNACE.

An exhaustive analysis of the performance of the Siemens Gas-furnaces at the Glass-works, Saint-Gobin, has been entered into by M. Kraus,[*] from which the following abstract is derived.

The system of Dr. Siemens is there applied to the furnaces for melting glass. The following description and analysis have reference to one of these furnaces.

The chamber to be heated is 28 feet long, 11¼ feet wide, and 6·56 feet high inside to the crown of the arch overhead: —making a volume of, say, 1,412 cubic feet. It contains

[*] "Etude sur le Four à Gaz et à Chaleur Régénérée," in the *Annales du Génie Civil*, 1874, page 36, &c.

20 melting-pots, weighing 440 lbs. each, and holding
770 lbs. of material; say, in all 8,800 pounds of pottery, and
15,400 pounds of material to be melted. This furnace,
previously to the employment of the Siemens system, consumed 12 tons of coal in 24 hours, holding the same charges.

There are five generators in the open air, four of which
are kept in action, and are sufficient for keeping up the
supply of heat to the furnace. The step-grate is inclined at
an angle of 50°. The brick slope is about 5 feet long,
nearly one-half longer than the grate, which is 3·61 feet
in length. The bars are hollow and kept cool by a current
of water. The floor of the gazogene is 8·20 feet below
the surface of the ground, and the charging box is 20 or
24 inches above the level. The air is admitted between five
bars only, within a vertical height of about 18 inches. The
average width is 6 feet, and the area is (6 × 1·5 =) 9
square feet of open grate for one furnace ; or, for four furnaces, 36 square feet. The quantity of coal consumed is
7 tons in 24 hours, and the area for admission of air amounts
to 4¾ square feet per ton consumed per hour, being about
20 lbs. of coal per square foot per hour :—a rate of consumption which answers very well. The fuel consists of a
mixture of one-fourth of dry coal or of coke, and three-
fourths of richly bituminous coal from Anzin. The generators act quietly, the gas is thick and plentiful, and under
good pressure.'

Passing from the generator, the gas travels along a horizontal passage, whence it rises into a chimney or vertical
passage 23 feet high ; thence into the cooling tube, 3 feet
9 inches in diameter, and 26 feet 3 inches long, from which
it descends by a chimney similar and parallel to the first,
for a depth of 29 feet 6 inches, into an underground passage
towards the valves. All these passages are easily accessible
by doorways, through which they may be cleaned with
promptitude. The underground passage is 3·28 feet by

8·44 feet, and is 197 feet in length up to the valves ; it is at a level of from 16 to 18 inches below the grate of the generator. The valves are circular, 24 inches in diameter. The passage to the regenerators is 24 inches high, and 32 inches wide.

The brick chambers, or regenerators, are 8·28 feet wide, and 7·88 feet high, and 11·48 feet long ; making a capacity of 278 cubic feet. The bricks are laid in 12 tiers, and they are uniformly traversed by the gases. The chimney is near the generators, and is placed 197 feet from the furnaces. It is 3·28 feet square, and stands 49 feet high. The gases are discharged into it by a flue 3·28 feet by 3·44 feet, fitted with a damper. The area of admission into the chimney is from ¼th to ⅓th of the section of the flue. Under these conditions, the furnace works advantageously. The temperature in the furnace is very high and very regular; whilst the temperature in the chimney is very low. The economy of combustible effected by the introduction of the regenerative furnaces amounts to 42 per cent.

In the following calculation, it is assumed that the specific heats are constant. The sense of the calculations would not be materially affected, if the actual specific heats at high temperatures were known and taken into account.

FRENCH AND ENGLISH MEASURES.

1 Kilogramme	. . . .	2·205 lbs.
1 Tonne	. . . .	0·9842 ton.
1° Centigrade	. . . .	1°·4 Fahrenheit.

Temperature by Centigrade, equal to (temperature by Fahrenheit — 32) × ⅝.
Temperature by Fahrenheit, equal to (temperature by Centigrade × ⅝) + 32.

1 Calorie	. . . .	3·968 English heat-units.

The fuel supplied to the generators, from which the following analyses were made, consisted of a mixture of ¾ths of bituminous coal, and ¼th of dry coal. Water was not supplied to the generator. The composition of the gases produced was as follows :—

	Volumes.	
Carbonic Oxide . .	24·2 ⎫	
Hydrogen . . .	8·2 ⎬	34·6 Combustible
Carburetted Hydrogen	2·2 ⎭	
Carbonic Acid . .	4·2 ⎫	
Nitrogen . . .	61·2 ⎬	65·4 Non-Combustible
	100·0	100·0

Here may be added, parenthetically, the composition of the combustible gases according to M. Boistel, in a paper read by him to the "Société des Ingenieurs Civils," in 1867 :—

	Per cent.
Carbonic Oxide	21½ to 24
Carbonic Acid	4 to 6
Nitrogen	60 to 64
Hydrogen	5·2 to 9·5
Hydro-carbons . . .	1·3 to 2·6
	100

Multiplying the quantities given by M. Kraus by the respective densities of the gases, the composition by weight is obtained :—

	Volumes.		Specific density.		Weights.		Per cent.
Carbonic Oxide . .	24·2	×	·9674	=	23·639	or	25·89
Hydrogen . . .	8·2	×	·0692	=	·567	„	·62
Carburetted Hydrogen	2·2	×	·5527	=	1·216	„	1·33
Carbonic Acid . .	4·2	×	1·5290	=	6·421	„	7·04
Nitrogen . . .	61·2	×	·9713	=	59·444	„	65·12
	100·0	×	·9128	=	91·277	or	100·00

The three combustible gases together amount to 27·84 per cent., and the two non-combustible gases to 72·16 per cent. A small quantity of aqueous vapour is mixed with these gases, and is condensed. From the composition of the gases, the average composition of the coal is deduced, as follows, omitting the ash :—

Per cent.

Carbon	.	.	.	.	84·38
Hydrogen	.	.	.	.	6·17
Oxygen	.	.	.	.	6·90
Nitrogen, &c.	.	.	.	.	2·55
					100·00

The quantities of gas produced per 100 kilogrammes of this coal are—

Kilogramme.

Carbonic Oxide	.	.	.	.	155·87
Carbonic Acid	.	.	.	.	42·39
Carburetted Hydrogen		.	.		8·02
Hydrogen	.	.	.	.	3·74
Nitrogen	.	.	.	.	392·15
					602·17

The water condensed in the passages, together with the tars, soot, &c., amounted to 3·79 kilogrammes.

To find the total capacity for heat of these products, multiply them respectively by their specific heats:—

	Kilogrammes.		Specific heat.		Calories.
Carbonic Oxide . .	155·87	×	·2479	=	38·66
Carbonic Acid . .	42·39	×	·2164	=	9·17
Carburetted Hydrogen	8·02	×	·5929	=	4·76
Hydrogen . . .	3·74	×	3·4046	=	12·73
Nitrogen . . .	392·15	×	·2440	=	95·68
100 Kilogrammes of fuel	602·17	×	·2673	=	161·00

Showing that the specific heat of the gases, taken together, is ·2673, and that the 602 kilogrammes of gases produced from 100 kilogrammes of the fuel absorb 161 calories for each degree of temperature.

The quantity of air necessary to burn these 602 kilogrammes of gases, allowing a surplus of 20 per cent. over the quantity chemically consumed in forming carbonic acid and aqueous vapour with the carbon and hydrogen respectively, amounts, according to M. Kraus, to 788·69 kilogrammes, comprising 181·81 kilogrammes of oxygen and 607·88 kilogrammes of nitrogen. The volume of this quantity of air at

62° Fahr., at the rate of 13·14 cubic feet per pound weight, or 28·97 cubic feet per kilogramme, is 22,850 cubic feet. The volume of the 602 kilogrammes of gases as produced, is equal to (602·17 ÷ ·9128 × 28·97 =) 19,110 cubic feet at 62° Fahr.

The total capacity for heat of this quantity of air, of which the specific heat is ·238, is equal to (788·69 × ·238 =) 187·76 calories per 1° C. of temperature.

By the conversion of the carbon into carbonic oxide and acid, and of a part of the hydrogen into water, there are generated 273,584 calories, or French heat-units; thus :—

Kilogrammes.		Calories.
66·803	Carbon into Carbonic Oxide . .	165,671
11·560	,, ,, Acid . .	93,405
·421	Hydrogen into Water . . .	14,568
		273,584 (A)

The total heating power of 100 kilogrammes of coal amounts to 862,961 calories :—

	Calories.
84·38 Carbon into Carbonic Acid . . .	681,790
6·17 Hydrogen into Water	181,171
	862,961 (B)

Dividing the quantity A, the heat disengaged in the generator, by the quantity B, the whole heat of combustion, it is found that the former is 31·7 per cent. of the latter.

Heat disengaged by the complete conversion of the gases in the Furnace.—Dr. Siemens allows 20 per cent. excess of air in the furnace above that which is chemically consumed in combustion. The gases of 100 kilogrammes of coal yield, in burning, 587,391 calories, thus :—

Kilogrammes.		Calories.
155·874	Carbonic Oxide into Carbonic Acid .	374,098
8·023	Carburetted Hydrogen into Carbonic Acid and Water	104,804
3·741	Hydrogen into Water . . .	108,489
		587,391

The products of combustion, including 20 per cent. excess of air, are—

	Kilogrammes.
Carbonic Acid	309·393
Water	47·205
Nitrogen	999·527
Atmospheric Oxygen	30·218
	1386·343

Of which the capacity for heat is found by multiplying the items respectively by their specific heats :—

	Kilogrammes.		Specific heat.		Calories.
Carbonic Acid	309·393	×	·2164	=	66·952
Water	47·205	×	·4750	=	22·430
Nitrogen	999·527	×	·2440	=	243·885
Oxygen	30·218	×	·2182	=	6·588
	1386·343	×	·2452	=	339·845

That is to say, the products of combustion absorb 339·845 units of heat for 1° C. of temperature, and the rise of temperature by combustion therefore amounts to $\left(\frac{587,391}{339·845} = \right)$ 1738° C., equivalent to 3128° Fahr.

Distribution of the Heat.—For the purpose of estimating the distribution of the heat produced, M. Kraus analyses the performance of the heating-furnaces at the Sougland iron-works, in the department of l'Aisne, France, constructed on the Siemens system. This furnace heats 14 piles of iron, weighing 50 kilogrammes, or 110 lbs. each. From a quantity of iron equal to 9,000 kilogrammes, or about 9 tons, heated in this furnace, 5,600 kilogrammes, or about 5·6 tons of blooms, exclusive of ends and wasters, were produced for the manufacture of plates. The quantity of Charleroi coal consumed was 2,000 kilogrammes, or about 2 tons per day of 24 hours—at the rate of 4·44 cwt. per ton of iron heated. It is assumed that the composition of the gases is the same as that of the gases at Saint Gobain.

The 2,000 kilogrammes of coal contain 10 per cent. of

ash; leaving 1,800 kilogrammes of net coal. The valves were reversed every half-hour; and in each interval 37¼ kilogrammes of coal were consumed, developing (587,391 ÷ 37·5 =) 220,272 calories.

Loss by the Chimney.—Suppose that the temperature in the chimney be taken at 100° C., or 212° Fahr., which is easily obtained. The burnt gases of 100 kilogrammes of coal, weighing 1,386 kilogrammes, have a capacity for 339·85 calories per 1° C., and for 37½ kilogrammes of coal, the gases, weighing 520 kilogrammes, have a capacity for 127·50 calories per 1° C. For 100° C., the total heat carried off would amount to (127·500 × 100 =) 12,750 calories. If the gases be not cooled below 200° C., they would carry off 25,500 calories by the chimney. Allow that the actual loss amounts to, say, 25,000 calories.

Loss by transmission through the Walls of the Regenerator.—M. Kraus assumes that the gases leave the furnace at the temperature of the welding-heat of iron, which he takes at 1,600 C., or 2,912° Fahr. His calculation, in which he employs the formulas of MM. Dulong and Petit, conducts him to the conclusion that the loss of heat by radiation and conduction through the walls amounts to 27,750 calories for each half-hour.

Quantity of Heat absorbed by the Iron.—To heat 9,000 kilogrammes of iron to, say, 1,600 C., or 2,900° Fahr., in 24 hours: the specific heat is ·185, and the heat thus absorbed is (9,000 × 1,600° ÷ ·185 =) 2,664,000 calories in 24 hours, or 55,500 calories in each half-hour.[*]

Quantity of Heat lodged in the Furnace.—To maintain the necessary temperature, and compensate for losses by transmission through the enclosure, the heat required

---

[*] M. Kraus assumed for calculation the ordinarily assigned specific heat of iron, ·114; but M. Ponsard has shown good reason for adopting the value ·185 for very high temperatures, and this value is employed in the text.

is 112,766 calories, an amount which is obtained by difference.

SUMMARY OF THE DISTRIBUTION OF THE HEAT OF 37½ KILOGRAMMES
OF COAL.

	Calories.	Per cent.
For the conversion of the fuel into gas . .	102,593 or	31·7

	Calories.	Per cent.	
Loss by chimney . . .	25,000	or 7·7	
Loss through walls of Regenerator	27,750	or 8·6	
Absorbed by the iron . .	55,500	or 16·5	
Lodged in the furnace and loss through walls . .	112,766	or 35·5	221,016 or 68·3
			323,609 or 100·0

The net heat utilised in heating the iron is shown to amount to 16½ per cent. of the total heat of combustion, or to 25 per cent. of the heat which is generated in the furnace.

Quantity of Heat intercepted by the Regenerators.—The burnt gases, on leaving the furnace, have a temperature of 1,600° C., and carry off (127·50 × 1,600 =) 204,000 calories in half an hour. In this expression, 127·50 is the capacity for heat of the burnt gases from 37½ kilogrammes of coal. Of these 204,000 calories, the chimney takes off 25,000 ; the walls of the regenerators traversed by the flames take (27,750 ÷ 2 =) 13,876 ; and the remainder, 165,125 calories, are retained by the regenerators, and are available for heating the air and the gases approaching the furnace. Supposing that the compartments of the regenerator for gas and for air are of equal capacity, then, in each there are (165,125 ÷ 2 =) 82,562 calories stored up. The quantity of the gases produced from 37½ kilogrammes of coal is (602 × 37·5 ÷ 100 =) 225·75 kilogrammes ; of which the capacity for heat is (161 × 225·75 ÷ 602 =) 60·38 units per 1° C. of temperature. Then, the gases are raised, by the absorption of 82,562 calories, to the average temperature (82,562 ÷ 60·38 =) 1,367 C., or 2,492° Fahr.

The capacity for heat of the air necessary to burn the

gases, including 20 per cent. excess, was calculated to be 187·76 calories per 1° C., per 100 kilogrammes of coal. For 37½ lbs. it is (187·76 × 37½ ÷ 100), or (187·76 × 225·75 ÷ 602 =) 70·41 calories per 1 C.; and the air is raised to the temperature (82,562 ÷ 70·41 =) 1,172 C., or 2,109 Fahr.

It is desirable that the air and the gas should be raised to the same temperature, and, for this object, the air-compartment is made larger than the gas-compartment. The mean temperature of the air and the gas, if heated to the same degree, would be $\frac{165,125}{(60·38 + 70·41)} = \frac{165,125}{130·79} = 1262$ C., or 2,272 Fahr.

The mixture of air and gas should, then, have a mean temperature of 1,262 C. before combustion. After combustion, the 165,125 calories are distributed amongst the burnt gases, of which the capacity for heat is a little less than that of the gases and the air together before combustion, being 127·50 instead of 130·79. The temperature given by the regenerator to the products of combustion is, therefore, (165,125 ÷ 127·50 =) 1,295 C., or 2,363 Fahr., which are added to the temperature generated by combustion. The combustion of the gases produces 1728° C. of heat, and the sum of this and the heat supplied from the regenerator, namely, 1,295° C., is equal to 3,023° C., or 5,477° Fahr.

Dr. Siemens recently stated,[*] with reference to the performance of the regenerative furnace, that a ton of iron could be heated to the welding point with a consumption of 7 cwts. of coal,—a quantity considerably greater than the quantity given by M. Kraus, page 352; and that a ton of steel could be melted with a consumption of 12 cwts.

Dr. Siemens,[†] several years ago, made numerous applica-

---

[*] In a lecture "On the Utilisation of Heat and other Material Forces," delivered at Glasgow, March 4th, 1878; page 19.

[†] In his paper on "A Steam Jet," in the *Proceedings of the Institution of Mechanical Engineers*, 1872; page 109.

tions of a jet of steam, peculiarly designed for efficiency, arranged as a blower, drawing and forcing air for accelerating the distillation of fuel in his gas-producers. The blower is built into the side wall of the producer, and the combined current of air and steam delivered by it issues through an opening into the space underneath the grate, which is closed by doors. The small proportion of steam that enters together with the air is just sufficient to assist beneficially in the production of the combustible gas, by being converted into carbonic oxide and hydrogen. The advantages found to result from applying these blowers to gas-producers are, that coal-dust of the most inferior description can be used, and that the production of gas in each producer, in consuming small fuel, is raised from 1½ tons to 3 tons, in 24 hours.

CHAPTER XXII.

THE PONSARD GAS-FURNACE, WITH RECUPERATOR.

THE Ponsard gas-furnace, Figs. 135 to 138, like the Siemens furnace, recovers by means of a "recuperator" a large proportion of the heat carried from the furnace by the burnt gases. Gazogenes of different forms are employed, according to the nature of the combustible. There are ordinary gazogenes, constructed with grates, in the manner already described, supplied with ordinary atmospheric air; and there are gazogenes which are supplied with heated air,—*gazogènes surchauffés*,—which are supplied with air previously heated by the recuperator.*

The "superheated gazogene" is very different in form from the ordinary gazogene: it has not any grate, the use of which, traversed by highly heated air, at a temperature of from 1,500° to 1,800° Fahr., is impracticable, as an iron grate would be quickly destroyed by the heat of the air passing through it. The generator consists simply of a close rectangular chamber, below the surface of the ground, having a level floor, and arched over. It is supplied with fuel through a longitudinal opening from above, which is filled up until it blocks and closes the openings, so that there is no direct communication with the external air. The current

---

* An excellent account of the Ponsard furnace, contributed by M. Sylvain Périssé, is published in the "Mémoires de la Société des Ingénieurs Civils," 1874; page 752.

of heated air enters the chamber at one flank of the heap of
fuel, and traverses it; whilst the gas which is generated by

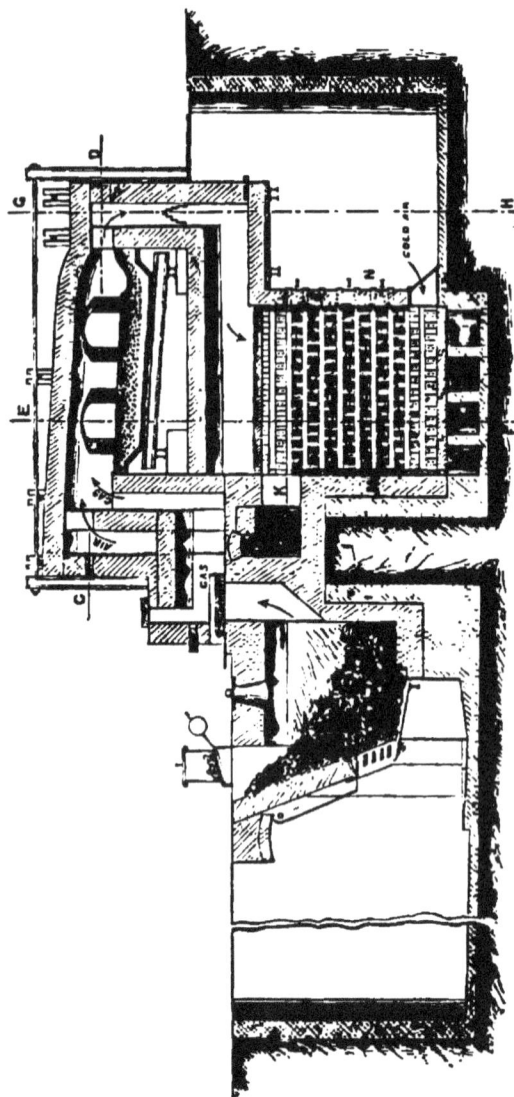

Fig. 135.—The Ponsard Gas-Furnace.—Longitudinal Section.

this means passes off through a passage from the other flank
of the heap.   Ash and slag settle down on the floor of the

chamber, and are cleared out from time to time, by doorways
in the sides of the chamber, which are usually closed up.

The recuperator is constructed of firebrick. A rectangular chamber is divided by a number of vertical diaphragms
into compartments, occupied alternately by the burnt gases
and the air for the furnace. The heat of the gases on one
side passes laterally through the diaphragms to the air on
the other side, and, that the air may freely circulate from
one compartment to another, the air-compartments are connected by numerous hollowed bricks, through which the
currents may pass, and which serve at the same time to

Fig. 136.—The Ponsard Gas-Furnace.—Superheated Gazogene.

increase the heating surface, and consequently also the com-
pactness and efficiency of the apparatus. In the figures it is
shown that the hot burnt gases enter the recuperator at the
upper part, descend through the compartments b, b, &c., and
pass out at the lower part. towards the chimney; whilst,
on the contrary, the cold air enters at the lower part, passes
upwards through the compartments a, a, &c., and passes out
at the upper part. By this arrangement, each of the strata
or sheets of air is enclosed between two envelopes of hot
gas, and absorbs heat continuously from the brick diaphragms
between which it is enclosed, for the whole height of the
traverse. The transverse, or bonding bricks, as they may be
called, are so alternated, vertically as well as laterally, that

they do not interfere with such variable expansion and con-traction as may be brought into play, and are therefore the less likely to cause dislocation. The admission of the air to the recuperator is regulated by a damper.

The hot gases, in descending, naturally occupy every

Fig. 137.—The Ponsard Gas-Furnace.—Longitudinal Section through Recuperator.

portion of the space into which they enter, and through which they pass. This is a peculiarity of a descending current,—to be found also in the Siemens regenerator ; whilst, again, the ascending air, expanding as it goes, natur-ally, though by another kind of effort, occupies every portion of the passage through which it ascends.

When it is required to divide the supply of air, so as to draw off a portion of it for the supply of the gazogene, the current may readily be split, by building up a solid iron partition in each air-space at right angles to the diaphragms. The separated current, controlled by a damper, may thus be led off apart from the main body of the current of air destined for the supply of the furnace. There are, in this case, two distinct recuperators within one chamber.

A recuperator constructed of bricks of the ordinary sizes, presents a total surface of 190 square feet per cubic yard of volume, calculated from outside dimensions; of which one half is employed in cooling the burnt gases, and the other half in heating the air. The weight per cubic yard is about 1,500 pounds, and the cost is 60s. per cubic yard, when bricks cost 4s. 4d. per cwt. The bricks are formed with grooves to receive the fire-clay with which the joints are made, and to insure air-tight jointing. But leakages, when they do happen, are harmless for explosion, since the elements in presence of each other are not explodible. It is true, nevertheless, that in the construction of the recuperator a greater degree of care is required than in that of the regenerator.

M. Périssé deduces from the results of his observations that the conductibility of fire-brick augments with the temperature. By numerous experiments with a water pyrometer, he has tested the temperature through which the air is raised, in passing through the recuperator, and he finds that it ranges between 1,000° and 1,100° C., or 1,800° and 2,000° Fahr.

Laboratory.—This is the name given by M. Ponsard to the combustion-chamber, which is placed above the levels of the gazogene and the recuperator. The gas and the heated air arrive, therefore, in the laboratory, with an ascensional force of about a fifth of an inch, more or less, and do not depend upon any mechanical means of creating a draught. The com-

R

bustion is, then, effected under pressure; and the relative proportions of the air and the gas, as well as the regulation of the draught, are regulated by means of three dampers, two of which control the elements, and the third is placed at the base of the chimney. The operation of the furnace is continuous. The combustible gases are led direct to the furnace without being cooled down, as in the Siemens furnace; and there is no deposit of tar or of hydro-carbons in the passages.

The excess of atmospheric air required for effecting complete combustion in the furnace, it is estimated, need not exceed 10 per cent. by weight of the products of combustion.

Ponsard Furnace at the Basacle Forge, Toulouse.—This furnace was constructed for heating, to a welding-heat, piles of old iron, for the manufacture of hoop iron, wheel-tyres, and the like. The piles, weighing from 100 lbs. to 180 lbs. each, are welded, and each pile is rolled into two billets from 20 to 80 inches in length. The billets are reheated in the furnace and then rolled into merchant iron.

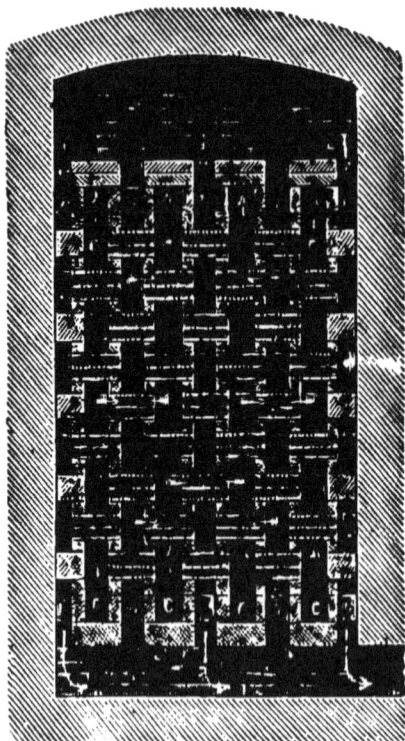

Fig. 138.—The Ponsard Gas-furnace.—Transverse Section of Recuperator.

Previously to the employment of the Ponsard furnace, which

was started in October, 1872, the work of the forge was done
with two ordinary furnaces, having a maximum width of
4 feet 7 inches, and a length of 7 feet 10¼ inches, with a
square fire-grate having 8·7 square feet of area. These have
been superseded by the Ponsard furnace, which alone produces 30 per cent. more iron than the other two together.
The hearth of the new furnace is 6·56 feet wide and 10¼ feet
long. There are two charging doorways, of which the
doorway farther from the gazogene is used for receiving the
piles; the piles are afterwards moved up towards the second
doorway, by which they are taken out of the furnace. The
billets are always introduced cold by the nearer door;
they arrive at a welding-heat in from ten to twenty minutes,
according to size, whilst the piles are being heated opposite
the farther door.

There is one gazogene, of the ordinary construction,
supplied with Decazeville coal of the size of hazel-nuts
(*petites noisette*), of which the constituents by destructive
distillation, after having been desiccated at 230° Fahr., are
as follows :—

Water	1·60
Ash	11·40
Coke (dull)	40·60
Volatilised matters	37·10
	300·00

The recuperator has a total volume of 381 cubic feet, or
14 cubic yards; with a total heating surface of 1,354 square
feet. The weight of special bricks amounts to 9¼ tons,
equivalent to 1,480 lbs. per cubic yard. The heating
surface amounts to 97 square feet per cubic yard, and as
much for the coaling of the burnt gases.

The quantity of iron charged into the furnace during a
given fortnight of uniform work—under unusually favourable conditions—amounted to above 20 tons in the 24
hours. During the fortnight previous to making a thorough

repair of the recuperator, when it was out of order, the quantity charged did not exceed 15 tons in the 24 hours.

The quantity of Decazeville coal consumed in 31 days of the month of May, 1874, was 119·60 tons, of which 58·10 tons were consumed by day and 61·50 tons by night. On Sundays, the fire was slackened, as the furnace was not at work. For six days in the week, the fuel was consumed at the rate of 4·20 tons; for the 24 hours of Sunday the rate was only 1·70 tons. For the whole week of seven days the consumption was, consequently, 27 tons; being, for the fortnight's performance already noticed, at the rate of 493 pounds, or 4·41 cwt., per ton of iron charged.

To compare with this performance, it may be stated that each of the old furnaces consumed nearly the same quantity of coal as the Ponsard furnace does now, whilst the new furnace produces about 30 per cent. more iron than the two old ones together. The resulting economy of combustible, therefore, amounts to about 60 per cent.; besides, the coal formerly consumed was from Carmaux, a dearer coal than the Decazeville coal.

The average results of the comparative performances of the Ponsard and the old furnaces are thus summarised by M. Pélégry :—

	Ponsard furnace.	Two old furnaces.
Iron charged into the furnace in 24 hours	17 tonnes	12·5 tonnes
Production per month of 24 working days, in piles subjected to two heats	200 do.	150 do.
Coal consumed, including days of rest (mixture of Decazeville and Carmaux for the old furnaces)	120 do.	180 do.
Coal, per tonne, per heat	662 lbs.	1,323 lbs.
Coal, per ton, per heat.	672 lbs. or 6 cwt.	1,344 lbs. or 12 cwt.

Here an economy of 50 per cent. is exhibited.   Seventeen

months after the Ponsard furnace was started, the recuperator was, for the first time, entirely rebuilt. The cooling-
down and the relighting of the recuperator occupied 4 days
each, and the total time during which the furnace was off
duty for this repair was 15 days. M. Pélégry estimates
that the furnace must be stopped for repair 15 days every
quarter. The grate is cleaned every 12 hours. A high
temperature in the gazogene is preferred, as it increases the
production of the furnace, and the consumption of fuel per
ton is less. The recuperator is cleaned out every week: the
gas-flues are cleared every 12 hours. The cleaning is easily
done, through cleaning-doors conveniently placed.

PONSARD FURNACE, SUPERHEATING, AT VIEUX-CONDÉ (NORD).

This furnace has been in operation since July, 1873.
It is at work from 6 A.M. to 6.30 P.M. every day,
except Sundays, for reheating bar-iron used for the manufacture of nuts and other small forgings. During the remaining 8 hours of rest, out of the 24, the gazogene is
kept alight by receiving a double charge, whilst the three
dampers are closed until 3 A.M., when the stoker arrives,
and restores the furnace to good working order by 6 A.M.
The hearth of the furnace is 28 inches wide, and 5 feet
4 inches long. The gazogene is 3·28 feet wide. The recuperator has a volume of 124 cubic feet, or 4·57 cubic yards,
with 3·20 tons of special bricks, weighing 1,568 lbs. per
cubic yard. The heating surface for the entering air is
430 square feet, and there is an equal surface for receiving
the heat from the burnt gases. There is 94 square feet of
surface, on each side, per cubic yard of the recuperator.

The work of the furnace varies, but it consists chiefly
in heating 10 bars at a time, for the manufacture of nuts for
bolts of from $\frac{5}{8}$ inch to $1\frac{1}{16}$ inches in diameter. These
bars are from 8 feet to 10 feet long: they are withdrawn,

one at a time, from the furnace, and taken at a white heat to the punching machine. They are reheated six times before they are exhausted. Cold bars 1·36 inch by 1 inch are heated to a white heat in 3½ minutes ; and bars 1·2 inches by ·80 inch are similarly heated in 2 minutes 40 seconds. These smaller bars are heated up from a red heat in 2 minutes 10 seconds. From 2 to 2½ tons of nuts of average size, corresponding to 10,000 or 12,000 in number, are manufactured per day of 11 hours. With the old furnace, only 1·57 tons per day were turned out.

At the Ponsard furnace, from 2,400 lbs. to 2,600 lbs. of bituminous coal were consumed per day, containing 24 per cent. of ash. The old furnace consumed twice as much per day, of the same coal ; and the heat of the burnt gases was employed to generate steam in an old boiler, sufficient to supply an engine of 12-horse power. The burnt gases from the Ponsard furnace are hot enough to be utilised in heating a Field boiler, which it was designed to put down. From the following comparative statement of the consumption of fuel, it appears that there is a clear saving of 50 per cent. for given weights of screws turned out :—

OLD FURNACE.

	lbs.
Coal per day	4961
Deduct for supplying 12-horse power, at 6·62 lbs. per hour per horse power, for 11 hours . .	880
Balance for 1·57 tons of screws	4081
Or, per ton of screws	23·2 cwt.

PONSARD FURNACE.

Coal per day, for 2·16 tons of screws . . .	2536 lbs.
Or, per ton of screws	10·5 cwt.

It is estimated that the waste of iron has been reduced from 5 per cent. by the old furnace, to from 2 to 3 per cent. by the gas-furnace.

The grate is cleaned three times a day. The recuperator is cleaned every 15 days, or every month.

PONSARD FURNACES, AT SAN GIOVANNI, IN THE VAL D'ARNO (TUSCANY).

The fuel used in these furnaces is exclusively lignite, containing, when mined, from 40 to 50 per cent. of water. When air-dried, it loses about one-half of the water. In its ordinary condition for use, large and small, it contains from 20 to 25 per cent. of water, and 15 per cent. of ash.

The reheating furnaces to which the Ponsard principle is applied, have rectangular hearths, 4 feet 9 inches wide, and 9 feet 10 inches long, with two doors. The recuperator has a volume of 295 cubic feet, or 10·92 cubic yards, and a heating surface of 1,032 square feet, or 94 square feet per cubic yard. It weighs 7·4 tons, or 1,954 lbs. per cubic yard. The area of grate for each gazogene is 28·4 square feet; and for each grate from 6·70 to 6·90 tons of lignite are consumed per day of 24 hours, being at the rate of about 22 lbs. per square foot of grate per hour.

The production depends on the hygrometric condition of the fuel, and the time during which the generator has been lighted. It is absolutely indispensable that the furnaces should be lighted 12 hours before the iron is charged. It is better to wait even 18 or 24 hours. The production does not arrive at its maximum until 5 or 6 days after the fire is lit, and even longer, if the lignite be not very dry. On several days during the winter of 1873-74, the lignite supplied for the furnaces contained 50 per cent., or more, of water. With such fuel, the production did not amount to more than 3 or 3½ tons of iron in 12 hours; whilst it easily reached 5 or 5½ tons per day, when the fuel did not contain more than 20 or 25 per cent. of water.

According to the results of an observation made on the

influence of the length of time since lighting the furnace, on the rate of production, one of the furnaces, after having been alight 12 hours, produced 3,733 lbs. of round merchant iron in 12 hours. Five days later, the production in the same time was increased to 7,361 lbs. of round iron, plus 2,198 lbs. of flat iron; in all, 9,559 lbs. Two days still later, the furnace produced 11,607 lbs. of flat iron; and four days after that, 13,631 lbs. Next day the fire was put out, and relighted ten days afterwards. After it had been again alight for 12 hours, the production only amounted to 7,478 lbs.

The waste of iron forms a higher per-centage when the production is lower. For instance, on the first day above mentioned, the waste amounted to 26 per cent.; that is, 126 lbs. of puddled iron were required for the production of 100 lbs. of finished iron. Four days later, it was reduced to 9 per cent.

It appears that the quality of iron treated in the lignite furnaces is much improved by the process. The flame is constantly free from oxygen. This freedom from oxygen has been proved by the melting of copper without oxidation. The oxidation of the iron is due entirely to the presence of aqueous vapour in the lignite-gases, which is decomposed.

During an exceptionally good day's work, in the manufacture of bars about $1\frac{1}{4}$ inch by $\frac{5}{10}$ inch thick, 31·80 tons were charged, in the two furnaces, and 29·67 tons were produced. The fuel consumed amounted to $11\frac{3}{4}$ cwt. per ton charged, and to 12 cwt. per ton produced. These were minimum consumptions. The consumption of fuel is rapidly augmented when the proportion of water is increased.

Puddling Furnaces.—The Ponsard system is applied also to puddling furnaces at San Giovanni. Lignite very well dried is required for these furnaces. When the fuel holds 40 or 50 per cent. of water, it is impossible to obtain a temperature sufficiently high to properly reduce the iron. The scoria remains pasty, and is mixed with the iron. The production of a puddling furnace averages 2 tons in 24 hours.

M. Périssé adds, in 1875,* that at the iron-works of MM. Harel & Co., at Pont-Evêque, a puddling furnace produces from 14 to 20 tons of rolled iron in the 24 hours, with the consumption of from 4 cwt. to 4·43 cwt. of coal per ton; whilst, for the same production, the ordinary furnaces consume from 10 cwt. to 12 cwt. per ton. The immediate economy, irrespective of the utilisation of heat for generating steam, amounts to upwards of 60 per cent.

Steel-heating Furnace.—M. Périssé adds that at the works at Seraing, the consumption of fuel for heating steel ingots for the manufacture of rails amounts to from 3·2 to 3·4 cwt., consisting of from 80 to 85 per cent. of coal, and from 15 to 20 per cent. of cinders.

Combustible Gases generated in the Ponsard Furnace.—M. Ponsard gives a comparative table showing approximately the composition of the combustible gases generated from various coals in his producers, averaged from several series of observations, under varying conditions:—

PONSARD GAZOGENES.—COMBUSTIBLE GASES.

Products of the Gazogene.	I.	II.	III.	IV.	V.	Observations.
	per cent	per cent	per cent	per cent	per cent	
Carbonic oxide . .	21·0	25·0	22·5	22·0	24·0	
Carbonic acid . .	6·0	4·0	4·5	4·4	4·0	
Nitrogen	61·0	60·0	57·0	52·0	55·0	Calculated.
Hydrogen and Hydro-carbons }	12·0	11·0	16·0	21·6	17·0	By difference.
By volume . .	100·0	100·0	100·0	100·0	100·0	

The values in the first and second lines are found directly by analysis on Orsat's method ; for the third line, nitrogen,

* " Memoires de la Societe des Ingenieurs Civils." 1875; page 292.

R 3

the oxygen required to form the oxide and acid in the first and second lines, was in the first place calculated. Then for the three columns I., II., III., corresponding to ordinary gazogenes, half the oxygen found in the coals was deducted; the other half being assumed to be in combination with hydrogen. For the Analysis III., the oxygen supplied by the water evaporated under the grate is also deducted. In Analyses IV. and V., pertaining to superheated gazogenes, the whole of the oxygen found in the coal has been deducted, and also the oxygen in the steam introduced in case IV. The amount of atmospheric oxygen having thus been ascertained, the quantities of nitrogen were easily calculated.

In Analysis I. the coal used was of the nature of coking coal; it contained from 9 to 18 per cent. of ash. Deducting ash, the coal, as pure coal, after having been desiccated at 230° Fahr., yielded—

Coke	79	to 81
Volatile products	21	to 19
	100	100

Cinders, representing about a fifth of the fuel, were charged with the coal. The gazogene was fed with cold air, and was worked very hot,—rather too lively,—owing to circumstances.

In Analysis II. the charge consisted of $\frac{4}{5}$ths of bituminous coal, of the type of gas-coal, and $\frac{1}{5}$ths of cinders. The gazogene was supplied with cold air; it was worked slowly, the temperature of the gases being 1,200° Fahr.

In Analysis III., for an ordinary gazogene, the charge consisted of smithy coal, yielding 73 per cent. of coke, and 27 per cent. of volatile products. The coal contained 20 per cent. of ash; no cinders were added; the gases escaped at the temperature 1,560° Fahr. Water was distributed as spray below the grate.

In Analysis IV., for a superheated gazogene, working at the temperature of melting copper (1,996° Fahr.), supplied with air from the recuperator, containing a very small quantity of vapour, gas-coal was used, which yielded 65 per cent. of coke, and 85 per cent. of volatile products. It contained from 15 to 20 per cent. of ash.

In Analysis V. the conditions were the same as the Analysis IV., except that the air was dry.

Comparing IV. with V., the effect of the addition of water to the air in IV. appears in the greater per-centage of hydrogen and its compounds; and it appears also in comparing I. and II. with III. The effect of quicker conversion, together with a smaller allowance of cinders, is seen by comparing I. with II:—a less proportion of carbonic oxide, and a greater proportion of carbonic acid.

CHAPTER XXIII.

GORMAN'S HEAT-RESTORING GAS-FURNACE.

Mr. W. Gorman, of Glasgow, whose furnace for burning coal under steam-boilers has already been noticed, introduced his "Heat-restoring Gas-furnace," about the year 1870, in the neighbourhood of Glasgow.* It is shown in vertical section, in Fig. 189. The gas producer is formed differently from others which have been described: having a horizontal grate, surmounted by a deep square chamber, in which a bed of coal to the depth of two feet is in combustion. The combustible gas meets and mingles with the heated air for combustion, introduced at the entrance to the heating chamber. The "heat-restorer," employed for heating the air, is placed beneath the heating chamber. It consists of a number of fire-clay pipes laid horizontally, through which the cold air circulates, and from which it absorbs the heat brought down by the hot gases which traverse the pipes externally, thence passing to the chimney. The bed of the furnace above illustrated, is $5\frac{1}{2}$ feet wide, and $7\frac{1}{4}$ feet long. In such a furnace, employed for heating air, at the Mossend iron-works, the production of rolled iron required a surplus of $8\frac{1}{4}$ per cent. of iron charged; whilst, with an ordinary furnace, the production required a surplus of $12\frac{1}{4}$ per cent. At Coatbridge, the con-

* "On the Heat-restoring Gas-furnace" a paper by Mr. W. Gorman, in the *Transactions of the Institution of Engineers in Scotland*, 1871; p. 245.

sumption of coal per ton of iron charged and heated,
amounted to 8·80 per cent. in the old furnaces, and to 4·44
per cent. in the gas-furnace.

By the adoption of the principle of direct transmission of

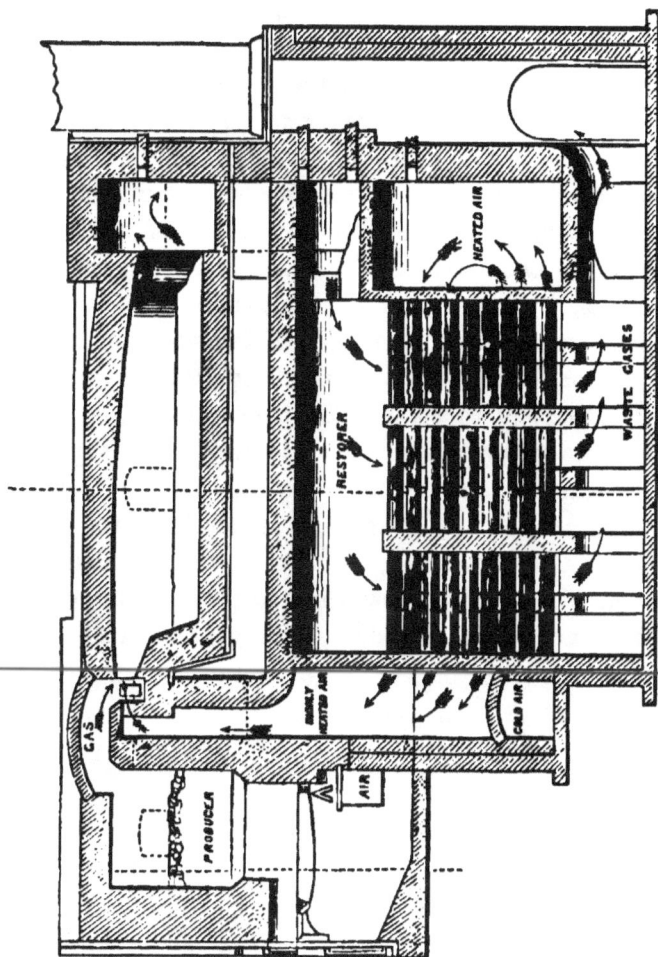

Fig. 139.—Gorman's Heat-restoring Gas-furnace.

heat in the furnace, from the hot gases to the entering air,
through comparatively thin clay pipes, a considerable degree
of "heat-restoring" capacity is developed within compact
limits.

CHAPTER XXIV.

WATER-GAS GENERATORS FOR HEATING PURPOSES.

SOME years ago, Mr. Joshua Kidd introduced a mode of gasifying fuel in conjunction with water. The fuel and water were completely gasified, and the results obtained were sufficiently encouraging to induce others, according to Mr. S. W. Davies, to work out the system to a practical issue.*

The fundamental principle of the generator described by Mr. Davies, is that of the gazogene employed in manufacturing operations. A blast of air is propelled, by the inductive action of a jet of superheated steam, into and through the fuel. Whilst the air is transformed into carbonic oxide and nitrogen, the steam is decomposed into its elements, hydrogen and oxygen; and the oxygen is converted into carbonic oxide. The decomposition of the steam is complete, and the resulting gaseous mixture consists entirely of carbonic oxide, hydrogen, nitrogen, and the inevitable proportion of carbonic acid. The supply of steam is generated in the fire-chamber itself; and as the whole apparatus is self-contained, the whole of the steam required for the purpose of the blast can be utilised as fuel, seeing that it is wholly transformed in passing through the fire. This method of procedure,—the generation of water-

* See an excellent paper by Mr. S. W. Davies, on "A new method for producing cheap Heating Gas for Domestic and Manufacturing Purposes," in the *Journal of the Society of Arts*, April 12, 1878, page 444.

gas—is substantially the same as Dr. Siemens employed some years ago for the supply of air to his gazogene.*

The generator of Mr. Davies' apparatus consists of an upright cylinder of cast-iron or wrought-iron, resting on and opening into a smaller cylinder, containing the fuel which rests on a grate near the lower end of the cylinder. The bottom of the lower cylinder is air-tight, and the blast is delivered into it under the grate. The steam-generator consists of a coil of thick wrought-iron pipe, which is placed within the upper cylinder, and is supported by it. The lower end of the coil is protected from the direct heat by a coating of gannister. The two ends of the coil are turned out through the side of the cylinder; the lower end being connected to a cistern or an accumulator under pressure, for the supply of water, and the upper end with a steam-pipe of smaller diameter, which is led down outside the apparatus, and is terminated by a small steam-tap immediately in front of the blast-pipe. A hopper for the supply of fuel is adapted to the upper end of the upper cylinder; and is fitted with a heavy gas-tight valve, by opening which at intervals, fuel is introduced into the generator. By another opening, the gas is conducted from the generator, and through a third opening, the state of the interior may be ascertained by inspection.

The supply of water to the coil for conversion into steam in a self-acting manner, by means of the steam-tap, is regulated so that the water is not only wholly evaporated, but is also superheated, in the upper part of the coil. The back pressure of the steam, prevented from escaping too rapidly, necessarily limits the space in the coil occupied by the water. An accumulator 2 feet high, 9 inches in diameter, is sufficient, even with so high a pressure as 60 lbs. on the square inch, for the service of a generator yielding 4,000 cubic feet of gases per hour. By a few strokes of the pump every quarter of an hour, the pressure was maintained, and the coil was

* Page 355 *ante.*

supplied with water. The accumulator consists simply of an
upright cylindrical vessel, provided with a force-pump and a
pressure-gauge, and containing air, by the compression of
which the required pressure may be obtained. With a
water-pressure of 15 lbs. per square inch in the accumulator, a gas-pressure of above 1 inch of water is obtained in
the generator. With 40 lbs. pressure, the gas acquired a
pressure of 2½ inches of water—a good working pressure.

The composition of the gases produced in the generator,
from peat-charcoal and coals, has been determined by analysis,
as follows :—

COMPOSITION OF GASES.

Fuel.	Pressure of Water.	Composition of Gases by volume.
	lbs. per square inch.	per cent.
Peat-charcoal	15 lbs.	CO . . . 28·6 H . . . 14·6 CO_2 . . 4·0 N . . . 53·0 100·2
Anthracite .	15 lbs.	CO . . 22·6 H . . . 10·0 CH_4 . . 4·9 CO_2 . . 4·5 N . . . 58·0 100·0
Anthracite and steam-coal (equal parts) . .	30 lbs.	CO . . 28·3 H . . . 9·3 CH_4 . . 5·2 CO_2 . . 6·2 N . . . 51·3 100·3
Anthracite .	60 lbs.	CO . . 26·4 H . . . 13·5 CH_4 . . 1·4 CO_2 . . 3·9 N . . . 51·8 100 0

The quantity of the mixed gases produced has been carefully determined. The gas from the generator was passed through 100 feet of 3-inch pipe, of which more than half was exposed in the open air; and then passed through a large meter. Before it reached the meter, the temperature had been reduced to 60° or 70° Fahr.

QUANTITY OF GAS PRODUCED PER POUND OF FUEL.

Fuel.	Pressure of Water.	Volume of Gas.
	lbs. per square inch.	cubic feet.
1. Anthracite	15 lbs.	69·5
2. Anthracite and steam-coal (equal parts) .	20 ,,	85·2
3. do. do. do. .	25 ,,	88·84
4. do. do. do. .	30 ,,	94·5
5. Anthracite	40 ,,	over 100

It is shown that the quantity of gas produced increases with the pressure; varying from 155,680 cubic feet to 224,000 cubic feet per ton of coal.

It appears from the results of very carefully conducted trials, when the gases were burnt in the open air, that the volume of this water-gas required to generate the same quantity of heat, is as 5 to 1 of ordinary coal-gas. The heat was measured by raising the temperature of a given weight of water through a given number of degrees.

To deduce from this datum the comparative cost of water-gas and ordinary coal-gas:—With a No. 1 generator, an average quantity of 1,000 cubic feet of gas can be produced per hour, consuming 10 lbs. of coal. For the working day of 10 hours, the quantity of coal consumed would be about 1 cwt.; and the gas produced, 10,000 cubic feet. The cost of its production is estimated as follows:—

	s.	d.
1 cwt. anthracite	1	1
Wages of attendant	4	0
Common coal and wood, for lighting the fire . .	0	2
Total cost . . .	5	3

The cost for the equivalent of 2,000 cubic feet of ordinary gas is from 7s. to 8s.   But the saving is shown to be much greater, when larger quantities are generated. Thus, a No. 2 generator consumes about 35 lbs. of coal per hour, and produces in that time 3,500 cubic feet of gas.   To produce 35,000 cubic feet of gas in 10 hours, the cost is as follows :

	s.	d.
3½ cwt. anthracite	3	6
Wages of attendant	4	0
Common coals and wood, for lighting the fire . .	0	4
Total cost . .	7	10

The cost for the equivalent quantity of common gas, 7,000 cubic feet, would be from 24s. 6d. to 28s., or more than three times as much as the cost for generator-gas.

CHAPTER XXV.

POWDERED FUEL.

It appears that Mr. John Bourne was the first publicly to advocate, in his patent of 1857, the use of coal or other fuels in the form of dust, for the generation of heat in furnaces. Aware that the more intimately and equally the mixture of fuel and the air for combustion can be effected and regulated ; and that, necessarily, the smaller the constituent particles of the fuel can be rendered, the more effectively and the more promptly can the desired mixture be accomplished, he says, in the edition of his " Treatise on the Steam Engine" published in 1861, page 858, "It appears to us that the fuel and the air must be fed in simultaneously, and the most feasible way of accomplishing this object seems to be in reducing the coal to dust, and blowing it into a chamber lined with fire-brick, so that the coal-dust may be ignited, by coming into contact with red-hot surfaces," &c.

Mr. T. R. Crampton, as early as the year 1868, instituted a long course of experimental investigation into the best means of generating and applying heat from the combustion of powdered coal. " It is not only necessary," says Mr. Crampton,* "to have the means of bringing together at will the proper equivalents of air and coal to insure perfect combustion ; but it is essential that the size of the coal

---

* See a paper by Mr. Crampton "On the Combustion of Powdered Fuel," in the *Journal of the Iron and Steel Institute*, 1873; page 91.

should be determined, and that it should, during its flotation
through the furnace, be so conducted that any overcharged
and undercharged currents of air and coal should be continually re-intermixed until the whole of the carbon is
consumed ; otherwise, were not this attended to, there would
be deposits of coal on one part of the furnace, and free
oxygen on others." In delivering a mixed current of air
and coal-dust through a pipe, it was discovered by Mr.
Crampton that, at times, although an absolute mixture
entered the pipes, conveying the coal and the air to the
furnace, yet, under certain circumstances, the materials
became separated,—more particularly when they had to pass
through bends—the coal being carried, by its superior

Fig. 140.—Crampton's Powdered Fuel Furnace.

momentum, to the outer part of the interior of the bend, and
being thus led to issue from the pipe into the furnace, as a
close stream, unmixed with air. To compensate for this
tendency to separate, Mr. Crampton, in his experiments
with fixed combustion-chambers, Fig. 140, introduced the
mixed currents by several inlets, inclined downwards so as
to strike the floor of the chamber, and to impinge upon each
other. The air and the fuel, playing over the floor, became
re-admixed, and were at once carried over the bridge into
the heating chamber.

The coal was reduced to powder sufficiently fine to pass
through a 80-sieve, at a cost, for labour, of 6d. per ton ; or,
including all charges, 1s. per ton. The air-current was produced by a fan-blast

In the application of the new system to puddling furnaces, at the Royal Gun Factories, Woolwich, Mr. Crampton constructed the furnace, shown in Fig. 141, in two compartments, a combustion-chamber A, and a puddling chamber B, opening into each other. It is not the purpose of the author to enter into details of construction. Suffice it to add, that the furnace was cylindrical and rotary, and that it was kept cool externally by a water-jacket; that the air was injected into the chamber A by an annular jet C, shown in section,

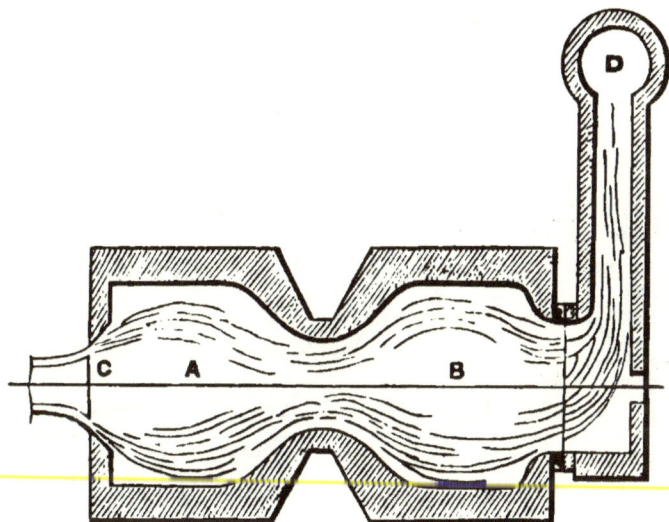

Fig. 141.—Crampton's Powdered Fuel Revolving Puddling Furnace.

into which the coal-dust was continuously delivered, and which drew the dust with it into the chamber. The annular current was inclined outwards, and in expanding conically, it struck the chamber on all sides; then, converging, it entered and passed through the puddling chamber, and thence to the chimney D.

The temperature which could be produced in this furnace was so high that wrought iron was melted in it without difficulty: 60 lbs. of wrought iron could be melted in 3½ hours.

Mr. Crampton stated that, in such a furnace, scrap-iron could be raised to a welding-heat, with the consumption of from 5 cwt. to 6 cwt. of coal ; whilst, from a subsequent report, it appeared that for puddling, from 17 cwt. to 20 cwt. of coal was consumed per ton of puddled bar.

Mr. Crampton subsequently made his revolving furnace as

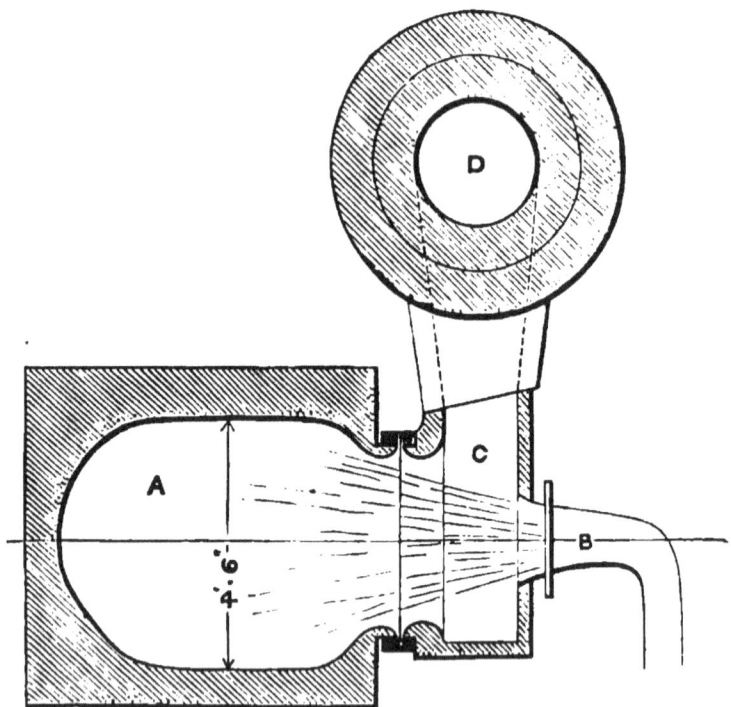

Fig. 142.—Crampton's Powdered Fuel Revolving Puddling Furnace—Single Chamber.

a single chamber, A, Fig. 142, 4 feet 6 inches in diameter, made to revolve at a speed not exceeding 15 turns per minute.   The mixed current from the pipe B is delivered across the flue-piece c, into the puddling chamber A, where combustion is completed.   That the current may be delivered in a state of uniform mixture, the bend of the delivery-pipe is divided longitudinally by diaphragms, as shown, which divide the current into sheets, and keep together the components of each sheet. The products of combustion wind their way out to the chimney D. The effect of the substitution of a single chamber in which the combustion should be effected in contact with the work, for the double chamber already noticed, was, as Mr. Crampton reports,* most satisfactory. He found that it did not have any injurious effect, and that a more intense heat was produced, with a considerably less consumption of fuel. Brickwork was eliminated from the construction of the furnace; the lining consisted exclusively of fettling. With 6½ cwt. charges of cold pig-iron, working night and day, the puddled bar was produced by a consumption of 14 cwt. of coal per ton of bar. Twenty tons of puddled iron was produced at the rate of 5 cwt. per hour, or 6 tons for 24 hours. According to the results of another series of trials, when the charges of pig were varied from 5 cwt. to 10 cwt., the coal consumed per ton of puddled bar amounted to 11¼ cwt., and it was concluded that the furnace was capable of turning out 5 tons of puddled bloom, with large charges, per shift of 12 hours, with a consumption of 10 cwt. of coal per ton. The quality of the iron produced, it may be added, has been generally acknowledged to be superior, both in purity and in strength.

The comparative results of Mr. Crampton's experiments with the double chamber and the single chamber, are conclusive as evidence of the first-rate importance of bringing the heat of combustion to bear by close and direct action upon the subject treated, for producing the greatest degree of efficiency.

The results arrived at by Mr. Crampton were corroborated by the results of the observations of M. Lavalley, recorded in a paper on the Crampton furnace, read by him at the Institution of Civil Engineers in France, in 1875.† He

* See Mr. Crampton's paper "On Crampton's Revolving Furnace and its Products," in the *Journal of the Iron and Steel Institute*, 1874; page 391.

† "Mémoires et Compte Rendu de la Société des Ingénieurs Civils, 1875;" page 266.

states that, as the result of long and numerous trials made at
Woolwich, charging cold pig-iron, the consumption of coal
did not exceed 600 kilogrammes per tonne, or 12 cwt. per
ton, of iron produced; and that this consumption could be
considerably diminished, if the iron were charged already
melted, or if the waste heat was utilised for heating up the
pig on a dandy. The following are the chief results of M.
Lavalley's observations :—

112 charges of cold pig-iron, 1,097 cwt., or 9·80 cwt. per
charge.

181½ hours, or 1 hour 23 minutes per charge.

1,301 cwt. of iron produced, being 204 cwt. or 18·6 per
cent. in excess of the weight of charges, derived from the
fettling.

1,057 cwt. of fettling consumed, of which the iron produced from it constituted about 20 per cent.

791 cwt. of coal consumed, or 12·2 cwt. per ton of iron
produced.

EMPLOYMENT OF POWDERED FUEL IN STEAM BOILERS.

Mr. Crampton stated, in his first paper, that he had made an
experiment with powdered coal as fuel in a large marine
boiler, containing 1,500 square feet of heating surface. The
fire-bars were taken out, and the furnaces lined with brick-
work. The combined current of air and coal was injected
into the furnaces. During a trial which lasted 24 hours, the
temperature in the smoke-box only varied within the narrow
range of from 380° to 400° Fahr.; whilst, according to
Mr. Crampton, the water was evaporated at the rate of from
10 lbs. to 11 lbs. per pound of coal. "It is impossible," he
says, "to produce smoke, as the whole of the volatile matter
is produced in the first instance. The consequence is that
the uptake of the boilers can have no flame in them."

Messrs. Whelpley and Storer devised a system of burning
powdered coal. The lump coal is first reduced to the state

of impalpable powder; it is then fed, together with air, through a conduit, from which it is drawn by a fan and discharged into the front of the furnace through an air-tight aperture. In 1876, the system was tried by Mr. Isherwood, for the American Government, under an externally fixed cylindrical boiler, 40 inches in diameter and 10 feet long, with flat ends, having 74 flue-tubes $2\frac{1}{4}$ inches in diameter, for the return draft. The draft was continued round the sides of the boiler, and the heating surface amounted to 442 square feet. The air for combustion was delivered into the closed ash-pit through a 5-inch vertical pipe. The powdered coal was delivered through a 2-inch horizontal pipe. The boiler was tried with powdered anthracite, on the new system, and with lump anthracite burned on an ordinary grate. A brick arch which was used with the dust-fuel, was removed to make way for the trial with lump-fuel, making an addition of 15 square feet of heating surface. Otherwise, the boiler remained in the same condition. The results, taken generally, of comparative trials, showed that semi-bituminous coal gave the same economic evaporation, whether it was consumed wholly in the lump state, or partly in lump and partly pulverised, or wholly pulverised. Distinguishing these three conditions as 1, 2, 3:—

	(1.) lbs.	(2, 3.) lbs.
Coal per square foot of grate per hour . .	11·11	11·35
Water evaporated per lb. of the combustible portion of the coal from and at 212 Fahr.	10.124	10·192
Temperature of Gases on leaving the boiler .	383·3°F.	381·8° F.

Here, no economy has been effected by the substitution of powdered fuel. It is probable that much of the coal-dust escaped unconsumed, as appears to have happened in 1868 and 1869 at Boston, in the course of experiments with powdered anthracite, when, it is recorded, pure carbon, to the amount of 38 per cent. of the fuel administered, was gradually accumulated in the furnace. The precipitation of this large per-centage of carbon points to the incompleteness of the mixture, and the want of time for completing the combustion of the residual coke-particles which remained to be burned, after the volatile portions of the fuel were burned off in flame.

A system of burning coal-dust for generating steam, introduced by Mr. G. K. Stevenson, of Valparaiso, was applied experimentally to a Cornish boiler, in Blackfriars, where it was at work in 1877. The boiler, Figs. 143, 144, was one of two, precisely alike, placed side by side : 5 feet 8 inches in diameter, and 28 feet 8 inches long, with a fire-tube 3 feet 6 inches in diameter. A fire-clay retort, 8 feet long, A, was placed within the fire-tube : it was numerously perforated

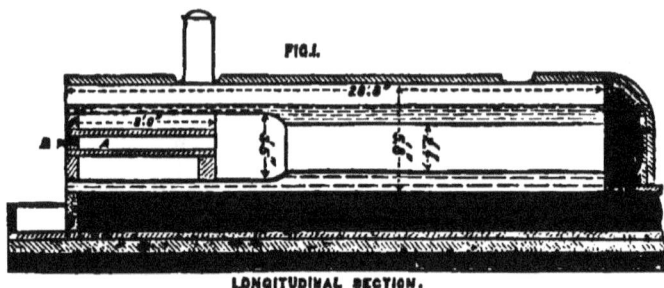

FIG.I.

LONGITUDINAL SECTION.

Fig. 143.—Stevenson's Coal-Dust Furnace, applied to a Cornish boiler.

with ¼-inch holes. The fuel was not finely pulverised, it was like a coarse powder. The air and the powder together, in measured quantities proportioned automatically, were driven together through the brick pipe B, into the retort, and there inflamed. A few fire-bricks placed in the flue, behind the bridge, acted as a bridge. From the results of comparative experiments,* it appears that, under the same circumstances, the boiler evaporated 8·8 lbs. of water per pound of powdered coal, against 6·5 lbs. per pound of lump coal. It is remarkable that the quantity of air delivered for combustion, per pound of powdered fuel, was just 12 lbs.,

* Reported in *The Engineer*, May 18, 1877 ; page 336.

or exactly such as was chemically consumed in effecting complete combustion. Yet, it is said, no smoke was produced.

It may be gathered from the foregoing experimental results

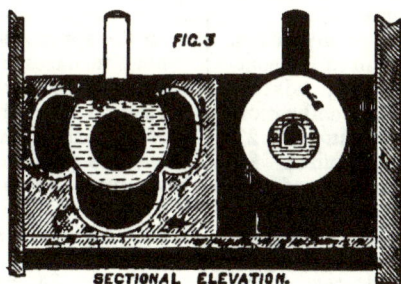

Fig. 144.—Stevenson's Coal-Dust Furnace.

with steam-boilers, that a non-conducting retort or enclosure was necessary for effecting the proper mixture of the coal and the air, and completing the combustion of the powdered fuel.

INDEX.

ABERAMAN coal, evaporative power of, 248

Abouchoff, experiments on the evaporative performance of stationary boiler at, 270

Air, atmospheric:—composition, 13; quantity required for combustion, 15, 17, 171; mixture with coal-gas for combustion, 21; supply of air in jets, 27, 111; quantity required for combustion of the coke of coal, 38; and of the gaseous portion, 39; area of orifices for supply to gaseous portion, 40; situation of orifices, 43, 58; mechanical agency for mixing gases and air, 43; forced blast, 46, 83, 84; regulation of the supply of air to the gas, 48; Mr. Josiah Parkes's system, 48; report of Sir Robert Kane and Dr. R. B. Brett, 49; their experiments with coal, 51; and with coke, 52; equalising supplies of gas and air, 56; Mr. John Dewrance's experiments, 59; admission of air through orifices in the fireplace, 60 (See *Furnaces*.)

Anderson, W., experiments on the evaporative performance of coal, wood, and peat, 267, 270

Argand furnaces, C. W. Williams's:—for land boilers, 63; for marine boilers, 69, 70, 71, 72, 87, 88; for a locomotive boiler, 92

Argand lamp, as an example of combustion of coal-gas, 24; imitation of, 93

Asphalte, 265

Aydon on liquid fuels, 274

BELL, J. LOTHIAN, on Price's retort puddling furnace, 320; on the appropriation of heat in blast furnaces, 327

Bi-carburetted hydrogen:—composition, 12; chemical process of its combustion, 16

Bicheroux puddling furnace, 311

Blaenavon, Newport furnace at, 313

Blast, forced:—gas furnace at Treveray, 45; blast of smoke and air, 83, 84; mechanical draught, 135, 136

Blast-furnaces, 218, 327; composition of the gases, 218

Boetius heating-furnace, 308

Boilers, steam:—principles of construction, 28; heating surface, 94; efficiency of flue-system, 94; the marine tubular boiler, 73, 95, 141; circulation of water, 99; priming, 108; locomotive boilers, 139

Boilers, locomotive: — Argand furnace in, 91; circulation of water and evaporation in, 139; coal-burning in, 256; combustion of coke in, 262; wood-burning in, 267

Boilers, marine:—furnace arrangements, 66 to 89; boiler of the *Llewellyn*, 85; circulation of water in flue-boilers, 116; flue-boiler of the *Liverpool* or *Great Liverpool*, 117, 130; flue-boiler of the *Great Britain*, 128; draught in flue-boilers, 128; circulation of water in tubular boilers, 139; working condition of tubular boilers, 142; disadvantages of the tubular boiler, 146; tubular boiler of the *Leeds*, 146; tubular boiler of the *Royal William*, with conductor-pins, 161; plate-surface boiler (Lamb and Summers') of the *Pacha*, 161; experimental tubular boilers, at Newcastle, 174, 245; practice of stokers on the Mersey, 239

Boiler, stationary:—principles of construction, 28; temperature in the flues, 30; grate-bar surface, 30; fire-chamber, 35; flame-bed, 36; furnaces for smoke-prevention, 42, 43, 64, 65, 79, 80, 82, 83, 84; draught in chimneys, 126; split draught for boilers, 132; mechanical draught, 134; Lancashire and Galloway boilers at Wigan, evaporative performance of coals in, 250; double-flue multitubular boilers at Abouchoff and at Erith, evaporative performances of coal, wood, and peat in, 267, 270; Cornish boiler, evaporative performance of petroleum in, 274

Bourne on the use of powdered fuel, 379

Brunton's revolving grate, 90

CADDICK and Mabery's iron furnaces, 324

Candle, flame and combustion of, 23, 24, 26

Carbon, as a constituent of coal, 7, 8; weight, 11; combustion of, 14, 38; carbon in flame, 166; carbon in smoke, 169, 172

Carbonic acid :—composition, 14; formation of, 17, 18

Carbonic oxide, formation of, 18, 19

Carburetted hydrogen :—composition, 12; chemical process of its combustion, 15, 170.

Casson-Dormoy puddling furnace, 313

Chamber of furnace, 35; high temperature in, 96

Chanter's furnace, 80

Clark, D. K., on rapidity of draught, 134; system of smoke prevention, 238

Coal, its constituents, gaseous and solid, 8, 10; gaseous elements, 10; their combination with air, 13, 27; the coke or solid portion, 38

Coal, combustion of, 194; regulation of supply of air, 195

Coal, evaporative power of :— Aberaman coal, 248; Hartley coals, evaporative performance of, 174, 245; evaporative power of, 248; South Lancashire and Cheshire coals, evaporative performance of, in Lancashire and Galloway boilers, at Wigan, 250; Hindley-yard coal, trials of, 250; evaporative performance of coal, wood, and peat in stationary boilers at Abouchoff and at Erith, 270

Coal-gas :—its formation in the furnace, 8, 9; combustion, 12; mixture of air with it for combustion, 21; Professor Daniell's opinion, 23; supply of air in jets, 27, 41; situation of orifices, 43; mechanical agency for mixing air and gases, 43; forced blast, 45; its continuous generation, 53; temperature and length of flame in the flues, 54; development of coal-gas and coke-gas, 55, 56; equalising supplies of gas and air, 56; explosive mixtures, 97

Coke:—quantity yielded by coals, 260; composition, 261; weight and bulk, 262; combustion in locomotives, 262; analysis of the gaseous products, 263; solid portion of coal in combustion, 38; air required for its combustion, 38

Coke, Durham, performance of, in blast furnaces, 327

Combustibility, 10

Combustion of coal, process of, 13; quantity of air required, 17; distillatory or gas-generating process, 29; conditions for complete combustion, 96; products of combustion, 169 (See *Coal-gas*.)

Combustion of coke, 262

Conductor-pins, increasing the heat - transmitting power of plate-surface by, 154; experimental results, 158; boiler of the *Royal William*, 160

Cotton-stalks, evaporative performance of, 273

Crampton's powdered-fuel puddling furnace, 379; use of powdered fuel for steam-boilers, 384

DANIELL, PROFESSOR, his opinion on the process of mixture of air and coal-gas in the furnace, 22

Davies, S. W., on the water-gas generator, 374

Davy, Sir Humphrey, on the essentials for combustion, 97; on heating combustible gases, 149

Decazeville coal, composition of, 363

Dewrance, John, his experiments on admitting air in divided streams, 59, 63; Argand furnace applied to a locomotive boiler, 91

Draught in iron furnaces, 189; action of the chimney, 190

Draught:—of furnaces, 124; in flues, 125; in chimneys, 126; in the boilers of the *Great Britain*, 128; split draught, 132; D. K. Clark on rapid draught, 134; mechanical draught, 45, 83, 84, 135; mechanical versus ordinary draught, 136

Dublin, City of, Steam Packet Company, report by Mr. Josiah Parkes to, 48; report of Sir Robert Kane and Dr. R. H. Brett on C. W. Williams's system, 49; report of Mr.

Joseph Clarke on the boilers of the *Llewellyn*, 85

EBELMEN, first proposal of gas-furnaces by, 278; his analysis of charcoal gas, 293; of peat gas, 294

Ebelmen and Sauvage, experiments by, on the combustion of coke, 263

Erith, experiments on the evaporative performances of coal, wood, and peat in stationary boiler at, 270

FAN draught, 136

Fichet, experiments by, on gas-furnaces, 282, 286

Flame:—its contact with the boiler should be avoided, 36; length of flame in flues, 54; how it may be extinguished, 97

Flame-bed, 36, 37

Fletcher, Lavington E., experiments on evaporative performance of South Lancashire and Cheshire coals conducted by, 250

Flues of boilers, temperature in, 30, 54, 98

Flues, multitubular. (See *Heating Surface*.)

Flues, split, 132

Forcing the fire, 144

French boiler, application of gas-furnace to, 286

Fuels:—chemical composition of, 229; air consumed in their combustion, 230; quantity of the gaseous products of combustion, 230; heat evolved, 231, temperature of combustion; 232; coal, 233; composition of British coals, 234: combustion of coal, 236; coke, 260; lignite, asphalte, and wood, 265; peat, 268; tan, straw, and cotton-stalks, 272; petroleum, 274; total heat of combustion of fuels, 276

Fuel, decomposition of, in gazogenes:—coal, 292; coke and charcoal, 293; peat, 294

Fuel, economy of, 189

Fuel, powdered, 379; first advocated by John Bourne, 379; T. R. Crampton's experiments, 379; his powdered-fuel furnace, 380; his revolving puddling furnace, 381, 382; Lavalley's experiments, 383

Used in steam-boilers, by Mr. Crampton, 384; by Whelpley and Storer, 384; Mr. Isherwood's experiments, 385; G. K. Stevenson's system, 386

Furnace:—grate-bars, 30, 37; chamber, 35; air through orifices in the fire-place, 60; various arrangements of furnaces, 63; C. W. Williams's Argand furnace, 63, 69, 70, 71, 72, 87, 88, 92, 239, 240, 246; ordinary marine furnace, 66, 73, 76, 87, 174; Parkes's split-bridge, 66; split-bridge modified, 67; split-bridge and air at the grate, 68; air at the bridge, 68; air at the bridge, and small grate, 69; air-box at bridge, 72, 73, 74, 75; perforated plate at bridge, 77; furnace with supplementary grate, 78; hot-air expedient and split-bridge, 79; Chanter's furnace, 80; hot air at the bridge, 80, 81; hot air from the flues, 82; blast of smoke and air, 83, 84; proper firing, 89; improper firing, 89; Argand furnace in a locomotive boiler, 91, 92; Robson's system, 177; Hobson and Hopkinson's system, 179; Stoney's system, 183

Furnaces, Gas. (See Gas Furnaces.)

Furnaces:—Ivison's system, with steam jets, 238; Clark's, D. K., system, with steam-inducted air-currents, 238; Combes's experiments on C. W. Williams's furnace, 240; Fairbairn's experiments on C. W. Williams's furnace, 240; Baker's system of undulated flues, with semi-elliptical chambers, 241; Wicksteed's experiments on Baker's system, 241; Gorman's furnace, 241; step-grate, 242; Marsilly's observations on the step-grate, 243

(See Gas-furnaces and Iron Furnaces.)

GALLOWAY boiler at Wigan, evaporative performance of, 250

Gas, manufacture of, application of gas-furnaces to, 282

Gaseous portion of coal in the furnace. (See Coal-gas.)

Gas-furnaces: — gas-furnace at Treveray, 45; function and operation, 277; first proposed by Ebelmen, 278; Gorman on gas-fuel, 278; Dr. Siemens on the gazogene, 279; manufacture of gas at Montreuil, 282; gas-furnaces for steam-boilers, 286; decomposition of fuel in the gazogene: coals, 292; coke, 293; charcoal, 293; peat, 294; Boetius heating furnace, 308; Bicheroux's heating furnace, 309; Smith-Casson's puddling furnace, 315; Siemens regenerative gas-furnace, 330; Ponsard gas-furnace with recuperator, 357; Gorman's heat-restoring gas-furnace, 371

Gazogenes, action of, 279

Glass-works at Saint-Gobin, Siemens' regenerative furnace at, 346

Gorman, on the best method of supplying air to ordinary furnaces, 241; on gas-fuel, 278; his heat-restoring gas-furnace, 372; performance at Mossend iron works, 372; at Coatbridge, 373

Grate-bar surface, proportions of, 30; depth below crown of furnace, 37 /1 0⁻

Grate, supplementary, 69, 78; Chanter's system, 80

Great Britain, boilers of, draught in, 128

Great Liverpool, boilers of, circulation of water in, and durability of plates, 117; draught, 130

Green's fuel-economiser, 251

Gun Factory, Royal, ordinary puddling furnaces at, consumption of fuel in, 296; Price's retort puddling furnace at, 316

HARTLEY coals, evaporative performance of, 174, 245; evaporative power of, 248

Haswell on the comparative evaporative performance of wood and coal in locomotives, 267

Head, Jeremiah, experiments by, on the Newport furnace, 312

Head, John, on the evaporative performance of straw and cotton-stalks, 272

Heat, total, of combustion of fuels, 276

Heated air:—at the bridge of the furnace, 80, 81; its supposed value, 147

Heating power of gas flames, 45

Heating surface of boilers, 94; absorbent power, 115, 122; durability of the plate-surface, 116; multitubular flues of marine boilers, objections to, 95, 97, 139, 144, 153; multitubular flues of locomotive boilers, 139; increase of absorbing power, by conductor-pins, 154; corrugated plate-surface, 160

Hindley Yard coal, evaporative performance of, 250

Hobson and Hopkinson's furnace, results of performance of, 176, 180

Hot air from the flues for the furnace, 82

Houldsworth's experiments on temperature in the flues, 30

Hydrogen, as a constituent of coal, 7, 8; weight and bulk, 11; combustion of, 14; influence of water generated by its combustion, 151; experiment condensing the vapour, 151

IRON furnaces: — draught, 189; proportion of heat utilised in common furnaces, 191; using compressed air, 192, 202, 208; influence of high wind no draught, 193; use of slack as fuel, 199; supplying heated air, 200, 209; economy by increasing the temperature of furnace, 204; observed temperature in furnaces, 209; defects of the old puddling furnace, 213; proposed new puddling furnace, 216; construction of ordinary iron furnaces, 295; distribution of heat generated, 297; utilising waste heat of ordinary iron-furnaces by generating steam, 301; utilising waste heat of furnaces by heating the air, 304; Boetius heating furnace, 308; Bicheroux's heating furnace, 311; the Newport puddling furnace, 311; the Casson-Dormoy puddling furnace, 313; Smith-Casson's gas-furnace, 315; Price's retort furnace, 316; Caddick and Mabery's furnace, 324; Crampton's furnace, 326, 380; Siemens' regenerative gas-furnace, 330; Ponsard gas-furnace, with recuperator, 357; Gorman's heat-restoring gas-furnace, 371

Iron, on the manufacture of, 213

Isherwood's experiments on powdered-fuel furnaces for steam-boilers, 385

JETS of air, combustion of gas with, 27

Jukes's moving bars, 90

KANE, Sir Robert, and Dr. R. B. Brett, report on C. W. Williams's furnace, 49; their experiments with coal, 51; and with coke, 52

Keerayef, experiments by, on the evaporative performance of coal, wood, and peat in stationary boilers, 270

Kidd, J., his mode of gasifying fuel, 374

Kraus on the performance of the Siemens gas-furnace at Saint-Gobin, 346

LAMB AND SUMMERS' plate-surface boiler, 161, 164

Lancashire boilers at Wigan, evaporative performance of, 250

Lavalley, his experiments with Crampton's powdered-fuel furnaces, 384

Leblanc, analysis of coke gases, 293

Leeds, boilers of the, 145, 146

Lignite, 265

Liquid fuel:—petroleum, its composition, 274; evaporative performance, 275

Liverpool and Manchester Railway, Mr. Dewrance's experiments on admitting air to furnace in divided streams, 59, 62; Argand furnace applied to a locomotive boiler, 91

Llewellyn steam packet, boilers of, 86

Locomotive boilers. (See *Boilers, Locomotive.*)

Locomotives, coal burning in, 256

Longridge, Armstrong, and Richardson, report of, on experimental performance of a marine boiler, to the Steam Collieries Association, Newcastle-on-Tyne, 174, 245

MACAR, J. de, comparison of furnaces by, 311

Marine boilers. (See *Boilers, Marine.*)

Montreuil, gas works at, employment of gas-furnaces, 282

Muller and Eichelbrenner, employment of gas-furnaces by, in the manufacture of gas, 282

NEWPORT furnace, 311

OLEFIANT GAS. (See *Bi-carburetted hydrogen.*)

PACHA, boiler of, with Lamb and Summers' plate-surface boilers, 161

Parkes's split-bridge, 58, 66, 67, 68

Peat:—composition, weight, and bulk, 268; evaporative performance in stationary boilers at Abouchoff and Erith, 270

Péclet on draught in flues, 125, 132; on mechanical draught, 134; on the heating power of tan, 272

Pélégry on the comparative performance of the Ponsard and old furnaces, 364

Petroleum, composition and evaporative performance of, 274

Piedbouf and Bisenius, puddling furnaces at their iron works, 311

Ponsard, analysis of gases in his gazogene, 294; on the economy of waste heat by generating steam, 301

Ponsard gas-furnace, 357; superheated gazogene, 359; the recuperator, 359; the laboratory, 361; comparative performance of the Ponsard and old furnaces, 364, 366; Ponsard furnace at Vieux - Condé, 365; at San Giovanni, 367; at Pont Eveque, 369; at Seraing, 369; gases generated, 369

Price's retort puddling furnace at the Royal Gun Factory, 316; Mr. Whitham's experiments, 322

Priming in boilers, source of, 108

Pyrometer, Houldsworth's, 32

REGENERATIVE stove, Siemens - Cowper, at Barrow works, 329

Reynolds on coal burning in locomotives, 256

Robson's furnace, results of performance of, 176, 178

Round Oak Iron-works, ordinary puddling furnace at, consumption of fuel in, 296; gas-producer at, 315

Royal William, boilers of, with conductor-pins, 160

SAW-DUST as fuel, 342

Siemens, Dr., on the functions of the gazogene, 279

Siemens' generator, analysis of coke gases in, 293

Siemens - Cowper regenerative stove for blast-furnaces, 329

Siemens' regenerative gas-furnace, 330; Professor Faraday's description, 330; the gazogene, 335; charging the gazogene, 338; temperature in the gazogene, 340; regenerators, 343; composition of firebricks, 344; generation and distribution of heat, by M. Kraus, 346; heat disengaged by the complete conversion of the gases in the furnace, 351; employment of a jet of steam as a blower, 355

Slack, use of, as fuel for iron-furnaces, 199

Smith-Casson, gas-furnace by, 315

Smoke:—it cannot be burned, 6, 166; generation of it, 167; its composition, 168, 169, 171; cause of it, 171, 173; weight of carbon in smoke, 172

South Lancashire and Cheshire coals, evaporative performance of, 250

Split-bridge, Mr. Parkes's system, 48, 58

Split-flues, 132

Stanley's self-feeding apparatus, 90

Steam, its composition, formation, 14 (See *Hydrogen*.)

Steam-boilers, gas-furnace for, 286

Stevenson's powdered-fuel furnace for steam-boilers, 386

Stoney's furnace, 182

Straw, evaporative performance of, 273

TAN, evaporative performance of, 272

Temperature in flues of boilers, 30, 54, 98

Treveray, gas-furnace at, 45

URE, Dr., on the Argand furnace, 63; on circulation of water by heat, 99; on evaporative efficacy of corrugated plates, 160

WATER, circulation of, in steam-boilers, 99; experimental evidence, 99; ebullition, 102, 109; relation of circulation to durability of the plates, 116; the *Great Liverpool*, 117; marine boilers, 139, 144

Water-gas generators for heating purposes by S. W. Davies, 374; composition of the gases, 376; cost, 378

West's report, 80, 82, 85

Whelpley and Storer's system of burning powdered fuel for steam-boilers, 384

Whitham's experiments on Price's retort furnace, 322

Williams, C. W., his Argand furnace, 49, 63 to 92; experimental performance of a marine boiler at Newcastle-on-Tyne on his system, 176, 181; on the practice of stokers on the Mersey, 239

Wise, Field, and Aydon, system of burning petroleum by, 275

Wood:—composition, weight, and bulk, 266; evaporative performance, 266, 270

LONDON: PRINTED BY VIRTUE AND CO., LIMITED, CITY ROAD.

𝕬𝖊𝖆𝖑𝖊'𝖘 𝕽𝖚𝖉𝖎𝖒𝖊𝖓𝖙𝖆𝖗𝖞 𝕾𝖊𝖗𝖎𝖊𝖘.

PHILADELPHIA, 1876.
THE PRIZE MEDAL
Was awarded to the Publishers for
Books : Rudimentary Scientific,
"WEALE'S SERIES," ETC.

A NEW LIST OF
WEALE'S SERIES
RUDIMENTARY SCIENTIFIC, EDUCATIONAL, AND CLASSICAL.

LONDON, 1862.
THE PRIZE MEDAL
Was awarded to the Publishers of
"WEALE'S SERIES."

These popular and cheap Series of Books, now comprising nearly Three Hundred distinct works in almost every department of Science, Art, and Education, are recommended to the notice of Engineers, Architects, Builders, Artisans, and Students generally, as well as to those interested in Workmen's Libraries, Free Libraries, Literary and Scientific Institutions, Colleges, Schools, Science Classes, &c., &c.

N.B.—In ordering from this List it is recommended, as a means of facilitating business and obviating error, to quote the numbers affixed to the volumes, as well as the titles and prices.

*** The prices quoted are for limp cloth ; but the volumes marked with a ‡ may be had strongly bound in cloth boards for 6d. extra.

RUDIMENTARY SCIENTIFIC SERIES.

ARCHITECTURE, BUILDING, ETC.

No.
16. *ARCHITECTURE—ORDERS*—The Orders and their Æsthetic Principles. By W. H. LEEDS. Illustrated. 1s. 6d.
17. *ARCHITECTURE—STYLES*—The History and Description of the Styles of Architecture of Various Countries, from the Earliest to the Present Period. By T. TALBOT BURY, F.R.I.B.A., &c. Illustrated. 2s.
*** ORDERS AND STYLES OF ARCHITECTURE, in One Vol., 3s. 6d.
18. *ARCHITECTURE—DESIGN*—The Principles of Design in Architecture, as deducible from Nature and exemplified in the Works of the Greek and Gothic Architects. By E. L. GARBETT, Architect. Illustrated. 2s.
*** *The three preceding Works, in One handsome Vol., half bound, entitled* "MODERN ARCHITECTURE," *price 6s.*
22. *THE ART OF BUILDING*, Rudiments of. General Principles of Construction, Materials used in Building, Strength and Use of Materials, Working Drawings, Specifications, and Estimates. By E. DOBSON, 2s.‡
23. *BRICKS AND TILES*, Rudimentary Treatise on the Manufacture of; containing an Outline of the Principles of Brickmaking. By EDW. DOBSON, M.R.I.B.A. With Additions by C. TOMLINSON, F.R.S. Illustrated, 3s.‡

☞ *The ‡ indicates that these vols. may be had strongly bound at 6d. extra.*

CROSBY LOCKWOOD AND CO., 7, STATIONERS' HALL COURT, E.C.

Architecture, Building, etc., *continued.*

25. *MASONRY AND STONECUTTING*, Rudimentary Treatise
on; in which the Principles of Masonic Projection and their application to
the Construction of Curved Wing-Walls, Domes, Oblique Bridges, and
Roman and Gothic Vaulting, are concisely explained. By EDWARD DOBSON,
M.R.I.B.A., &c. Illustrated with Plates and Diagrams. 2s. 6d.‡

44. *FOUNDATIONS AND CONCRETE WORKS*, a Rudimentary
Treatise on; containing a Synopsis of the principal cases of Foundation
Works, with the usual Modes of Treatment, and Practical Remarks on
Footings, Planking, Sand, Concrete, Béton, Pile-driving, Caissons, and
Cofferdams. By E. DOBSON, M.R.I.B.A., &c. Fourth Edition, revised by
GEORGE DODD, C.E. Illustrated. 1s. 6d.

42. *COTTAGE BUILDING*. By C. BRUCE ALLEN, Architect.
Eighth Edition, revised and enlarged. Numerous Illustrations. 1s. 6d.

45. *LIMES, CEMENTS, MORTARS, CONCRETES, MASTICS,*
PLASTERING, &c. By G. R. BURNELL, C.E. Eleventh Edition. 1s. 6d.

57. *WARMING AND VENTILATION*, a Rudimentary Treatise
on; being a concise Exposition of the General Principles of the Art of Warming and Ventilating Domestic and Public Buildings, Mines, Lighthouses,
Ships, &c. By CHARLES TOMLINSON, F.R.S., &c. Illustrated. 3s.

83**. *CONSTRUCTION OF DOOR LOCKS*. Compiled from the
Papers of A. C. HOBBS, Esq., of New York, and Edited by CHARLES TOM-
LINSON, F.R.S. To which is added, a Description of Fenby's Patent Locks,
and a Note upon IRON SAFES by ROBERT MALLET, M.I.C.E. Illus. 2s. 6d.

111. *ARCHES, PIERS, BUTTRESSES, &c.*: Experimental Essays
on the Principles of Construction in; made with a view to their being useful
to the Practical Builder. By WILLIAM BLAND. Illustrated. 1s. 6d.

116. *THE ACOUSTICS OF PUBLIC BUILDINGS;* or, The
Principles of the Science of Sound applied to the purposes of the Architect and
Builder. By T. ROGER SMITH, M.R.I.B.A., Architect. Illustrated. 1s. 6d.

124. *CONSTRUCTION OF ROOFS*, Treatise on the, as regards
Carpentry and Joinery. Deduced from the Works of ROBISON, PRICE, and
TREDGOLD. Illustrated. 1s. 6d.

127. *ARCHITECTURAL MODELLING IN PAPER*, the Art of.
By T. A. RICHARDSON, Architect. Illustrated. 1s. 6d.

128. *VITRUVIUS—THE ARCHITECTURE OF MARCUS
VITRUVIUS POLLO.* In Ten Books. Translated from the Latin by
JOSEPH GWILT, F.S.A., F.R.A.S. With 23 Plates. 5s.

130. *GRECIAN ARCHITECTURE*, An Inquiry into the Principles
of Beauty in; with a Historical View of the Rise and Progress of the Art in
Greece. By the EARL OF ABERDEEN. 1s.

*** *The two preceding Works in One handsome Vol., half bound, entitled* "ANCIENT
ARCHITECTURE," *price 6s.*

132. *DWELLING-HOUSES*, a Rudimentary Treatise on the Erection
of. By S. H. BROOKS, Architect. New Edition, with Plates. 2s. 6d.‡

156. *QUANTITIES AND MEASUREMENTS*, How to Calculate and
Take them in Bricklayers', Masons', Plasterers', Plumbers', Painters', Paper-
hangers', Gilders', Smiths', Carpenters', and Joiners' Work. By A. C.
BEATON, Architect and Surveyor. New and Enlarged Edition. Illus. 1s. 6d.

175. *LOCKWOOD & CO.'S BUILDER'S AND CONTRACTOR'S*
PRICE BOOK, for 1879, containing the latest Prices of all kinds of Builders'
Materials and Labour, and of all Trades connected with Building: Lists of
the Members of the Metropolitan Board of Works, of Districts, District
Officers, and District Surveyors, and the Metropolitan Bye-laws. Edited by
FRANCIS T. W. MILLER, Architect and Surveyor. 3s. 6d.; half bound, 4s.

182. *CARPENTRY AND JOINERY*—THE ELEMENTARY PRIN-
CIPLES OF CARPENTRY. Chiefly composed from the Standard Work of
THOMAS TREDGOLD, C.E. With Additions from the Works of the most
Recent Authorities, and a TREATISE ON JOINERY by E. WYNDHAM
TARN, M.A. Numerous Illustrations. 3s. 6d.‡

The ‡ indicates that these vols. may be had strongly bound at 6d. extra.

LONDON : CROSBY LOCKWOOD AND CO.,

Architecture, Building, etc., *continued.*

182*. *CARPENTRY AND JOINERY. ATLAS* of 35 Plates to accompany the foregoing book. With Descriptive Letterpress. 4to. 6s.; cloth boards, 7s. 6d.

187. *HINTS TO YOUNG ARCHITECTS.* By George Wightwick. New, Revised, and enlarged Edition. By G. Huskisson Guillaume, Architect. With numerous Woodcuts. 3s. 6d.‡

188. *HOUSE PAINTING, GRAINING, MARBLING, AND SIGN WRITING:* A Practical Manual of. With 9 Coloured Plates of Woods and Marbles, and nearly 150 Wood Engravings. By Ellis A. Davidson. Second Edition, carefully revised. 5s. cloth limp; 6s. cloth boards.

189. *THE RUDIMENTS OF PRACTICAL BRICKLAYING.* In Six Sections: General Principles; Arch Drawing, Cutting, and Setting; Pointing; Paving, Tiling, Materials; Slating and Plastering; Practical Geometry, Mensuration, &c. By Adam Hammond. Illustrated. 1s. 6d.

191. *PLUMBING.* A Text-Book to the Practice of the Art or Craft of the Plumber. With Chapters upon House Drainage, embodying the latest Improvements. Containing about 300 Illustrations. By W. P. Buchan, Sanitary Engineer. 3s.‡

192. *THE TIMBER IMPORTER'S, TIMBER MERCHANT'S,* and BUILDER'S STANDARD GUIDE; comprising copious and valuable Memoranda for the Retailer and Builder. By Richard E. Grandy. Second Edition, Revised. 3s.‡

CIVIL ENGINEERING, ETC.

13. *CIVIL ENGINEERING,* the Rudiments of; for the Use of Beginners, for Practical Engineers, and for the Army and Navy. By Henry Law, C.E. Including a Section on Hydraulic Engineering, by George R. Burnell, C.E. 5th Edition, with Notes and Illustrations by Robert Mallet, A.M., F.R.S. Illustrated with Plates and Diagrams. 5s.‡

29. *THE DRAINAGE OF DISTRICTS AND LANDS.* By G. Drysdale Dempsey, C.E. New Edition, enlarged. Illustrated. 1s. 6d.

30. *THE DRAINAGE OF TOWNS AND BUILDINGS.* By G. Drysdale Dempsey, C.E. New Edition. Illustrated. 2s. 6d.
. With "Drainage of Districts and Lands," in One Vol., 3s. 6d.

31. *WELL-DIGGING, BORING, AND PUMP-WORK.* By John George Swindell, Assoc. R.I.B.A. New Edition, revised by G. R. Burnell, C.E. Illustrated. 1s. 6d.

35. *THE BLASTING AND QUARRYING OF STONE,* for Building and other Purposes. With Remarks on the Blowing up of Bridges. By Gen. Sir John Burgoyne, Bart., K.C.B. Illustrated. 1s. 6d.

43. *TUBULAR AND OTHER IRON GIRDER BRIDGES.* Particularly describing the Britannia and Conway Tubular Bridges. With a Sketch of Iron Bridges, and Illustrations of the Application of Malleable Iron to the Art of Bridge Building. By G. D. Dempsey, C.E. New Edition, with Illustrations. 1s. 6d.

62. *RAILWAY CONSTRUCTION,* Elementary and Practical Instruction on. By Sir Macdonald Stephenson, C.E. New Edition, enlarged by Edward Nugent, C.E. Plates and numerous Woodcuts. 3s.

80*. *EMBANKING LANDS FROM THE SEA,* the Practice of. Treated as a Means of Profitable Employment for Capital. With Examples and Particulars of actual Embankments, and also Practical Remarks on the Repair of old Sea Walls. By John Wiggins, F.G.S. New Edition, with Notes by Robert Mallet, F.R.S. 2s.

81. *WATER WORKS,* for the Supply of Cities and Towns. With a Description of the Principal Geological Formations of England as influencing Supplies of Water; and Details of Engines and Pumping Machinery for raising Water. By Samuel Hughes, F.G.S., C.E. New Edition, revised and enlarged, with numerous Illustrations. 4s.‡

82**. *GAS WORKS,* and the Practice of Manufacturing and Distributing Coal Gas. By Samuel Hughes, C.E. New Edition, revised by W. Richards, C.E. Illustrated. 3s. 6d.‡

☞ *The ‡ indicates that these vols. may be had strongly bound at 6d. extra.*

Civil Engineering, etc., *continued.*

117. *SUBTERRANEOUS SURVEYING*, an Elementary and Practical Treatise on. By Thomas Fenwick. Also the Method of Conducting Subterraneous Surveys without the Use of the Magnetic Needle, and other Modern Improvements. By Thomas Baker, C.E. Illustrated. 2s. 6d.‡

118. *CIVIL ENGINEERING IN NORTH AMERICA*, a Sketch of. By David Stevenson, F.R.S.E., &c. Plates and Diagrams. 3s.

121. *RIVERS AND TORRENTS.* With the Method of Regulating their Courses and Channels. By Professor Paul Frisi, F.R.S. To which is added, AN ESSAY ON NAVIGABLE CANALS. Translated by Major-General J. Garstin, of the Bengal Engineers. Plates. 5s. Cloth boards.

197. *ROADS AND STREETS (THE CONSTRUCTION OF)*, in two Parts: I. The Art of Constructing Common Roads, by Henry Law, C.E., revised and condensed by D. Kinnear Clark, C.E.; II. Recent Practice, including pavements of Stone, Wood, and Asphalte, by D. K. Clark, M.I.C.E., with numerous Illustrations. 4s. 6d.‡ [*Just published.*

MECHANICAL ENGINEERING, ETC.

33. *CRANES*, the Construction of, and other Machinery for Raising Heavy Bodies for the Erection of Buildings, and for Hoisting Goods. By Joseph Glynn, F.R.S., &c. Illustrated. 1s. 6d.

34. *THE STEAM ENGINE*, a Rudimentary Treatise on. By Dr. Lardner. Illustrated. 1s. 6d.

59. *STEAM BOILERS:* their Construction and Management. By R. Armstrong, C.E. Illustrated. 1s. 6d.

63. *AGRICULTURAL ENGINEERING:* Farm Buildings, Motive Power, Field Machines, Machinery, and Implements. By G. H. Andrews, C.E. Illustrated. 3s.

67. *CLOCKS, WATCHES, AND BELLS*, a Rudimentary Treatise on. By Sir Edmund Beckett (late Edmund Beckett Denison, LL.D., Q.C.). A New, Revised, and considerably Enlarged Edition (the 6th), with very numerous Illustrations. 4s. 6d. cloth limp; 5s. 6d. cloth boards, gilt.

82. *THE POWER OF WATER*, as applied to drive Flour Mills, and to give motion to Turbines and other Hydrostatic Engines. By Joseph Glynn, F.R.S., &c. New Edition, Illustrated. 2s.‡

98. *PRACTICAL MECHANISM*, the Elements of; and Machine Tools. By T. Baker, C.E. With Remarks on Tools and Machinery, by J. Nasmyth, C.E. Plates. 2s. 6d.‡

114. *MACHINERY*, Elementary Principles of, in its Construction and Working. Illustrated by numerous Examples of Modern Machinery for different Branches of Manufacture. By C. D. Abel, C.E. 1s. 6d.

115. *ATLAS OF PLATES.* Illustrating the above Treatise. By C. D. Abel, C.E. 7s. 6d.

139. *THE STEAM ENGINE*, a Treatise on the Mathematical Theory of, with Rules at length, and Examples for the Use of Practical Men. By T. Baker, C.E. Illustrated. 1s. 6d.

162. *THE BRASS FOUNDER'S MANUAL;* Instructions for Modelling, Pattern-Making, Moulding, Turning, Filing, Burnishing, Bronzing, &c. With copious Receipts, numerous Tables, and Notes on Prime Costs and Estimates. By Walter Graham. Illustrated. 2s. 6d.‡

164. *MODERN WORKSHOP PRACTICE*, as applied to Marine, Land, and Locomotive Engines, Floating Docks, Dredging Machines, Bridges, Cranes, Ship-building, &c., &c. By J. G. Winton. Illustrated. 3s.‡

165. *IRON AND HEAT*, exhibiting the Principles concerned in the Construction of Iron Beams, Pillars, and Bridge Girders, and the Action of Heat in the Smelting Furnace. By J. Armour, C.E. 2s. 6d ‡

166. *POWER IN MOTION:* Horse-Power, Motion, Toothed-Wheel Gearing, Long and Short Driving Bands, Angular Forces. By James Armour, C.E. With 73 Diagrams. 2s. 6d.‡

167. *THE APPLICATION OF IRON TO THE CONSTRUCTION* OF BRIDGES, GIRDERS, ROOFS, AND OTHER WORKS. By Francis Campin, C.E. Second Edition, revised and corrected. 2s. 6d.‡

☞ *The ‡ indicates that these vols. may be had strongly bound at 6d. extra.*

LONDON : CROSBY LOCKWOOD AND CO.,

Mechanical Engineering, etc., *continued.*

171. THE WORKMAN'S MANUAL OF ENGINEERING DRAWING. By JOHN MAXTON, Engineer, Instructor in Engineering Drawing, Royal Naval College, Greenwich. Third Edition. Illustrated with 7 Plates and nearly 350 Woodcuts. 3s. 6d.‡

190. STEAM AND THE STEAM ENGINE, Stationary and Portable. Being an extension of Mr. John Sewell's "Treatise on Steam." By D. KINNEAR CLARK, M.I.C.E., Author of "Railway Machinery," &c., &c. With numerous Illustrations. 3s. 6d.‡

200. FUEL, its Combustion and Economy; consisting of Abridgments of "Treatise on the Combustion of Coal and the Prevention of Smoke," by C. W. WILLIAMS, A.I.C.E., and "The Economy of Fuel," by T. SYMES PRIDEAUX. With extensive additions on Recent Practice in the Combustion and Economy of Fuel—Coal, Coke, Wood, Peat, Petroleum, &c.—by the Editor, D. KINNEAR CLARK, M.I.C.E. With numerous Illustrations. 4s. 6d.‡
[Just Published.

202. LOCOMOTIVE ENGINES, a Rudimentary Treatise on. Comprising an Historical Sketch and Description of the Locomotive Engine by G. D. DEMPSEY, C.E.; with large additions treating of the Modern Locomotive, by D. KINNEAR CLARK, M.I.C.E. With numerous Illustrations. 3s.‡
[Just Published.

SHIPBUILDING, NAVIGATION, MARINE ENGINEERING, ETC.

51. NAVAL ARCHITECTURE, the Rudiments of; or an Exposition of the Elementary Principles of the Science, and their Practical Application to Naval Construction. Compiled for the Use of Beginners. By JAMES PEAKE, School of Naval Architecture, H.M. Dockyard, Portsmouth. Fourth Edition, corrected, with Plates and Diagrams. 3s. 6d.‡

53*. SHIPS FOR OCEAN AND RIVER SERVICE, Elementary and Practical Principles of the Construction of. By HAKON A. SOMMERFELDT, Surveyor of the Royal Norwegian Navy. With an Appendix. 1s.

53**. AN ATLAS OF ENGRAVINGS to Illustrate the above. Twelve large folding plates. Royal 4to, cloth. 7s. 6d.

54. MASTING, MAST-MAKING, AND RIGGING OF SHIPS, Rudimentary Treatise on. Also Tables of Spars, Rigging, Blocks; Chain, Wire, and Hemp Ropes, &c., relative to every class of vessels. Together with an Appendix of Dimensions of Masts and Yards of the Royal Navy of Great Britain and Ireland. By ROBERT KIPPING, N.A. Fourteenth Edition. Illustrated. 2s.‡

54*. IRON SHIP-BUILDING. With Practical Examples and Details for the Use of Ship Owners and Ship Builders. By JOHN GRANTHAM, Consulting Engineer and Naval Architect. 5th Edition, with Additions. 4s.

54**. AN ATLAS OF FORTY PLATES to Illustrate the above. Fifth Edition. Including the latest Examples, such as H.M. Steam Frigates "Warrior," "Hercules," "Bellerophon;" H.M. Troop Ship "Serapis," Iron Floating Dock, &c., &c. 4to, boards. 38s.

55. THE SAILOR'S SEA BOOK: a Rudimentary Treatise on Navigation. I. How to Keep the Log and Work it off. II. On Finding the Latitude and Longitude. By JAMES GREENWOOD, B.A., of Jesus College, Cambridge. To which are added, Directions for Great Circle Sailing; an Essay on the Law of Storms and Variable Winds; and Explanations of Terms used in Ship-building. Ninth Edition, with several Engravings and Coloured Illustrations of the Flags of Maritime Nations. 2s.

80. MARINE ENGINES, AND STEAM VESSELS, a Treatise on. Together with Practical Remarks on the Screw and Propelling Power, as used in the Royal and Merchant Navy. By ROBERT MURRAY, C.E., Engineer-Surveyor to the Board of Trade. With a Glossary of Technical Terms, and their Equivalents in French, German, and Spanish. Seventh Edition, revised and enlarged. Illustrated. 3s.‡

☞ *The ‡ indicates that these vols. may be had strongly bound at 6d. extra.*

7, STATIONERS' HALL COURT, LUDGATE HILL, E.C.

Shipbuilding, Navigation, etc., *continued.*

83*bis.* *THE FORMS OF SHIPS AND BOATS:* Hints, Experimentally Derived, on some of the Principles regulating Ship-building. By W.
BLAND. Seventh Edit on, revised, with numerous Illustrations and Models. 1s. 6d.

99. *NAVIGATION AND NAUTICAL ASTRONOMY,* in Theory
and Practice. With Attempts to facilitate the Finding of the Time and the
Longitude at Sea. By J. R. YOUNG, formerly Professor of Mathematics in
Belfast College. Illustrated. 2s. 6d.

100*. *TABLES* intended to facilitate the Operations of Navigation and
Nautical Astronomy, as an Accompaniment to the above Book. By J. R.
YOUNG. 1s. 6d.

106. *SHIPS' ANCHORS,* a Treatise on.   By GEORGE COTSELL,
N.A. Illustrated. 1s. 6d.

149. *SAILS AND SAIL-MAKING,* an Elementary Treatise on.
With Draughting, and the Centre of Effort of the Sails. Also, Weights
and Sizes of Ropes ; Masting, Rigging, and Sails of Steam Vessels, &c., &c.
Tenth Edition, enlarged, with an Appendix. By ROBERT KIPPING, N.A.,
Sailmaker, Quayside, Newcastle. Illustrated. 2s. 6d.‡

155. *THE ENGINEER'S GUIDE TO THE ROYAL AND*
MERCANTILE NAVIES. By a PRACTICAL ENGINEER. Revised by D.
F. M'CARTHY, late of the Ordnance Survey Office, Southampton. 3s.

PHYSICAL SCIENCE, NATURAL PHILO-
SOPHY, ETC.

1. *CHEMISTRY,* for the Use of Beginners.   By Professor GEORGE
FOWNES, F.R.S.   With an Appendix, on the Application of Chemistry to
Agriculture. 1s.

2. *NATURAL PHILOSOPHY,* Introduction to the Study of; for
the Use of Beginners. By C. TOMLINSON, Lecturer on Natural Science in
King's College School, London. Woodcuts. 1s. 6d.

4. *MINERALOGY,* Rudiments of; a concise View of the Properties
of Minerals. By A. RAMSAY, Jun. Woodcuts and Steel Plates. 3s.‡

6. *MECHANICS,* Rudimentary Treatise on; being a concise Exposition of the General Principles of Mechanical Science, and their Applications. By CHARLES TOMLINSON, Lecturer on Natural Science in King's
College School, London. Illustrated. 1s. 6d.

7. *ELECTRICITY;* showing the General Principles of Electrical
Science, and the purposes to which it has been applied. By Sir W. SNOW
HARRIS, F.R.S., &c.   With considerable Additions by R. SABINE, C.E.,
F.S.A. Woodcuts. 1s. 6d.

7*. *GALVANISM,* Rudimentary Treatise on, and the General Prin-
ciples of Animal and Voltaic Electricity. By Sir W. SNOW HARRIS. New
Edition, revised, with considerable Additions, by ROBERT SABINE, C.E.,
F.S A. Woodcuts. 1s. 6d.

8. *MAGNETISM;* being a concise Exposition of the General Prin-
ciples of Magnetical Science, and the Purposes to which it has been applied.
By Sir W. SNOW HARRIS. New Edition, revised and enlarged by H. M.
NOAD, Ph.D., Vice-President of the Chemical Society, Author of "A
Manual of Electricity," &c., &c.   With 165 Woodcuts. 3s. 6d.‡

11. *THE ELECTRIC TELEGRAPH;* its History and Progress;
with Descriptions of some of the Apparatus. By R. SABINE, C.E., F.S.A., &c.
Woodcuts. 3s.

12. *PNEUMATICS,* for the Use of Beginners.   By CHARLES
TOMLINSON. Illustrated. 1s. 6d.

72. *MANUAL OF THE MOLLUSCA ;* a Treatise on Recent and
Fossil Shells. By Dr. S. P. WOODWARD, A.L.S. With Appendix by
RALPH TATE, A.L.S., F.G.S.   With numerous Plates and 300 Woodcuts.
6s. 6d. Cloth boards, 7s. 6d.

The ‡ indicates that these vols. may be had strongly bound at 6d. extra.

LONDON : CROSBY LOCKWOOD AND CO.,

Physical Science, Natural Philosophy, etc., *continued.*

79**. *PHOTOGRAPHY*, Popular Treatise on; with a Description of the Stereoscope, &c. Translated from the French of D. VAN MONCKHOVEN, by W. H. THORNTHWAITE, Ph.D. Woodcuts. 1s. 6d.

96. *ASTRONOMY.* By the Rev. R. MAIN, M.A., F.R.S., &c. New Edition, with an Appendix on "Spectrum Analysis." Woodcuts. 1s. 6d.

97. *STATICS AND DYNAMICS*, the Principles and Practice of; embracing also a clear development of Hydrostatics, Hydrodynamics, and Central Forces. By T. BAKER, C.E. 1s. 6d.

138. *TELEGRAPH*, Handbook of the; a Manual of Telegraphy, Telegraph Clerks' Remembrancer, and Guide to Candidates for Employment in the Telegraph Service. By R. BOND. Fourth Edition, revised and enlarged: to which is appended, QUESTIONS on MAGNETISM, ELECTRICITY, and PRACTICAL TELEGRAPHY, for the Use of Students, by W. McGREGOR, First Assistant Superintendent, Indian Gov. Telegraphs. Woodcuts. 3s.‡

143. *EXPERIMENTAL ESSAYS.* By CHARLES TOMLINSON. I. On the Motions of Camphor on Water. II. On the Motion of Camphor towards the Light. III. History of the Modern Theory of Dew. Woodcuts. 1s.

173. *PHYSICAL GEOLOGY*, partly based on Major-General PORTLOCK's "Rudiments of Geology." By RALPH TATE, A.L.S., &c. Woodcuts. 2s.

174. *HISTORICAL GEOLOGY*, partly based on Major-General PORTLOCK's "Rudiments." By RALPH TATE, A.L.S., &c. Woodcuts. 2s. 6d.

173 & 174. *RUDIMENTARY TREATISE ON GEOLOGY*, Physical and Historical. Partly based on Major-General PORTLOCK's "Rudiments of Geology." By RALPH TATE, A.L.S., F.G.S., &c., &c. Numerous Illustrations. In One Volume. 4s. 6d.‡

183 & 184. *ANIMAL PHYSICS*, Handbook of. By Dr. LARDNER, D.C.L., formerly Professor of Natural Philosophy and Astronomy in University College, Lond. With 520 Illustrations. In One Vol. 7s. 6d., cloth boards.

*** *Sold also in Two Parts, as follows :—*

183. ANIMAL PHYSICS. By Dr. LARDNER. Part I., Chapters I—VII. 4s.
184. ANIMAL PHYSICS. By Dr. LARDNER. Part II., Chapters VIII—XVIII. 3s.

MINING, METALLURGY, ETC.

117. *SUBTERRANEOUS SURVEYING*, Elementary and Practical Treatise on, with and without the Magnetic Needle. By THOMAS FENWICK, Surveyor of Mines, and THOMAS BAKER, C.E. Illustrated. 2s. 6d.‡

133. *METALLURGY OF COPPER ;* an Introduction to the Methods of Seeking, Mining, and Assaying Copper, and Manufacturing its Alloys. By ROBERT H. LAMBORN, Ph.D. Woodcuts. 2s. 6d.‡

134. *METALLURGY OF SILVER AND LEAD.* A Description of the Ores; their Assay and Treatment, and valuable Constituents. By Dr. R. H. LAMBORN. Woodcuts. 2s. 6d.‡

135. *ELECTRO-METALLURGY;* Practically Treated. By ALEXANDER WATT, F.R.S.S.A. New Edition, enlarged. Woodcuts. 2s. 6d.‡

172. *MINING TOOLS*, Manual of. For the Use of Mine Managers, Agents, Students, &c. Comprising Observations on the Materials from, and Processes by, which they are manufactured ; their Special Uses, Applications, Qualities, and Efficiency. By WILLIAM MORGANS, Lecturer on Mining at the Bristol School of Mines. 2s. 6d.‡

172*. *MINING TOOLS, ATLAS* of Engravings to Illustrate the above, containing 235 Illustrations of Mining Tools, drawn to Scale. 4to. 4s. 6d.; cloth boards, 6s.

176. *METALLURGY OF IRON*, a Treatise on the. Containing History of Iron Manufacture, Methods of Assay, and Analyses of Iron Ores, Processes of Manufacture of Iron and Steel, &c. By H. BAUERMAN, F.G.S. Fourth Edition, enlarged, with numerous Illustrations. 4s. 6d.‡

☞ *The ‡ indicates that these vols. may be had strongly bound at 6d. extra.*

Mining, Metallurgy, etc., *continued.*

180. *COAL AND COAL MINING,* A Rudimentary Treatise on.
By WARINGTON W. SMYTH, M.A., F.R.S., &c., Chief Inspector of the
Mines of the Crown and of the Duchy of Cornwall. New Edition, revised
and corrected. With numerous Illustrations. 3s. 6d.‡

195. *THE MINERAL SURVEYOR AND VALUER'S COM-*
PLETE GUIDE, with new Traverse Tables, and Descriptions of Improved
Instruments; also the Correct Principles of Laying out and Valuing Mineral
Properties. By WILLIAM LINTERN, Mining and Civil Engineer. With
four Plates of Diagrams, Plans, &c. 3s. 6d.‡   [*Just published.*

AGRICULTURE, GARDENING, ETC.

29. *THE DRAINAGE OF DISTRICTS AND LANDS.* By
G. DRYSDALE DEMPSEY, C.E. Illustrated. 1s. 6d.
. *With "Drainage of Towns and Buildings," in One Vol., 3s. 6d.*

63. *AGRICULTURAL ENGINEERING:* Farm Buildings, Motive
Powers and Machinery of the Steading, Field Machines, and Implements.
By G. H. ANDREWS, C.E. Illustrated. 3s.

66. *CLAY LANDS AND LOAMY SOILS.* By Professor
DONALDSON. 1s.

131. *MILLER'S, MERCHANT'S, AND FARMER'S READY*
RECKONER, for ascertaining at sight the value of any quantity of Corn,
from One Bushel to One Hundred Quarters, at any given price, from £1 to
£5 per Qr. With approximate values of Millstones, Millwork, &c. 1s.

140. *SOILS, MANURES, AND CROPS.* (Vol. 1. OUTLINES OF
MODERN FARMING.) By R. SCOTT BURN. Woodcuts. 2s.

141. *FARMING AND FARMING ECONOMY,* Notes, Historical
and Practical, on. (Vol. 2. OUTLINES OF MODERN FARMING.) By R. SCOTT
BURN. Woodcuts. 3s.

142. *STOCK; CATTLE, SHEEP, AND HORSES.* (Vol. 3.
OUTLINES OF MODERN FARMING.) By R. SCOTT BURN. Woodcuts. 2s. 6d.

145. *DAIRY, PIGS, AND POULTRY,* Management of the. By
R. SCOTT BURN. With Notes on the Diseases of Stock. (Vol. 4. OUTLINES
OF MODERN FARMING.) Woodcuts. 2s.

146. *UTILIZATION OF SEWAGE, IRRIGATION, AND*
RECLAMATION OF WASTE LAND. (Vol. 5. OUTLINES OF MODERN
FARMING.) By R. SCOTT BURN. Woodcuts. 2s. 6d.

. *Nos. 140-1-2-5-6, in One Vol., handsomely half-bound, entitled "OUTLINES OF
MODERN FARMING." By ROBERT SCOTT BURN. Price 12s.*

177. *FRUIT TREES,* The Scientific and Profitable Culture of. From
the French of DU BREUIL, Revised by GEO. GLENNY. 187 Woodcuts. 3s. 6d.‡

198. *SHEEP; THE HISTORY, STRUCTURE, ECONOMY, AND*
DISEASES OF. By W. C. SPOONER, M.R.V.C., &c. Fourth Edition,
considerably enlarged; with numerous fine engravings, including some
specimens of New and Improved Breeds. 366 pp. 3s. 6d.‡ [*Just published.*

201. *KITCHEN GARDENING MADE EASY.* Showing how to
prepare and lay out the ground, the best means of cultivating every known
Vegetable and Herb, with cultural directions for the management of them
all the year round. By GEORGE M. F. GLENNY, Editor of "Glenny's Illustrated Garden Almanack," and Author of "Floriculture," &c. 1s. 6d.‡
[*Just Published.*

☞ *The ‡ indicates that these vols. may be had strongly bound at 6d. extra.*

LONDON : CROSBY LOCKWOOD AND CO.,

FINE ARTS.

20. *PERSPECTIVE FOR BEGINNERS.* Adapted to Young Students and Amateurs in Architecture, Painting, &c. By GEORGE PYNE, Artist. Woodcuts. 2s.

40 *GLASS STAINING ;* or, Painting on Glass, The Art of. Com-
& prising Directions for Preparing the Pigments and Fluxes, laying them upon
41. the Glass, and Firing or Burning in the Colours. From the German of Dr. GESSERT. To which is added, an Appendix on THE ART OF ENAMELLING, &c., with THE ART OF PAINTING ON GLASS. From the German of EMANUEL OTTO FROMBERG. In One Volume. 2s. 6d.

69. *MUSIC,* A Rudimentary and Practical Treatise on. With numerous Examples. By CHARLES CHILD SPENCER. 2s. 6d.

71. *PIANOFORTE,* The Art of Playing the. With numerous Exercises and Lessons. Written and Selected from the Best Masters, by CHARLES CHILD SPENCER. 1s. 6d.

181. *PAINTING POPULARLY EXPLAINED,* including Fresco, Oil, Mosaic, Water Colour, Water-Glass, Tempera, Encaustic, Miniature, Painting on Ivory, Vellum, Pottery, Enamel, Glass, &c. With Historical Sketches of the Progress of the Art by THOMAS JOHN GULLICK, assisted by JOHN TIMBS, F.S.A. Fourth Edition, revised and enlarged, with Frontispiece and Vignette. 5s.‡

186. *A GRAMMAR OF COLOURING,* applied to Decorative Painting and the Arts. By GEORGE FIELD. New Edition, enlarged and adapted to the Use of the Ornamental Painter and Designer. By ELLIS A. DAVIDSON, Author of " Drawing for Carpenters," &c. With two new Coloured Diagrams and numerous Engravings on Wood. 3s.‡

ARITHMETIC, GEOMETRY, MATHEMATICS, ETC.

32. *MATHEMATICAL INSTRUMENTS,* a Treatise on; in which their Construction and the Methods of Testing, Adjusting, and Using them are concisely Explained. By J. F. HEATHER, M.A., of the Royal Military Academy, Woolwich. Original Edition, in 1 vol., Illustrated. 1s. 6d.

₊ *In ordering the above, be careful to say, " Original Edition," or give the number in the Series* (32) *to distinguish it from the Enlarged Edition in 3 vols.* (*Nos.* 168-9-70.)

60. *LAND AND ENGINEERING SURVEYING,* a Treatise on; with all the Modern Improvements. Arranged for the Use of Schools and Private Students; also for Practical Land Surveyors and Engineers. By T. BAKER, C.E. New Edition, revised by EDWARD NUGENT, C.E. Illustrated with Plates and Diagrams. 2s.‡

61*. *READY RECKONER FOR THE ADMEASUREMENT OF LAND.* By ABRAHAM ARMAN, Schoolmaster, Thurleigh, Beds. To which is added a Table, showing the Price of Work, from 2s. 6d. to £1 per acre, and Tables for the Valuation of Land, from 1s. to £1,000 per acre, and from one pole to two thousand acres in extent, &c., &c. 1s. 6d.

76. *DESCRIPTIVE GEOMETRY,* an Elementary Treatise on ; with a Theory of Shadows and of Perspective, extracted from the French of G. MONGE. To which is added, a description of the Principles and Practice of Isometrical Projection ; the whole being intended as an introduction to the Application of Descriptive Geometry to various branches of the Arts. By J. F. HEATHER, M.A. Illustrated with 14 Plates. 2s.

178. *PRACTICAL PLANE GEOMETRY:* giving the Simplest Modes of Constructing Figures contained in one Plane and Geometrical Construction of the Ground. By J. F. HEATHER, M.A. With 215 Woodcuts. 2s.

179. *PROJECTION :* Orthographic, Topographic, and Perspective : giving the various Modes of Delineating Solid Forms by Constructions on a Single Plane Surface. By J. F. HEATHER, M.A. [*In preparation.*

₊ *The above three volumes will form a* COMPLETE ELEMENTARY COURSE OF MATHEMATICAL DRAWING.

☞ *The ‡ indicates that these vols. may be had strongly bound at 6d. extra.*

7, STATIONERS' HALL COURT, LUDGATE HILL, E.C.

Arithmetic, Geometry, Mathematics, etc., *continued.*

83. *COMMERCIAL BOOK-KEEPING.* With Commercial Phrases and Forms in English, French, Italian, and German. By JAMES HADDON, M.A., Arithmetical Master of King's College School, London. 1s. 6d.

84. *ARITHMETIC,* a Rudimentary Treatise on: with full Explanations of its Theoretical Principles, and numerous Examples for Practice. For the Use of Schools and for Self-Instruction. By J. R. YOUNG, late Professor of Mathematics in Belfast College. New Edition, with Index. 1s. 6d.

84*. A KEY to the above, containing Solutions in full to the Exercises, together with Comments, Explanations, and Improved Processes, for the Use of Teachers and Unassisted Learners. By J. R. YOUNG. 1s. 6d.

85. *EQUATIONAL ARITHMETIC,* applied to Questions of Interest,
85*. Annuities, Life Assurance, and General Commerce; with various Tables by which all Calculations may be greatly facilitated. By W. HIPSLEY. 2s.

86. *ALGEBRA,* the Elements of. By JAMES HADDON, M.A., Second Mathematical Master of King's College School. With Appendix, containing miscellaneous Investigations, and a Collection of Problems in various parts of Algebra. 2s.

86*. A KEY AND COMPANION to the above Book, forming an extensive repository of Solved Examples and Problems in Illustration of the various Expedients necessary in Algebraical Operations. Especially adapted for Self-Instruction. By J. R. YOUNG. 1s. 6d.

88. *EUCLID,* THE ELEMENTS OF: with many additional Propositions
89. and Explanatory Notes: to which is prefixed, an Introductory Essay on Logic. By HENRY LAW, C.E. 2s. 6d.‡

Sold also separately, viz.:—

88. EUCLID, The First Three Books. By HENRY LAW, C.E. 1s.
89. EUCLID, Books 4, 5, 6, 11, 12. By HENRY LAW, C.E. 1s. 6d.

90. *ANALYTICAL GEOMETRY AND CONIC SECTIONS,* a Rudimentary Treatise on. By JAMES HANN, late Mathematical Master of King's College School, London. A New Edition, re-written and enlarged by J. R. YOUNG, formerly Professor of Mathematics at Belfast College. 2s.‡

91. *PLANE TRIGONOMETRY,* the Elements of. By JAMES HANN, formerly Mathematical Master of King's College, London. 1s.

92. *SPHERICAL TRIGONOMETRY,* the Elements of. By JAMES HANN. Revised by CHARLES H. DOWLING, C.E. 1s.
• Or with "The Elements of Plane Trigonometry," in One Volume, 2s.

93. *MENSURATION AND MEASURING,* for Students and Practical Use. With the Mensuration and Levelling of Land for the Purposes of Modern Engineering. By T. BAKER, C.E. New Edition, with Corrections and Additions by E. NUGENT, C.E. Illustrated. 1s. 6d.

94. *LOGARITHMS,* a Treatise on; with Mathematical Tables for facilitating Astronomical, Nautical, Trigonometrical, and Logarithmic Calculations; Tables of Natural Sines and Tangents and Natural Cosines. By HENRY LAW, C.E. Illustrated. 2s. 6d.‡

101. *MEASURES, WEIGHTS, AND MONEYS OF ALL NATIONS,* and an Analysis of the Christian, Hebrew, and Mahometan Calendars. By W. S. B. WOOLHOUSE, F.R.A.S., &c. 1s. 6d.

102. *INTEGRAL CALCULUS,* Rudimentary Treatise on the. By HOMERSHAM COX, B.A. Illustrated. 1s.

103. *INTEGRAL CALCULUS,* Examples on the. By JAMES HANN, late of King's College, London. Illustrated. 1s.

101. *DIFFERENTIAL CALCULUS,* Examples of the. By W. S. B. WOOLHOUSE, F.R.A.S., &c. 1s. 6d.

104. *DIFFERENTIAL CALCULUS,* Examples and Solutions of the. By JAMES HADDON, M.A. 1s.

☞ *The ‡ indicates that these vols. may be had strongly bound at 6d. extra.*

LONDON : CROSBY LOCKWOOD AND CO.,

Arithmetic, Geometry, Mathematics, etc., *continued.*

105. *MNEMONICAL LESSONS.* — GEOMETRY, ALGEBRA, AND TRIGONOMETRY, in Easy Mnemonical Lessons. By the Rev. THOMAS PENYNGTON KIRKMAN, M.A. 1s. 6d.

136. *ARITHMETIC,* Rudimentary, for the Use of Schools and Self-Instruction. By JAMES HADDON, M.A. Revised by ABRAHAM ARMAN. 1s. 6d.

137. A KEY TO HADDON'S RUDIMENTARY ARITHMETIC. By A. ARMAN. 1s. 6d.

158. *THE SLIDE RULE, AND HOW TO USE IT;* containing full, easy, and simple Instructions to perform all Business Calculations with unexampled rapidity and accuracy. By CHARLES HOARE, C.E. With a Slide Rule in tuck of cover. 3s.‡

168. *DRAWING AND MEASURING INSTRUMENTS.* Including—I. Instruments employed in Geometrical and Mechanical Drawing, and in the Construction, Copying, and Measurement of Maps and Plans. II. Instruments used for the purposes of Accurate Measurement, and for Arithmetical Computations. By J. F. HEATHER, M.A., late of the Royal Military Academy, Woolwich, Author of "Descriptive Geometry," &c., &c. Illustrated. 1s. 6d.

169. *OPTICAL INSTRUMENTS.* Including (more especially) Telescopes, Microscopes, and Apparatus for producing copies of Maps and Plans by Photography. By J. F. HEATHER, M.A. Illustrated. 1s. 6d.

170. *SURVEYING AND ASTRONOMICAL INSTRUMENTS.* Including—I. Instruments Used for Determining the Geometrical Features of a portion of Ground. II. Instruments Employed in Astronomical Observations. By J. F. HEATHER, M.A. Illustrated. 1s. 6d.

The above three volumes form an enlargement of the Author's original work, "Mathematical Instruments: their Construction, Adjustment, Testing, and Use," the Eleventh Edition of which is on sale, price 1s. 6d. (See No. 32 in the Series.)

168.⎫ *MATHEMATICAL INSTRUMENTS.* By J. F. HEATHER,
169.⎬ M.A. Enlarged Edition, for the most part entirely re-written. The 3 Parts as
170.⎭ above, in One thick Volume. With numerous Illustrations. 4s. 6d.‡

185. *THE COMPLETE MEASURER;* setting forth the Measurement of Boards, Glass, &c., &c.; Unequal-sided, Square-sided, Octagonal-sided, Round Timber and Stone, and Standing Timber. With a Table showing the solidity of hewn or eight-sided timber, or of any octagonal-sided column. Compiled for Timber-growers, Merchants, and Surveyors, Stonemasons, Architects, and others. By RICHARD HORTON. Third Edition, with valuable additions. 4s.; strongly bound in leather, 5s.

196. *THEORY OF COMPOUND INTEREST AND ANNUITIES;* with Tables of Logarithms for the more Difficult Computations of Interest, Discount, Annuities, &c. By FÉDOR THOMAN, of the Société Crédit Mobilier, Paris. 4s.‡ [*Just published.*

199. *INTUITIVE CALCULATIONS;* or, Easy and Compendious Methods of Performing the various Arithmetical Operations required in Commercial and Business Transactions; together with Full Explanations of Decimals and Duodecimals, several Useful Tables, and an Examination and Discussion of the best Schemes for a Decimal Coinage. By DANIEL O'GORMAN. Twenty-fifth Edition, corrected and enlarged by J. R. YOUNG, formerly Professor of Mathematics in Belfast College. 3s.‡ [*Just published.*

MISCELLANEOUS VOLUMES.

36. *A DICTIONARY OF TERMS used in ARCHITECTURE, BUILDING, ENGINEERING, MINING, METALLURGY, ARCHÆOLOGY, the FINE ARTS, &c.* By JOHN WEALE. Fifth Edition. Revised by ROBERT HUNT, F.R.S., Keeper of Mining Records. Numerous Illustrations. 5s. cloth limp; 6s. cloth boards.

50. *THE LAW OF CONTRACTS FOR WORKS AND SERVICES.* By DAVID GIBBONS. Third Edition, revised and considerably enlarged. 3s.‡

☞ *The ‡ indicates that these vols. may be had strongly bound at 6d. extra.*

Miscellaneous Volumes, *continued.*

112. *MANUAL OF DOMESTIC MEDICINE.* By R. GOODING, B.A.. M.D. Intended as a Family Guide in all Cases of Accident and Emergency. 2s.‡

112-. *MANAGEMENT OF HEALTH.* A Manual of Home and Personal Hygiene. By the Rev. JAMES BAIRD, B.A. 1s.

150. *LOGIC,* Pure and Applied.   By S. H. EMMENS.   1s. 6d.

152. *PRACTICAL HINTS FOR INVESTING MONEY.* With an Explanation of the Mode of Transacting Business on the Stock Exchange. By FRANCIS PLAYFORD, Sworn Broker. 1s. 6d.

153. *SELECTIONS FROM LOCKE'S ·ESSAYS ON THE HUMAN UNDERSTANDING.* With Notes by S. H. EMMENS. 2s.

154. *GENERAL HINTS TO EMIGRANTS.* Containing Notices of the various Fields for Emigration. With Hints on Preparation for Emigrating, Outfits, &c., &c. With Directions and Recipes useful to the Emigrant. With a Map of the World. 2s.

157. *THE EMIGRANT'S GUIDE TO NATAL.* By ROBERT JAMES MANN, F.R.A.S., F.M.S. Second Edition, carefully corrected to the present Date. Map. 2s.

193. *HANDBOOK OF FIELD FORTIFICATION,* intended for the Guidance of Officers Preparing for Promotion, and especially adapted to the requirements of Beginners. By Major W. W. KNOLLYS, F.R.G.S., 93rd Sutherland Highlanders, &c. With 163 Woodcuts. 3s.‡

194. *THE HOUSE 'MANAGER :* Being a Guide to Housekeeping. Practical Cookery, Pickling and Preserving, Household Work, Dairy Management, the Table and Dessert, Cellarage of Wines, Home-brewing and Wine-making, the Boudoir and Dressing-room, Travelling, Stable Economy, Gardening Operations, &c. By AN OLD HOUSEKEEPER. 3s. 6d.‡

194. *HOUSE BOOK* (*The*). Comprising :—I. THE HOUSE MANAGER.
112.   By an OLD HOUSEKEEPER. II. DOMESTIC MEDICINE. By RALPH GOODING,
&   M.D. III. MANAGEMENT OF HEALTH. By JAMES BAIRD. In One Vol.,
112*.   strongly half-bound. 6s.   [*Just published.*

EDUCATIONAL AND CLASSICAL SERIES.

HISTORY.

1. **England, Outlines of the History of;** more especially with reference to the Origin and Progress of the English Constitution. A Text Book for Schools and Colleges. By WILLIAM DOUGLAS HAMILTON, F.S.A., of Her Majesty's Public Record Office. Fourth Edition, revised. Maps and Woodcuts. 5s. ; cloth boards, 6s.

5. **Greece, Outlines of the History of;** in connection with the Rise of the Arts and Civilization in Europe. By W. DOUGLAS HAMILTON, of University College, London, and EDWARD LEVIEN, M.A., of Balliol College, Oxford. 2s. 6d. ; cloth boards, 3s. 6d.

7. **Rome, Outlines of the History of:** from the Earliest Period to the Christian Era and the Commencement of the Decline of the Empire. By EDWARD LEVIEN, of Balliol College, Oxford. Map, 2s. 6d. ; cl. bds. 3s. 6d.

9. **Chronology of History, Art, Literature, and Progress,** from the Creation of the World to the Conclusion of the Franco-German War. The Continuation by W. D. HAMILTON, F.S.A., of Her Majesty's Record Office. 3s. ; cloth boards, 3s. 6d.

50. **Dates and Events. in English History,** for the use of Candidates in Public and Private Examinations. By the Rev. E. RAND. 1s.

The ‡ indicates that these vols. may be had strongly bound at 6d. extra.

LONDON : CROSBY LOCKWOOD AND CO.,

ENGLISH LANGUAGE AND MISCEL-LANEOUS.

11. **Grammar of the English Tongue,** Spoken and Written. With an Introduction to the Study of Comparative Philology. By HYDE CLARKE, D.C.L. Third Edition. 1s.

11*. **Philology:** Handbook of the Comparative Philology of English, Anglo-Saxon, Frisian, Flemish or Dutch, Low or Platt Dutch, High Dutch or German, Danish, Swedish, Icelandic, Latin, Italian, French, Spanish, and Portuguese Tongues. By HYDE CLARKE, D.C.L. 1s.

12. **Dictionary of the English Language,** as Spoken and Written. Containing above 100,000 Words. By HYDE CLARKE, D.C.L. 3s. 6d.; cloth boards, 4s. 6d.; complete with the GRAMMAR, cloth bds., 5s. 6d.

48. **Composition and Punctuation,** familiarly Explained for those who have neglected the Study of Grammar. By JUSTIN BRENAN. 16th Edition. 1s.

49. **Derivative Spelling-Book:** Giving the Origin of Every Word from the Greek, Latin, Saxon, German, Teutonic, Dutch, French, Spanish, and other Languages; with their present Acceptation and Pronunciation. By J. ROWBOTHAM, F.R.A.S. Improved Edition. 1s. 6d.

51. **The Art of Extempore Speaking:** Hints for the Pulpit, the Senate, and the Bar. By M. BAUTAIN, Vicar-General and Professor at the Sorbonne. Translated from the French. Sixth Edition, carefully corrected. 2s. 6d.

52. **Mining and Quarrying,** with the Sciences connected therewith. First Book of, for Schools. By J. H. COLLINS, F.G.S., Lecturer to the Miners' Association of Cornwall and Devon. 1s.

53. **Places and Facts in Political and Physical Geography,** for Candidates in Public and Private Examinations. By the Rev. EDGAR RAND, B.A. 1s.

54. **Analytical Chemistry,** Qualitative and Quantitative, a Course of. To which is prefixed a Brief Treatise upon Modern Chemical Nomenclature and Notation. By WM. W. PINK, Practical Chemist, &c., and GEORGE E. WEBSTER, Lecturer on Metallurgy and the Applied Sciences, Nottingham. 2s.

THE SCHOOL MANAGERS' SERIES OF READING BOOKS,

Adapted to the Requirements of the New Code. Edited by the Rev. A. R. GRANT, Rector of Hitcham, and Honorary Canon of Ely; formerly H.M. Inspector of Schools.

INTRODUCTORY PRIMER,

	s. d.		s. d.
FIRST STANDARD	0 6	FOURTH STANDARD	1 2
SECOND „	0 10	FIFTH „	1 6
THIRD „	1 0	SIXTH „	1 6

LESSONS FROM THE BIBLE. Part I. Old Testament. 1s.
LESSONS FROM THE BIBLE. Part II. New Testament, to which is added THE GEOGRAPHY OF THE BIBLE, for very young Children. By Rev. C. THORNTON FORSTER. 1s. 2d. *** Or the Two Parts in One Volume. 2s.

FRENCH.

24. **French Grammar.** With Complete and Concise Rules on the Genders of French Nouns. By G. L. STRAUSS, Ph.D. 1s. 6d.

25. **French-English Dictionary.** Comprising a large number of New Terms used in Engineering, Mining, on Railways, &c. By ALFRED ELWES. 1s. 6d.

26. **English-French Dictionary.** By ALFRED ELWES. 2s.

25,26. **French Dictionary** (as above). Complete, in One Vol., 3s.; cloth boards, 3s. 6d. *** Or with the GRAMMAR, cloth boards, 4s. 6d.

French, *continued.*

47. **French and English Phrase Book:** containing Introductory Lessons, with Translations, for the convenience of Students; several Vocabularies of Words, a Collection of suitable Phrases, and Easy Familiar Dialogues. 1s.

GERMAN.

39. **German Grammar.** Adapted for English Students, from Heyse's Theoretical and Practical Grammar, by Dr. G. L. STRAUSS. 1s.

40. **German Reader:** A Series of Extracts, carefully culled from the most approved Authors of Germany; with Notes, Philological and Explanatory. By G. L. STRAUSS, Ph.D. 1s.

41. **German Triglot Dictionary.** By NICHOLAS ESTERHAZY, S. A. HAMILTON. Part I. English-German-French. 1s.

42. **German Triglot Dictionary.** Part II. German-French-English. 1s.

43. **German Triglot Dictionary.** Part III. French-German-English. 1s.

41-43. **German Triglot Dictionary** (as above), in One Vol., 3s.; cloth boards, 4s. *₊* Or with the GERMAN GRAMMAR, cloth boards, 5s.

ITALIAN.

27. **Italian Grammar,** arranged in Twenty Lessons, with a Course of Exercises. By ALFRED ELWES. 1s.

28. **Italian Triglot Dictionary,** wherein the Genders of all the Italian and French Nouns are carefully noted down. By ALFRED ELWES. Vol. 1. Italian-English-French. 2s.

30. **Italian Triglot Dictionary.** By A. ELWES. Vol. 2. English-French-Italian. 2s.

32. **Italian Triglot Dictionary.** By ALFRED ELWES. Vol. 3. French-Italian-English. 2s.

28,30, **Italian Triglot Dictionary** (as above). In One Vol., 6s.
32.   cloth boards, 7s. 6d. *₊* Or with the ITALIAN GRAMMAR, cloth bds., 8s. 6d.

SPANISH AND PORTUGUESE.

34. **Spanish Grammar,** in a Simple and Practical Form. With a Course of Exercises. By ALFRED ELWES. 1s. 6d.

35. **Spanish-English and English-Spanish Dictionary.** Including a large number of Technical Terms used in Mining, Engineering, &c., with the proper Accents and the Gender of every Noun. By ALFRED ELWES. 4s.; cloth boards, 5s. *₊* Or with the GRAMMAR, cloth boards, 6s.

55. **Portuguese Grammar,** in a Simple and Practical Form. With a Course of Exercises. By ALFRED ELWES, Author of "A Spanish Grammar," &c. 1s. 6d. [*Just published.*]

HEBREW.

46*. **Hebrew Grammar.** By Dr. BRESSLAU. 1s. 6d.

44. **Hebrew and English Dictionary,** Biblical and Rabbinical; containing the Hebrew and Chaldee Roots of the Old Testament Post-Rabbinical Writings. By Dr. BRESSLAU. 6s. *₊* Or with the GRAMMAR, 7s.

46. **English and Hebrew Dictionary.** By Dr. BRESSLAU. 3s.

44,46. **Hebrew Dictionary** (as above), in Two Vols., complete, with 46*. the GRAMMAR, cloth boards, 12s.

LONDON : CROSBY LOCKWOOD AND CO.,

LATIN.

19. **Latin Grammar.** Containing the Inflections and Elementary Principles of Translation and Construction. By the Rev. Thomas Goodwin, M.A., Head Master of the Greenwich Proprietary School. 1s.

20. **Latin–English Dictionary.** By the Rev. Thomas Goodwin, M.A. 2s.

22. **English–Latin Dictionary;** together with an Appendix of French and Italian Words which have their origin from the Latin. By the Rev. Thomas Goodwin, M.A. 1s. 6d.

20,22. **Latin Dictionary** (as above). Complete in One Vol., 3s. 6d.; cloth boards, 4s. 6d. •‚• Or with the Grammar, cloth boards, 5s. 6d.

LATIN CLASSICS. With Explanatory Notes in English.

1. **Latin Delectus.** Containing Extracts from Classical Authors, with Genealogical Vocabularies and Explanatory Notes, by H. Young. 1s.

2. **Cæsaris Commentarii de Bello Gallico.** Notes, and a Geographical Register for the Use of Schools, by H. Young. 2s.

12. **Ciceronis Oratio pro Sexto Roscio Amerino.** Edited, with an Introduction, Analysis, and Notes Explanatory and Critical, by the Rev. James Davies, M.A. 1s.

13. **Ciceronis Orationes in Catilinam, Verrem, et pro Archia.** With, Introduction, Analysis, and Notes Explanatory and Critical, by Rev. T. H. L. Leary, D.C.L. formerly Scholar of Brasenose College, Oxford. 1s. 6d. *[Just published.]*

14. **Ciceronis Cato Major, Lælius, Brutus, sive de Senectute, de Amicitia, de Claris Oratoribus Dialogi.** With Notes by W. Brownrigg Smith, M.A., F.R.G.S. 2s.

3. **Cornelius Nepos.** With Notes. By H. Young. 1s.

6. **Horace;** Odes, Epode, and Carmen Sæculare. Notes by H. Young. 1s. 6d.

7. **Horace;** Satires, Epistles, and Ars Poetica. Notes by W. Brownrigg Smith, M.A., F.R.G.S. 1s. 6d.

21. **Juvenalis Satiræ.** With Prolegomena and Notes by T. H. S. Escott, B.A., Lecturer on Logic at King's College, London. 2s.

16. **Livy:** History of Rome. Notes by H. Young and W. B. Smith, M.A. Part 1. Books i., ii., 1s. 6d.

16*. ——— Part 2. Books iii., iv., v., 1s. 6d.

17. ——— Part 3 Books xxi., xxii., 1s. 6d.

8. **Sallustii Crispi Catalina et Bellum Jugurthinum.** Notes Critical and Explanatory, by W. M. Donne, B.A., Trin. Coll., Cam. 1s. 6d.

10. **Terentii Adelphi, Hecyra, Phormio.** Edited, with Notes, Critical and Explanatory, by the Rev. James Davies, M.A. 2s.

9. **Terentii Andria et Heautontimorumenos.** With Notes, Critical and Explanatory, by the Rev. James Davies, M.A. 1s. 6d.

11. **Terentii Eunuchus, Comœdia.** Notes, by Rev. J. Davies, M.A. 1s. 6d.

4. **Virgilii Maronis Bucolica et Georgica.** With Notes on the Bucolics by W. Rushton, M.A., and on the Georgics by H. Young. 1s. 6d.

5. **Virgilii Maronis Æneis.** With Notes, Critical and Explanatory, by H. Young. New Edition, revised and improved. With copious Additional Notes by Rev. T. H. L. Leary, D.C.L., formerly Scholar of Brasenose College, Oxford. 3s. *[Just published.]*

5*. ——— Part 1. Books i.—vi., 1s. 6d. *[Just published.]*

5** ——— Part 2. Books vii.—xii., 2s. *[Just published.]*

19. **Latin Verse Selections,** from Catullus, Tibullus, Propertius, and Ovid. Notes by W. B. Donne, M.A., Trinity College, Cambridge. 2s.

20. **Latin Prose Selections,** from Varro, Columella, Vitruvius, Seneca, Quintilian, Florus, Velleius Paterculus, Valerius Maximus Suetonius, Apuleius, &c. Notes by W. B. Donne, M.A. 2s.

GREEK.

14. **Greek Grammar,** in accordance with the Principles and Philological Researches of the most eminent Scholars of our own day. By HANS
CLAUDE HAMILTON. 1s. 6d.

15,17. **Greek Lexicon.** Containing all the Words in General Use, with
their Significations, Inflections, and Doubtful Quantities. By HENRY R.
HAMILTON. Vol. 1. Greek-English, 2s.; Vol. 2. English-Greek, 2s. Or the
Two Vols. in One, 4s.: cloth boards, 5s.

14.15. **Greek Lexicon** (as above). Complete, with the GRAMMAR, in
17. One Vol., cloth boards, 6s.

GREEK CLASSICS. With Explanatory Notes in English.

1. **Greek Delectus.** Containing Extracts from Classical Authors,
with Genealogical Vocabularies and Explanatory Notes, by H. YOUNG. New
Edition, with an improved and enlarged Supplementary Vocabulary, by JOHN
HUTCHISON, M.A., of the High School, Glasgow. 1s. 6d.

30. **Æschylus:** Prometheus Vinctus: The Prometheus Bound. From
the Text of DINDORF. Edited, with English Notes, Critical and Explanatory,
by the Rev. JAMES DAVIES, M.A. 1s.

32. **Æschylus:** Septem Contra Thebes: The Seven against Thebes.
From the Text of DINDORF. Edited, with English Notes, Critical and Explanatory, by the Rev. JAMES DAVIES, M.A. 1s.

40. **Aristophanes:** Acharnians. Chiefly from the Text of C. H.
WEISE. With Notes, by C. S. T. TOWNSHEND, M.A. 1s. 6d.

26. **Euripides:** Alcestis. Chiefly from the Text of DINDORF. With
Notes, Critical and Explanatory, by JOHN MILNER, B.A. 1s. 6d.

23. **Euripides:** Hecuba and Medea. Chiefly from the Text of DIN-
DORF. With Notes, Critical and Explanatory, by W. BROWNRIGG SMITH,
M.A., F.R.G.S. 1s. 6d.

4-17. **Herodotus,** The History of, chiefly after the Text of GAISFORD.
With Preliminary Observations and Appendices, and Notes, Critical and
Explanatory, by T. H. L. LEARY, M.A., D.C.L.
    Part 1. Books i., ii. (The Clio and Euterpe), 2s.
    Part 2. Books iii., iv. (The Thalia and Melpomene), 2s.
    Part 3. Books v.-vii. (The Terpsichore, Erato, and Polymnia), 2s.
    Part 4. Books viii., ix. (The Urania and Calliope) and Index, 1s. 6d.

5-12. **Homer,** The Works of. According to the Text of BAEUMLEIN.
With Notes, Critical and Explanatory, drawn from the best and latest
Authorities, with Preliminary Observations and Appendices, by T. H. L.
LEARY, M.A., D.C.L.

THE ILIAD:	Part 1. Books i. to vi., 1s.6d.	Part 3. Books xiii. to xviii., 1s. 6d.
	Part 2. Books vii. to xii., 1s.6d.	Part 4. Books xix. to xxiv., 1s. 6d.
THE ODYSSEY:	Part 1. Books i. to vi., 1s.6d	Part 3. Books xiii. to xviii., 1s. 6d.
	Part 2. Books vii. to xii., 1s.6d.	Part 4. Books xix. to xxiv., and Hymns, 2s.

4. **Lucian's Select Dialogues.** The Text carefully revised, with
Grammatical and Explanatory Notes, by H. YOUNG. 1s.

13. **Plato's Dialogues:** The Apology of Socrates, the Crito, and
the Phædo. From the Text of C. F. HERMANN. Edited with Notes, Critical
and Explanatory, by the Rev. JAMES DAVIES, M.A. 2s.

18. **Sophocles:** Œdipus Tyrannus. Notes by H. YOUNG. 1s.

20. **Sophocles:** Antigone. From the Text of DINDORF. Notes,
Critical and Explanatory, by the Rev. JOHN MILNER, B.A. 2s.

41. **Thucydides:** History of the Peloponnesian War. Notes by H.
YOUNG. Book 1. 1s.

2, 3. **Xenophon's Anabasis;** or, The Retreat of the Ten Thousand.
Notes and a Geographical Register, by H. YOUNG. Part 1. Books i. to iii.,
1s. Part 2. Books iv. to vii., 1s.

42. **Xenophon's Panegyric on Agesilaus.** Notes and Introduction by LL. F. W. JEWITT. 1s. 6d.

43. **Demosthenes.** The Oration on the Crown and the Philippics.
With English Notes. By Rev. T. H. L. LEARY, D.C.L., formerly Scholar of
Brasenose College, Oxford. 1s. 6d.   [*Just Published.*

LONDON, *April*, 1878.

A Catalogue of Books ·

INCLUDING MANY

NEW & STANDARD WORKS

IN

ENGINEERING, ARCHITECTURE, AGRICULTURE, MATHEMATICS, MECHANICS, SCIENCE, &c. &c.

PUBLISHED BY

CROSBY LOCKWOOD & CO.,

7, STATIONERS'-HALL COURT, LUDGATE HILL, E.C.

ENGINEERING, SURVEYING, &c.

Humber's New Work on Water-Supply.

A COMPREHENSIVE TREATISE on the WATER-SUPPLY of CITIES and TOWNS. By WILLIAM HUMBER, Assoc. Inst. C.E., and M. Inst. M.E. Author of "Cast and Wrought Iron Bridge Construction," &c. &c. Imp. 4to. Illustrated with 50 Double Plates, 2 Single Plates, Coloured Frontispiece, and upwards of 250 Woodcuts, and containing 400 pages of Text, elegantly and substantially half-bound in morocco. 6*l.* 6*s.*

List of Contents:—

I. Historical Sketch of some of the means that have been adopted for the Supply of Water to Cities and Towns.—II. Water and the Foreign Matter usually associated with it.—III. Rainfall and Evaporation.—IV. Springs and the water-bearing formations of various districts.—V. Measurement and Estimation of the Flow of Water.—VI. On the Selection of the Source of Supply.—VII. Wells.—VIII. Reservoirs.—IX. The Purification of Water.—X. Pumps.—XI. Pumping Machinery.—XII. Conduits.—XIII. Distribution of Water.—XIV. Meters, Service Pipes, and House Fittings.—XV. The Law and Economy of Water Works.—XVI. Constant and Intermittent Supply.—XVII. Description of Plates.—Appendices, giving Tables of Rates of Supply, Velocities, &c. &c., together with Specifications of several Works illustrated, among which will be found :—Aberdeen, Bideford, Canterbury, Dundee, Halifax, Lambeth, Rotherham, Dublin, and others.

OPINIONS OF THE PRESS.

"The most systematic and valuable work upon water supply hitherto produced in English, or in any other language."—*Engineer* (first notice).

"Mr. Humber's work is characterised almost throughout by an exhaustiveness much more distinctive of French and German than of English technical treatises."—*Engineer* (third notice).

"We can congratulate Mr. Humber on having been able to give so large an amount of information on a subject so important as the water supply of cities and towns. The plates, fifty in number, are mostly drawings of executed works, and alone would have commanded the attention of every engineer whose practice may lie in this branch of the profession."—*Builder.*

Humber's Modern Engineering.   First Series.

A RECORD of the PROGRESS of MODERN ENGINEER-
ING, 1863. Comprising Civil, Mechanical, Marine, Hydraulic,
Railway, Bridge, and other Engineering Works, &c. By WILLIAM
HUMBER, Assoc. Inst. C.E., &c.   Imp. 4to, with 36 Double
Plates, drawn to a large scale, and Photographic Portrait of John
Hawkshaw, C.E., F.R.S., &c.   3*l.* 3*s.* half morocco.

List of the Plates.

NAME AND DESCRIPTION.	PLATES.	NAME OF ENGINEER.
Victoria Station and Roof—L. B.& S. C. Rail	1 to 8	Mr. R. Jacomb Hood, C.E.
Southport Pier	9 and 10	Mr. James Brunlees, C.E.
Victoria Station and Roof—L. C. & D. & G. W. Railways	11 to 15A	Mr. John Fowler, C.E.
Roof of Cremorne Music Hall	16	Mr. William Humber, C.E.
Bridge over G. N. Railway	17	Mr. Joseph Cubitt, C.E.
Roof of Station—Dutch Rhenish Railway	18 and 19	Mr. Euschedi, C.E.
Bridge over the Thames—West London Extension Railway	20 to 24	Mr. William Baker, C.E.
Armour Plates	25	Mr. James Chalmers, C.E.
Suspension Bridge, Thames	26 to 29	Mr. Peter W. Barlow, C.E.
The Allen Engine	30	Mr. G. T. Porter, M.E.
Suspension Bridge, Avon	31 to 33	Mr. John Hawkshaw, C.E. and W. H. Barlow, C.E.
Underground Railway	34 to 36	Mr. John Fowler, C.E.

With copious Descriptive Letterpress, Specifications, &c.

" Handsomely lithographed and printed. It will find favour with many who desire
to preserve in a permanent form copies of the plans and specifications prepared for the
guidance of the contractors for many important engineering works."—*Engineer.*

Humber's Modern Engineering.   Second Series.

A RECORD of the PROGRESS of MODERN ENGINEER-
ING, 1864 ; with Photographic Portrait of Robert Stephenson,
C.E., M.P., F.R.S., &c.   3*l.* 3*s.* half morocco.

List of the Plates.

NAME AND DESCRIPTION.	PLATES.	NAME OF ENGINEER.
Birkenhead Docks, Low Water Basin	1 to 15	Mr. G. F. Lyster, C.E.
Charing Cross Station Roof—C. C. Railway.	16 to 18	Mr. Hawkshaw, C.E.
Digswell Viaduct—Great Northern Railway.	19	Mr. J. Cubitt, C.E.
Robbery Wood Viaduct—Great N. Railway.	20	Mr. J. Cubitt, C.E.
Iron Permanent Way	20a	
Clydach Viaduct—Merthyr, Tredegar, and Abergavenny Railway	21	Mr. Gardner, C.E.
Ebbw Viaduct ditto ditto ditto	22	Mr. Gardner, C.E.
College Wood Viaduct—Cornwall Railway	23	Mr. Brunel.
Dublin Winter Palace Roof	24 to 26	Messrs. Ordish & Le Feuvre.
Bridge over the Thames—L. C. & D. Railw.	27 to 32	Mr. J. Cubitt, C.E.
Albert Harbour, Greenock	33 to 36	Messrs. Bell & Miller.

With copious Descriptive Letterpress, Specifications, &c.

"A *résumé* of all the more interesting and important works lately completed in Great
Britain ; and containing, as it does, carefully executed drawings, with full working
details, it will be found a valuable accessory to the profession at large."—*Engineer.*

" Mr. Humber has done the profession good and true service, by the fine collection
of examples he has here brought before the profession and the public."—*Practical
Mechanics' Journal.*

Humber's Modern Engineering. Third Series.

A RECORD of the PROGRESS of MODERN ENGINEER-
ING, 1865. Imp. 4to, with 40 Double Plates, drawn to a large
scale, and Photo Portrait of J. R. M'Clean, Esq., late President
of the Institution of Civil Engineers. 3*l*. 3*s*. half morocco.

List of Plates and Diagrams.

MAIN DRAINAGE, METROPOLIS.

NORTH SIDE.

Plate 1. Map showing Interception of
Sewers.—2 and 3. Middle Level Sewer.
Sewer under Regent's Canal; and Junc-
tion with Fleet Ditch.—4, 5, and 6. Out-
fall Sewer. Bridge over River Lea.
Elevation and Details. — 7. Outfall
Sewer. Bridge over Marsh Lane, North
Woolwich Railway, and Bow and Barking
Railway Junction.—8, 9, and 10. Outfall
Sewer. Bridge over Bow and Barking
Railway. Elevation and Details.—
11 and 12. Outfall Sewer. Bridge over
East London Waterworks' Feeder. Ele-
vation and Details.—13 and 14. Outfall
Sewer. Reservoir. Plan and Section.—
15. Outfall Sewer. Tumbling Bay and
Outlet.—16. Outfall Sewer. Penstocks.

SOUTH SIDE.

Plates 17 and 18. Outfall Sewer. Ber-
mondsey Branch.—19, 20, 21, and 22.

MAIN DRAINAGE, METROPOLIS,

continued—

Outfall Sewer. Reservoir and Outlet.
Plan and Details.—23. Outfall Sewer.
Filth Hoist —24. Sections of Sewers
(North and South Sides).

THAMES EMBANKMENT.

Plate 25. Section of River Wall.—
26 and 27. Steam-boat Pier, Westminster.
Elevation and Details. — 28. Landing
Stairs between Charing Cross and Water-
loo Bridges.—29 and 30. York Gate.
Front Elevation. Side Elevation and
Details.—31, 32, and 33. Overflow and
Outlet at Savoy Street Sewer. Details ;
and Penstock.—34, 35, and 36. Steam-boat
Pier, Waterloo Bridge. Elevation and
Details.—37. Junction of Sewers. Plans
and Sections —38. Gullies. Plans and
Sections.—39. Rolling Stock.—40. Granite
and Iron Forts.

With copious Descriptive Letterpress, &c.

Humber's Modern Engineering. Fourth Series.

A RECORD of the PROGRESS of MODERN ENGINEER-
ING, 1866. Imp. 4to, with 36 Double Plates, drawn to a large
scale, and Photographic Portrait of John Fowler, Esq., President
of the Institution of Civil Engineers. 3*l*. 3*s*. half morocco.

List of the Plates and Diagrams.

NAME AND DESCRIPTION.	PLATES.	NAME OF ENGINEER.
Abbey Mills Pumping Station, Main Drainage, Metropolis.	1 to 4	Mr. Bazalgette, C.E.
Barrow Docks	5 to 9	Messrs. M'Clean & Stillman, [C. E.
Manquis Viaduct, Santiago and Valparaiso Railway	10, 11	Mr. W. Loyd, C.E.
Adams' Locomotive, St. Helen's Canal Railw.	12, 13	Mr. H. Cross, C.E.
Cannon Street Station Roof	14 to 16	Mr. J. Hawkshaw, C.E.
Road Bridge over the River Moka.	17, 18	Mr. H. Wakefield, C.E.
Telegraphic Apparatus for Mesopotamia	19	Mr. Siemens, C.E.
Viaduct over the River Wye, Midland Railw.	20 to 22	Mr. W. H. Barlow, C.E.
St. Germans Viaduct, Cornwall Railway	23, 24	Mr. Brunel, C.E.
Wrought-Iron Cylinder for Diving Bell	25	Mr. J. Coode, C.E.
Millwall Docks	26 to 31	Messrs. J. Fowler, C.E., an William Wilson, C.E.
Milroy's Patent Excavator	32	Mr. Milroy, C.E.
Metropolitan District Railway	33 to 38	Mr. J. Fowler, and Mr. T M. Johnson, C.E.
Harbours, Ports, and Breakwaters	A to C	

With Copious Descriptive Letterpress, Specifications, &c.

Humber's Great Work on Bridge Construction.

A COMPLETE and PRACTICAL TREATISE on CAST and WROUGHT-IRON BRIDGE CONSTRUCTION, including Iron Foundations. In Three Parts—Theoretical, Practical, and Descriptive. By WILLIAM HUMBER, Assoc. Inst. C. E., and M. Inst. M. E. Third Edition, revised and much improved, with 115 Double Plates (20 of which now first appear in this edition), and numerous additions to the Text. In 2 vols. imp. 4to, 6l. 16s. 6d. half-bound in morocco.

. "A very valuable contribution to the standard literature of civil engineering. In addition to elevations, plans, and sections, large scale details are given, which very much enhance the instructive worth of these illustrations. No engineer would willingly be without so valuable a fund of information."—*Civil Engineer and Architect's Journal.*

"Mr. Humber's stately volumes lately issued—in which the most important bridges erected during the last five years, under the direction of our most eminent engineers, are drawn and specified in great detail."—*Engineer.*

"A book—and particularly a large and costly treatise like Mr. Humber's—which has reached its third edition may certainly be said to have established its own reputation."—*Engineering.*

Strains, Formulæ & Diagrams for Calculation of.

A HANDY BOOK for the CALCULATION of STRAINS in GIRDERS and SIMILAR STRUCTURES, and their STRENGTH ; consisting of Formulæ and Corresponding Diagrams, with numerous Details for Practical Application, &c. By WILLIAM HUMBER, Assoc. Inst. C. E., &c. Second Edition. Fcap. 8vo, with nearly 100 Woodcuts and 3 Plates, 7s. 6d. cloth.

"The arrangement of the matter in this little volume is as convenient as it well could be. . . . . The system of employing diagrams as a substitute for complex computations is one justly coming into great favour, and in that respect Mr. Humber's volume is fully up to the times."—*Engineering.*

"The formulæ are neatly expressed, and the diagrams good."—*Athenæum.*

"Mr. Humber has rendered a great service to the architect and engineer by producing a work especially treating on the methods of delineating the strains on iron beams, roofs, and bridges by means of diagrams."—*Builder.*

Barlow on the Strength of Materials, enlarged.

A TREATISE ON THE STRENGTH OF MATERIALS, with Rules for application in Architecture, the Construction of Suspension Bridges, Railways, &c. ; and an Appendix on the Power of Locomotive Engines, and the effect of Inclined Planes and Gradients. By PETER BARLOW, F.R.S. A New Edition, revised by his Sons, P. W. BARLOW, F.R.S., and W. H. BARLOW, F.R.S., to which are added Experiments by HODGKINSON, FAIRBAIRN, and KIRKALDY ; an Essay (with Illustrations) on the effect produced by passing Weights over Elastic Bars, by the Rev. ROBERT WILLIS, M.A., F.R.S. And Formulæ for Calculating Girders, &c. The whole arranged and edited by W. HUMBER, Assoc. Inst. C.E., Author of "A Complete and Practical Treatise on Cast and Wrought-Iron Bridge Construction," &c. 8vo, 400 pp., with 19 large Plates, and numerous woodcuts, 18s. cloth.

"The book is undoubtedly worthy of the highest commendation."—*Mining Journal.*

"The best book on the subject which has yet appeared. . . . . We know of no work that so completely fulfils its mission."—*English Mechanic.*

"The standard treatise upon this particular subject."—*Engineer.*

Tramways and Tram-Traffic.

TRAMWAYS: their CONSTRUCTION and WORKING.
Containing a Comprehensive History of the System; an exhaustive Analysis of the Various Modes of Traction, including Horse Power, Steam, Heated Water, and Compressed Air; a Description of the varieties of Rolling Stock; and ample Details of Cost and Working Expenses, with Special reference to the Tramways of the United Kingdom. By D. KINNEAR CLARK, M. I. C. E., Author of 'Railway Machinery,' &c., in one vol. 8vo, with numerous illustrations and thirteen folding plates, 18*s.* cloth. [*Just published.*

Iron and Steel.

'IRON AND STEEL': a Work for the Forge, Foundry, Factory, and Office. Containing Ready, Useful, and Trustworthy Information for Ironmasters and their Stocktakers; Managers of Bar, Rail, Plate, and Sheet Rolling Mills; Iron and Metal Founders; Iron Ship and Bridge Builders; Mechanical, Mining, and Consulting Engineers; Architects, Contractors, Builders, and Professional Draughtsmen. By CHARLES HOARE, Author of 'The Slide Rule,' &c. Eighth Edition. Revised throughout and considerably enlarged. With folding Scales of "Foreign Measures compared with the English Foot," and "fixed Scales of Squares, Cubes, and Roots, Areas, Decimal Equivalents, &c." Oblong, 32mo, leather elastic-band, 6*s.*

"We cordially recommend this book to those engaged in considering the details of all kinds of iron and steel works. . . . . It has been compiled with care and accuracy. . . . . Many useful rules and hints are given for lessening the amount of arithmetical labour which is always more or less necessary in arranging iron and steel work of all kinds, and a great quantity of useful tables for preparing estimates of weights, dimensions, strengths of structures, costs of work, &c., will be found in Mr. Hoare's book.—*Naval Science.*

Weale's Engineer's Pocket-Book.

THE ENGINEER'S, ARCHITECT'S, and CONTRACTOR'S POCKET-BOOK (LOCKWOOD & Co.'s; formerly WEALE's). Published Annually. In roan tuck, gilt edges, with 10 Copper-Plates and numerous Woodcuts. 6*s.*

"A vast amount of really valuable matter condensed into the small dimensions of a book which is, in reality, what it professes to be—a pocket-book. . . . We cordially recommend the book.—*Colliery Guardian.*
"It contains a large amount of information peculiarly valuable to those for whose use it is compiled. We cordially commend it to the engineering and architectural professions generally."—*Mining Journal.*

Iron Bridges, Girders, Roofs, &c.

A TREATISE on the APPLICATION of IRON to the CONSTRUCTION of BRIDGES, GIRDERS, ROOFS, and OTHER WORKS; showing the Principles upon which such Structures are Designed, and their Practical Application. Especially arranged for the use of Students and Practical Mechanics, all Mathematical Formulæ and Symbols being excluded. By FRANCIS CAMPIN, C.E. Second Edition revised and corrected. With numerous Diagrams. 12mo, cloth boards, 3*s.*.

"Invaluable to those who have not been educated in mathematics."—*Colliery Guardian.*
"Remarkably accurate and well written."—*Artisan.*

Pioneer Engineering.

PIONEER ENGINEERING. A Treatise on the Engineering Operations connected with the Settlement of Waste Lands in New Countries. By EDWARD DOBSON, Assoc. Inst. C.E., Author of "The Art of Building," &c. With numerous Plates and Wood Engravings. Crown 8vo, 10s. 6d., cloth.   [*Just published*.

"A most useful handbook to engineering pioneers."—*Iron.*
"The author's experience has been turned to good account, and the book is likely to be of considerable service to pioneer engineers."—*Building News.*

New Iron Trades' Companion.

THE IRON AND METAL TRADES' COMPANION: Being a Calculator containing a Series of Tables upon a new and comprehensive plan for expeditiously ascertaining the value of any goods bought or sold by weight, from 1s. per cwt. to 112s. per cwt., and from one farthing per pound to one shilling per pound. Each Table extends from one pound to 100 tons; to which are appended Rules on Decimals, Square and Cube Root, Mensuration of Superficies and Solids, &c.; also Tables of Weights of Materials, and other Useful Memoranda. By THOMAS DOWNIE. Strongly bound in leather, 396 pp., 9s.

"A most useful set of tables, and will supply a want, for nothing like them before existed."—*Building News.*
"Will save the possessor the trouble of making numerous intricate calculations. Although specially adapted to the iron and metal trades, the tables contained in this handy little companion will be found useful in every other business in which merchandise is bought and sold by weight."—*Railway News.*

Sanitary Work.

SANITARY WORK IN THE SMALLER TOWNS AND IN VILLAGES. Comprising:—1. Some of the more Common Forms of Nuisance and their Remedies; 2. Drainage; 3. Water Supply. A useful book for Members of Local Boards and Rural Sanitary Authorities, Health Officers, Engineers, Surveyors, Builders, and Contractors. By CHARLES SLAGG, Assoc. Inst. C.E. Crown 8vo, 5s., cloth.

"Mr. Slagg has brought together much valuable information, and has a happy lucidity of expression; and he has been industrious in collecting data."—*Athenæum.*
"This is a very useful book, and may be safely recommended. . . . . The author, Mr. Charles Slagg, has had practical experience in the works of which he treats. There is a great deal of work required to be done in the smaller towns and villages, and this little volume will help those who are willing to do it."—*Builder.*

Sanitary Engineering.

WHOLESOME HOUSES: being an exposition of the Banner System of Sanitation. By EDWARD GREGSON BANNER, C.E. New and enlarged edition (25th thousand), illustrated with numerous Wood Engravings. Crown 8vo, sewed 6d., cloth 1s. [*Just published*.

Steam Engine.

STEAM AND THE STEAM ENGINE, Stationary and Portable, an Elementary Treatise on. Being an Extension of Mr. John Sewell's Treatise on Steam. By D. KINNEAR CLARK, C.E., M.I.C.E., Author of "Railway Locomotives," &c. With Illustrations. 12mo, 4s., cloth.

"Every essential part of the subject is treated of competently, and in a popular style."—*Iron.*

Strains.

THE STRAINS ON STRUCTURES OF IRONWORK;
with Practical Remarks on Iron Construction. By F. W. SHEILDS,
M. Inst. C.E. Second Edition, with 5 plates. Royal 8vo, 5s. cloth.

CONTENTS.—Introductory Remarks; Beams Loaded at Centre; Beams Loaded at
unequal distances between supports; Beams uniformly Loaded; Girders with triangu-
lar bracing Loaded at centre; Ditto, Loaded at unequal distances between supports;
Ditto, uniformly Loaded; Calculation of the Strains on Girders with triangular
Basings; Cantilevers; Continuous Girders; Lattice Girders; Girders with Vertical
Struts and Diagonal Ties; Calculation of the Strains on Ditto; Bow and String
Girders; Girders of a form not belonging to any regular figure; Plate Girders; Ap-
portionments of Material to Strain; Comparison of different Girders; Proportion of
Length to Depth of Girders; Character of the Work; Iron Roofs.

Construction of Iron Beams, Pillars, &c.

IRON AND HEAT, Exhibiting the Principles concerned in the
Construction of Iron Beams, Pillars, and Bridge Girders, and the
Action of Heat in the Smelting Furnace. By JAMES ARMOUR,
C.E. Woodcuts, 12mo, cloth boards, 3s. 6d. ; cloth limp, 2s. 6d.

"A very useful and thoroughly practical little volume, in every way deserving of
circulation amongst working men."—*Mining Journal.*
"No ironworker who wishes to acquaint himself with the principles of his own
trade can afford to be without it."—*South Durham Mercury.*

Power in Motion.

POWER IN MOTION : Horse Power, Motion, Toothed Wheel
Gearing, Long and Short Driving Bands, Angular Forces, &c.
By JAMES ARMOUR, C.E. With 73 Diagrams. 12mo, cloth
boards, 3s. 6d.

"Numerous illustrations enable the author to convey his meaning as explicitly as
it is perhaps possible to be conveyed. The value of the theoretic and practical knowledge imparted cannot well be over estimated."—*Newcastle Weekly Chronicle.*

Metallurgy of Iron.

A TREATISE ON THE METALLURGY OF IRON : con-
taining Outlines of the History of Iron Manufacture, Methods of
Assay, and Analyses of Iron Ores, Processes of Manufacture of
Iron and Steel, &c. By H. BAUERMAN, F.G.S., Associate of the
Royal School of Mines. With numerous Illustrations. Fourth
Edition, revised and much enlarged. 12mo, cloth boards, 5s. 6d.

"Carefully written, it has the merit of brevity and conciseness, as to less important
points, while all material matters are very fully and thoroughly entered into."—
Standard.

Trigonometrical Surveying.

AN OUTLINE OF THE METHOD OF CONDUCTING A
TRIGONOMETRICAL SURVEY, for the Formation of Geo
graphical and Topographical Maps and Plans, Military Recon-
naissance, Levelling, &c., with the most useful Problems in Geodesy
and Practical Astronomy, and Formulæ and Tables for Facilitating
their Calculation. By LIEUT-GENERAL FROME, R.E., late In-
spector-General of Fortifications, &c. Fourth Edition, Enlarged,
thoroughly Revised, and partly Re-written. By CAPTAIN CHARLES
WARREN, R.E., F.G.S. With 19 Plates and 115 Woodcuts,
royal 8vo, 16s. cloth.

Practical Tunnelling.

PRACTICAL TUNNELLING : Explaining in detail the Setting out of the Works, Shaft-sinking and Heading-Driving, Ranging the Lines and Levelling under Ground, Sub-Excavating, Timbering, and the Construction of the Brickwork of Tunnels with the amount of labour required for, and the Cost of, the various portions of the work. By FREDERICK WALTER SIMMS, M. Inst. C.E., author of "A Treatise on Levelling." Third Edition, Revised and Extended, with additional chapters illustrating the Recent Practice of Tunnelling as exemplified by the St. Gothard, Mont Cenis, and other modern works, by D. KINNEAR CLARK, M. Inst. C.E. Imp. 8vo, with 21 Folding Plates and numerous Wood Engravings, 30s., cloth.

"It is the only practical treatise on the great art of tunnelling. Mr. Clark's work brings the exigencies of tunnel enterprise up to our own time. The great length of modern tunnels has led to a new difficulty in the art, which the last generation was ignorant of, namely, the difficulty of ventilation. In Mr. Clark's supplement we find this branch of the subject has been fully considered. Mr. Clark's additional chapters on the Mont Cenis and St. Gothard Tunnels contain minute and valuable experiences and data relating to the method of excavation by compressed air, the heading operations, rock-boring machinery, process of enlargement, ventilation in course of construction by compressed air, labour and cost, &c."—*Building News.*
" The estimation in which Mr. Simms' book on tunnelling has been held for over thirty years cannot be more truly expressed than in the words of the late Professor Rankine :—' The best source of information on the subject of tunnels is Mr. F. W. Simms' work on " Practical Tunnelling."'—*The Architect.*

Levelling.

A TREATISE on the PRINCIPLES and PRACTICE of LEVELLING ; showing its Application to Purposes of Railway and Civil Engineering, in the Construction of Roads ; with Mr. TELFORD'S Rules for the same. By FREDERICK W. SIMMS, F.G.S., M. Inst. C.E. Sixth Edition, very carefully revised, with the addition of Mr. LAW'S Practical Examples for Setting out Railway Curves, and Mr. TRAUTWINE'S Field Practice of Laying out Circular Curves. With 7 Plates and numerous Woodcuts. 8vo, 8s. 6d. cloth.   *•* TRAUTWINE on Curves, separate, 5s.

"One of the most important text-books for the general surveyor, and there is scarcely a question connected with levelling for which a solution would be sought but that would be satisfactorily answered by consulting the volume."—*Mining Journal.*
" The text-book on levelling in most of our engineering schools and colleges."—*Engineer.*

The High-Pressure Steam Engine.

THE HIGH-PRESSURE STEAM ENGINE ; an Exposition of its Comparative Merits, and an Essay towards an Improved System of Construction, adapted especially to secure Safety and Economy. By Dr. ERNST ALBAN, Practical Machine Maker, Plau, Mecklenberg. Translated from the German, with Notes, by Dr. POLE, F.R.S., M. Inst. C.E., &c. &c. With 28 fine Plates, 8vo, 16s. 6d. cloth.

" A work like this, which goes thoroughly into the examination of the high-pressure engine, the boiler, and its appendages, &c., is exceedingly useful, and deserves a place a every scientific library."—*Steam Shipping Chronicle.*

Reynolds' Locomotive-Engine Driving.

LOCOMOTIVE-ENGINE DRIVING ; a Practical Manual for
Engineers in charge of Locomotive Engines. By MICHAEL
REYNOLDS, Inspector, Locomotive and Carriage Department,
London, Brighton, and South Coast Railway. Second Edition.
With Illustrations. Crown 8vo, 4s. 6d., cloth. [*Just published.*
" Mr. Reynolds has supplied a want, and has supplied it well. We can confidently
recommend the book not only to the practical driver, but to every one who takes an
interest in the performance of locomotive engines."—*The Engineer.*
"The work is as novel as it is useful, and if drivers and firemen will but take as
much pains in reading it as the author has in writing it, there can be no question as
to the benefit they will derive."—*English Mechanic.*
" Mr. Reynolds has opened a new chapter in the literature of the day. This
admirable practical treatise, of the practical utility of which we have to speak in
terms of warm commendation."—*Athenæum.*

Hydraulics.

HYDRAULIC TABLES, CO-EFFICIENTS, and FORMULÆ
for finding the Discharge of Water from Orifices, Notches, Weirs,
Pipes, and Rivers. With New Formulæ, Tables,. and General
Information on Rain-fall, Catchment-Basins, Drainage, Sewerage,
Water Supply for Towns and Mill Power. By JOHN NEVILLE,
Civil Engineer, M.R.I.A. Third Edition, carefully revised, with
considerable Additions. Numerous Illustrations. Cr. 8vo, 14s. cloth.
" Undoubtedly an exceedingly useful and elaborate compilation."—*Iron.*
" Alike valuable to students and engineers in practice."—*Mining Journal.*

Strength of Cast Iron, &c.

A PRACTICAL ESSAY on the STRENGTH of CAST IRON
and OTHER METALS. By the late THOMAS TREDGOLD, Mem.
Inst. C.E., Author of "Elementary Principles of Carpentry," &c.
Fifth Edition, Edited by EATON HODGKINSON, F.R.S. ; to
which are added EXPERIMENTAL RESEARCHES on the
STRENGTH and OTHER PROPERTIES of CAST IRON.
By the EDITOR. The whole Illustrated with 9 Engravings and
numerous Woodcuts. 8vo, 12s. cloth.
₊ HODGKINSON ON CAST IRON, separately. Price 6s. cloth.

Steam Boilers.

A TREATISE ON STEAM BOILERS : their Strength, Con-
struction, and Economical Working. By R. WILSON, A.I.C.E.
Fourth Edition. 12mo, 6s., cloth.
" The best work on boilers which has come under our notice."—*Engineering.*
" The best treatise that has ever been published on steam boilers."—*Engineer.*

Tables of Curves.

TABLES OF TANGENTIAL ANGLES and MULTIPLES
for setting out Curves from 5 to 200 Radius. By ALEXANDER
BEAZELEY, M. Inst. C.E. Printed on 48 Cards, and sold in a
cloth box, waistcoat-pocket size, 3s. 6d.
" Each table is printed on a small card, which, being placed on the theodolite, leaves
the hands free to manipulate the instrument—no small advantage as regards the rapidity
of work. They are clearly printed, and compactly fitted into a small case for the
pocket—an arrangement that will recommend them to all practical men."—*Engineer.*
" Very handy : a man may know that all his day's work must fall on two of these
cards, which he puts into his own card-case, and leaves the rest behind."—*Athenæum.*

Slate and Slate Quarrying.

A TREATISE ON SLATE AND SLATE QUARRYING, Scientific, Practical, and Commercial. By D. C. DAVIES, F.G.S., Mining Engineer, &c. With numerous Illustrations and Folding Plates. Crown 8vo, 6s., cloth. [*Just published.*

"A useful and practical hand-book on an important industry."—*Engineering.*
" A useful embodiment of practical information derived from original sources, combined with a digest of everything that has already appeared likely to be of interest."—*Building News.*
" There is no other book which contains so much information concerning the procedure observed in taking quarries, the processes employed in working them, and such full statistics of the present and past position of the great slate trade of Wales."—*The Architect.*

Earthwork.

EARTHWORK TABLES, showing the Contents in Cubic Yards of Embankments, Cuttings, &c., of Heights or Depths up to an average of 80 feet. By JOSEPH BROADBENT, C. E., and FRANCIS CAMPIN, C.E. Cr. 8vo, oblong, 5s. cloth.

"The way in which accuracy is attained, by a simple division of each cross section into three elements, two of which are constant and one variable, is ingenious."—*Athenæum.*
" Cannot fail to come into general use."—*Mining Journal.*

Surveying (Land and Marine).

LAND AND MARINE SURVEYING, In Reference to the Preparation of Plans for Roads and Railways, Canals, Rivers, Towns' Water Supplies, Docks and Harbours ; with Description and Use of Surveying Instruments. By W. DAVIS HASKOLL, C.E. 8vo, 12s. 6d. cloth, with 14 folding Plates, and numerous Woodcuts.

"A most useful and well arranged book for the aid of a student."—*Builder.*
" Cannot fail to prove of the utmost practical utility, and may be safely recommended to all students who aspire to become clean and expert surveyors."—*Mining Journal.*

Engineering Fieldwork.

THE PRACTICE OF ENGINEERING FIELDWORK, applied to Land and Hydraulic, Hydrographic, and Submarine Surveying and Levelling. Second Edition, revised, with considerable additions, and a Supplementary Volume on WATER-WORKS, SEWERS, SEWAGE, and IRRIGATION. By W. DAVIS HASKOLL, C.E. Numerous folding Plates. Demy 8vo, 2 vols. in one, cloth boards, 1l. 1s. (published at 2l. 4s.)

Mining, Surveying and Valuing.

THE MINERAL SURVEYOR AND VALUER'S COMPLETE GUIDE, comprising a Treatise on Improved Mining Surveying, with new Traverse Tables ; and Descriptions of Improved Instruments ; also an Exposition of the Correct Principles of Laying out and Valuing Home and Foreign Iron and Coal Mineral Properties. By WILLIAM LINTERN, Mining and Civil Engineer. With four Plates of Diagrams, Plans, &c., 12mo, 4s., cloth.

"Contains much valuable information given in a small compass, and which, as far as we have tested it, is thoroughly trustworthy."—*Iron and Coal Trades Review.*

*** The above, bound with THOMAN'S TABLES. (See page 22). Price 7s. 6d., cloth.

Fire Engineering.

FIRES, FIRE-ENGINES, AND FIRE BRIGADES. With a History of Fire-Engines, their Construction, Use, and Management ; Remarks on Fire-Proof Buildings, and the Preservation of Life from Fire ; Statistics of the Fire Appliances in English Towns ; Foreign Fire Systems ; Hints on Fire Brigades, &c., &c. By CHARLES F. T. YOUNG, C.E. With numerous Illustrations, handsomely printed, 544 pp., demy 8vo, 1*l.* 4*s.* cloth.

" We can most heartily commend this book. . . . . It is really the only English work we now have upon the subject."—*Engineering.*

" We strongly recommend the book to the notice of all who are in any way interested in fires, fire-engines, or fire-brigades."—*Mechanics' Magazine.*

Manual of Mining Tools.

MINING TOOLS. For the use of Mine Managers, Agents, Mining Students, &c. By WILLIAM MORGANS, Lecturer on Practical Mining at the Bristol School of Mines. Volume of Text. 12mo. With an Atlas of Plates, containing 235 Illustrations. 4to. Together, 9*s.* cloth boards.

" Students in the Science of Mining, and not only they, but subordinate officials in mines, and even Overmen, Captains, Managers, and Viewers may gain practical knowledge and useful hints by the study of Mr. Morgans' Manual."—*Colliery Guardian.*

" A very valuable work, which will tend materially to improve our mining literature."—*Mining Journal.*

Common Sense for Gas-Users.

COMMON SENSE FOR GAS-USERS : a Catechism of Gas-Lighting for Householders, Gasfitters, Millowners, Architects, Engineers, &c., &c. By ROBERT WILSON, C.E., Author of " A Treatise on Steam Boilers." Second Edition. Crown 8vo, sewed, with Folding Plates and Wood Engravings, 2*s.* 6*d.* [*Just published.*

Gas and Gasworks.

A TREATISE on GASWORKS and the PRACTICE of MANUFACTURING and DISTRIBUTING COAL GAS. By SAMUEL HUGHES, C.E. Fourth Edition, revised by W. RICHARDS, C.E. With 68 Woodcuts, 12mo, 4*s.*, cloth boards.

Waterworks for Cities and Towns.

WATERWORKS for the SUPPLY of CITIES and TOWNS, with a Description of the Principal Geological Formations of England as influencing Supplies of Water. By SAMUEL HUGHES, F.G.S., Civil Engineer. New and enlarged edition, 12mo, with numerous Illustrations, 5*s.*, cloth boards.

" One of the most convenient, and at the same time reliable works on a subject, the vital importance of which cannot be over-estimated."—*Bradford Observer.*

Coal and Coal Mining.

COAL AND COAL MINING : a Rudimentary Treatise on. By WARINGTON W. SMYTH, M.A., F.R.S., &c., Chief Inspector of the Mines of the Crown and of the Duchy of Cornwall. New edition, revised and corrected. 12mo, with numerous Illustrations, 4*s.* 6*d.*, cloth boards.

" Every portion of the volume appears to have been prepared with much care, and as an outline is given of every known coal-field in this and other countries, as well as of the two principal methods of working, the book will doubtless interest a very large number of readers."—*Mining Journal.*

Roads and Streets.

THE CONSTRUCTION OF ROADS AND STREETS. In Two Parts. I. The Art of Constructing Common Roads. By HENRY LAW, C.E. Revised and Condensed by D. KINNEAR CLARK, C.E.—II. Recent Practice in the Construction of Roads and Streets : including Pavements of Stone, Wood, and Asphalte. By D. KINNEAR CLARK, C.E., M.I.C.E., Author of "Railway Machinery," "A Manual of Rules, Tables, and Data," &c. With numerous Illustrations. 12mo, 5s., cloth.   [*Just published.*

" A book which every borough surveyor and engineer must possess, and which will be of considerable service to architects, builders, and property owners generally."— *Building News.*

" The volume is suggestive, and will be an acquisition not only to engineers but to the greater number of people in this country on whom devolves the administration of roads as a part of the system of local government."— *The Architect.*

" To highway and town surveyors this book will have the utmost value, and as containing the largest amount of information in the shortest space and at the lowest price, we may predict for it a wide circulation."— *Journal of Gas Lighting.*

Field-Book for Engineers.

THE ENGINEER'S, MINING SURVEYOR'S, and CONTRACTOR'S FIELD-BOOK. By W. DAVIS HASKOLL, C.E. Third Edition, enlarged, consisting of a Series of Tables, with Rules, Explanations of Systems, and Use of Theodolite for Traverse Surveying and Plotting the Work with minute accuracy by means of Straight Edge and Set Square only; Levelling with the Theodolite, Casting out and Reducing Levels to Datum, and Plotting Sections in the ordinary manner; Setting out Curves with the Theodolite by Tangential Angles and Multiples with Right and Left-hand Readings of the Instrument; Setting out Curves without Theodolite on the System of Tangential Angles by Sets of Tangents and Offsets; and Earthwork Tables to 80 feet deep, calculated for every 6 inches in depth. With numerous wood-cuts, 12mo, 12s. cloth.

" The book is very handy, and the author might have added that the separate tables of sines and tangents to every minute will make it useful for many other purposes, the genuine traverse tables existing all the same."— *Athenæum.*

" A very useful work for the practical engineer and surveyor."— *Railway News.*

" The work forms a handsome pocket volume, and cannot fail, from its portability and utility, to be extensively patronised by the engineering profession."— *Mining Journal.*

" We strongly recommend it to all classes of surveyors."— *Colliery Guardian.*

Earthwork, Measurement and Calculation of.

A MANUAL on EARTHWORK. By ALEX. J. S. GRAHAM, C.E., Resident Engineer, Forest of Dean Central Railway. With numerous Diagrams. 18mo, 2s. 6d. cloth.

" As a really handy book for reference, we know of no work equal to it ; and the railway engineers and others employed in the measurement and calculation of earthwork will find a great amount of practical information very admirably arranged, and available for general or rough estimates, as well as for the more exact calculations required in the engineers' contractor's offices."— *Artizan.*

Harbours.

THE DESIGN and CONSTRUCTION of HARBOURS : A Treatise on Maritime Engineering. By THOMAS STEVENSON, F.R.S.E., F.G.S., M.I.C.E. Second Edition, containing many additional subjects, and otherwise generally extended and revised. With 20 Plates and numerous Cuts. Small 4to, 15s. cloth.

Bridge Construction in Masonry, Timber, & Iron.

EXAMPLES OF BRIDGE AND VIADUCT CONSTRUC-
TION OF MASONRY, TIMBER, AND IRON ; consisting of
46 Plates from the Contract Drawings or Admeasurement of select
Works. By W. DAVIS HASKOLL, C.E.   Second Edition, with
the addition of 554 Estimates, and the Practice of Setting out Works,
illustrated with 6 pages of Diagrams.   Imp. 4to, 2*l.* 12*s.* 6*d.* half-
morocco.

" One of the very few works extant descending to the level of ordinary routine, and
treating on the common every-day practice of the railway engineer. . . . A work of
the present nature by a man of Mr. Haskoll's experience, must prove invaluable to
hundreds. The tables of estimates appended to this edition will considerably enhance
its value."—*Engineering.*

Mathematical Instruments, their Construction, &c.

MATHEMATICAL INSTRUMENTS : THEIR CONSTRUC-
TION, ADJUSTMENT, TESTING, AND USE; comprising
Drawing, Measuring, Optical, Surveying, and Astronomical Instruments.   By J. F. HEATHER, M.A., Author of " Practical Plane
Geometry," " Descriptive Geometry," &c.   Enlarged Edition, for
the most part entirely rewritten.   With numerous Wood-cuts.
12mo, 5*s.* cloth boards.

Drawing for Engineers, &c.

THE WORKMAN'S MANUAL OF ENGINEERING
DRAWING.   By JOHN MAXTON, Instructor in Engineering
Drawing, Royal Naval College, Greenwich, formerly of R. S. N. A.,
South Kensington.   Third Edition, carefully revised.   With upwards
of 300 Plates and Diagrams.   12mo, cloth, strongly bound, 4*s.* 6*d.*

" Even accomplished draughtsmen will find in it much that will be of use to them.
A copy of it should be kept for reference in every drawing office."—*Engineering.*
" Indispensable for teachers of engineering drawing."—*Mechanics' Magazine.*

Oblique Arches.

A PRACTICAL TREATISE ON THE CONSTRUCTION of
OBLIQUE ARCHES.   By JOHN HART.   Third Edition, with
Plates.   Imperial 8vo, 8*s.* cloth.

Oblique Bridges.

A PRACTICAL and THEORETICAL ESSAY on OBLIQUE
BRIDGES, with 13 large folding Plates.   By GEO. WATSON
BUCK, M. Inst. C.E.   Second Edition, corrected by W. H.
BARLOW, M. Inst. C.E.   Imperial 8vo, 12*s.* cloth.

" The standard text-book for all engineers regarding skew arches, is Mr. Buck's
treatise, and it would be impossible to consult a better."—*Engineer.*

Pocket-Book for Marine Engineers.

A POCKET BOOK FOR MARINE ENGINEERS.   Con-
taining useful Rules and Formulæ in a compact form.   By FRANK
PROCTOR, A.I.N.A.   Second Edition, revised and enlarged.
Royal 32mo, leather, gilt edges, with strap, 4*s.*

" We recommend it to our readers as going far to supply a long-felt want."—
Naval Science.
" A most useful companion to all marine engineers."—*United Service Gazette.*
" Scarcely anything required by a naval engineer appears to have been for-
gotten."—*Iron.*

Grantham's Iron Ship-Building, enlarged.

ON IRON SHIP-BUILDING; with Practical Examples and
Details. Fifth Edition. Imp. 4to, boards, enlarged from 24 to 40
Plates (21 quite new), including the latest Examples. Together
with separate Text, 12mo, cloth limp, also considerably enlarged.
By JOHN GRANTHAM, M. Inst. C.E., &c. 2l. 2s. complete.

"A thoroughly practical work, and every question of the many in relation to iron
shipping which admit of diversity of opinion, or have various and conflicting personal
interests attached to them, is treated with sober and impartial wisdom and good sense.
. . . . As good a volume for the instruction of the pupil or student of iron naval
architecture as can be found in any language."—*Practical Mechanics' Journal*.
"A very elaborate work. . . . It forms a most valuable addition to the history
of iron shipbuilding, while its having been prepared by one who has made the subject
his study for many years, and whose qualifications have been repeatedly recognised,
will recommend it as one of practical utility to all interested in shipbuilding."—*Army
and Navy Gazette*.
"Mr. Grantham's work is of great interest. . . . It is also valuable as a record
of the progress of iron shipbuilding. . . . It will, we are confident, command an
extensive circulation among shipbuilders in general. . . . By order of the Board
of Admiralty, the work will form the text-book on which the examination in iron ship-
building of candidates for promotion in the dockyards will be mainly based."—
Engineering.

Weale's Dictionary of Terms.

A DICTIONARY of TERMS used in ARCHITECTURE,
BUILDING, ENGINEERING, MINING, METALLURGY,
ARCHÆOLOGY, the FINE ARTS, &c. By JOHN WEALE.
Fifth Edition, revised and corrected by ROBERT HUNT, F.R.S.,
Keeper of Mining Records, Editor of "Ure's Dictionary of Arts,"
&c. 12mo, cloth boards, 6s.

"A book for the enlightenment of those whose memory is treacherous or education
deficient in matters scientific and industrial. The additions made of modern disco-
veries and knowledge are extensive. The result is 570 pages of concentrated essence
of elementary knowledge, admirably and systematically arranged, and presented in
neat and handy form."—*Iron*.
"The best small technological dictionary in the language."—*Architect*.
"A comprehensive and accurate compendium. Author, editor, and publishers deserve high commendations for producing such a useful work. We can warmly recommend such a dictionary as a standard work of reference to our subscribers. Every
ironmonger should procure it—no engineer should be without it—builders and architects must admire it—metallurgists and archæologists would profit by it."—*Ironmonger*.
"The absolute accuracy of a work of this character can only be judged of after
extensive consultation, and from our examination it appears very correct and very
complete."—*Mining Journal*.
"There is no need now to speak of the excellence of this work; it received the ap-
proval of the community long ago. Edited now by Mr. Robert Hunt, and published
in a cheap, handy form, it will be of the utmost service as a book of reference scarcely
to be exceeded in value."—*Scotsman*.

Steam.

THE SAFE USE OF STEAM: containing Rules for Unprofessional Steam Users. By an ENGINEER. Third Edition. 12mo.
Sewed, 6d.

N. B.—*This little work should be in the hands of every person
having to deal with a Steam Engine of any kind.*

"If steam-users would but learn this little book by heart, and then hand it to
their stokers to do the same, and see that the latter do it, boiler explosions would
become sensations by their rarity."—*English Mechanic*.

ARCHITECTURE, &c.

Construction.

THE SCIENCE of BUILDING: An Elementary Treatise on the Principles of Construction. By E. WYNDHAM TARN, M.A., Architect. With 47 Wood Engravings. Demy 8vo. 8s. 6d. cloth.

"A very valuable book, which we strongly recommend to all students."—*Builder.*
"No architectural student should be without this hand-book."—*Architect.*
"An able digest of information which is only to be found scattered through various works."—*Engineering.*

Beaton's Pocket Estimator.

THE POCKET ESTIMATOR FOR THE BUILDING TRADES, being an easy method of estimating the various parts of a Building collectively, more especially applied to Carpenters' and Joiners' work, priced according to the present value of material and labour. By A. C. BEATON, Author of 'Quantities and Measurements.' Second Edition. Carefully revised. 33 Woodcuts. Leather. Waistcoat-pocket size. 1s. 6d.

Beaton's Builders' and Surveyors' Technical Guide.

THE POCKET TECHNICAL GUIDE AND MEASURER FOR BUILDERS AND SURVEYORS: containing a Complete Explanation of the Terms used in Building Construction, Memoranda for Reference, Technical Directions for Measuring Work in all the Building Trades, &c., &c. By A. C. BEATON. With 19 Woodcuts. Leather. Waistcoat-pocket size. 1s. 6d.

Villa Architecture.

A HANDY BOOK of VILLA ARCHITECTURE; being a Series of Designs for Villa Residences in various Styles. With Detailed Specifications and Estimates. By C. WICKES, Architect, Author of "The Spires and Towers of the Mediæval Churches of England," &c. Complete in 1 vol. 61 Plates. 4to, 2l. 2s. half morocco.
. A SELECTION FROM THE ABOVE, containing 30 Designs, with Detailed Specifications, Estimates, &c. 21s. half morocco.
"The whole of the designs bear evidence of their being the work of an artistic architect, and they will prove very valuable and suggestive."—*Building News.*

House Painting.

HOUSE PAINTING, GRAINING, MARBLING, AND SIGN WRITING: a Practical Manual of. With 9 Coloured Plates of Woods and Marbles, and nearly 150 Wood Engravings. By ELLIS A. DAVIDSON, Author of 'Building Construction,' &c. Second Edition, carefully revised. 12mo, 6s. cloth boards.
"Contains a mass of information of use to the amateur and of value to the practical man."—*English Mechanic.*
"Deals with the practice of painting in all its parts, from the grinding of colours to varnishing and gilding."—*Architect.*

Wilson's Boiler and Factory Chimneys.

BOILER AND FACTORY CHIMNEYS; their Draught-power and Stability, with a chapter on Lightning Conductors. By ROBERT WILSON, A I.C.E., Author of "Treatise on Steam Boilers," &c., &c. Crown 8vo, 3s. 6d., cloth. [*Just published.*
"A most valuable book of its kind, full of useful information, definite in statement, and thoroughly practical in treatment."—*The Local Government Chronicle.*

A Book on Building.

A BOOK ON BUILDING, CIVIL AND ECCLESIASTICAL. By Sir EDMUND BECKETT, Bart., LL.D., Q.C., F.R.A.S., Author of "Clocks and Watches and Bells," &c. Crown 8vo, with Illustrations, 7s. 6d., cloth.

"A book which is always amusing and nearly always instructive. Sir E. Beckett will be read for the raciness of his style. We are able very cordially to recommend all persons to read it for themselves. The style throughout is in the highest degree condensed and epigrammatic."—*Times.*

"We commend the book to the thoughtful consideration of all who are interested in the building art."—*Builder.*

"There is hardly a subject connected with either building or repairing on which sensible and practical directions will not be found, the use of which is probably destined to prevent many an annoyance, disappointment, and unnecessary expense." —*Daily News*

Architecture, Ancient and Modern.

RUDIMENTARY ARCHITECTURE, Ancient and Modern. Consisting of VITRUVIUS, translated by JOSEPH GWILT, F.S.A., &c., with 23 fine copper plates; GRECIAN Architecture, by the EARL of ABERDEEN; the ORDERS of Architecture, by W. H. LEEDS, Esq.; The STYLES of Architecture of Various Countries, by T. TALBOT BURY; The PRINCIPLES of DESIGN in Architecture, by E. L. GARBETT. In one volume, half-bound (pp. 1,100), copiously illustrated, 12s.

* *Sold separately, in two vols., as follows—*

ANCIENT ARCHITECTURE. Containing Gwilt's Vitruvius and Aberdeen's Grecian Architecture. Price 6s. half-bound.

N.B.— *This is the only edition of VITRUVIUS procurable at a moderate price.*

MODERN ARCHITECTURE. Containing the Orders, by Leeds; The Styles, by Bury; and Design, by Garbett. 6s. half-bound.

The Young Architect's Book.

HINTS TO YOUNG ARCHITECTS. By GEORGE WIGHT-WICK, Architect, Author of "The Palace of Architecture," &c. &c. New Edition, revised and enlarged. By G. HUSKISSON GUILLAUME, Architect. Numerous illustrations. 12mo, cloth boards, 4s.

"Will be found an acquisition to pupils, and a copy ought to be considered as necessary a purchase as a box of instruments."—*Architect.*

"Contains a large amount of information, which young architects will do well to acquire, if they wish to succeed in the everyday work of their profession.'—*English Mechanic.*

Drawing for Builders and Students.

PRACTICAL RULES ON DRAWING for the OPERATIVE BUILDER and YOUNG STUDENT in ARCHITECTURE. By GEORGE PYNE, Author of a "Rudimentary Treatise on Perspective for Beginners." With 14 Plates, 4to, 7s. 6d. boards.

Builder's and Contractor's Price Book.

LOCKWOOD & CO.'S BUILDER'S AND CONTRACTOR'S PRICE BOOK for 1878, containing the latest prices of all kinds of Builders' Materials and Labour, and of all Trades connected with Building, &c., &c. The whole revised and edited by FRANCIS T. W. MILLER, Architect and Surveyor. Fcap. 8vo, strongly half-bound, 4s.

Handbook of Specifications.

THE HANDBOOK OF SPECIFICATIONS; or, Practical Guide to the Architect, Engineer, Surveyor, and Builder, in drawing up Specifications and Contracts for Works and Constructions. Illustrated by Precedents of Buildings actually executed by eminent Architects and Engineers. Preceded by a Preliminary Essay, and Skeletons of Specifications and Contracts, &c., &c. By Professor THOMAS L. DONALDSON, M.I.B.A. With A REVIEW OF THE LAW OF CONTRACTS. By W. CUNNINGHAM GLEN, of the Middle Temple. With 33 Lithographic Plates, 2 vols., 8vo, 2l. 2s.

"In these two volumes of 1,100 pages (together), forty-four specifications of executed works are given, including the specifications for parts of the new Houses of Parliament, by Sir Charles Barry, and for the new Royal Exchange, by Mr. Tite, M.P. Donaldson's Handbook of Specifications must be bought by all architects."—Builder.

Taylor and Cresy's Rome.

THE ARCHITECTURAL ANTIQUITIES OF ROME. By the late G. L. TAYLOR, Esq., F.S.A., and EDWARD CRESY, Esq. New Edition, thoroughly revised, and supplemented under the editorial care of the Rev. ALEXANDER TAYLOR, M.A. (son of the late G. L. Taylor, Esq.), Chaplain of Gray's Inn. This is the only book which gives on a large scale, and with the precision of architectural measurement, the principal Monuments of Ancient Rome in plan, elevation, and detail. Large folio, with 130 Plates, half-bound, 3l. 3s.

. Originally published in two volumes, folio, at 18l. 18s.

Specifications for Practical Architecture.

SPECIFICATIONS FOR PRACTICAL ARCHITECTURE: A Guide to the Architect, Engineer, Surveyor, and Builder; with an Essay on the Structure and Science of Modern Buildings. By FREDERICK ROGERS, Architect. With numerous Illustrations. Demy 8vo, 15s., cloth. (Published at 1l. 10s.)

. A volume of specifications of a practical character being greatly required, and the old standard work of Alfred Bartholomew being out of print, the author, on the basis of that work, has produced the above. He has also inserted specifications of works that have been erected in his own practice.

The House-Owner's Estimator.

THE HOUSE-OWNER'S ESTIMATOR; or, What will it Cost to Build, Alter, or Repair? A Price-Book adapted to the Use of Unprofessional People as well as for the Architectural Surveyor and Builder. By the late JAMES D. SIMON, A.R.I.B.A. Edited and Revised by FRANCIS T. W. MILLER, Surveyor. With numerous Illustrations. Second Edition, with the prices carefully revised to 1875. Crown 8vo, cloth, 3s. 6d.

"In two years it will repay its cost a hundred times over."—Field.
"A very handy book for those who want to know what a house will cost to build, alter, or repair."—English Mechanic.
"Especially valuable to non-professional readers."—Mining Journal.

Useful Text-Book for Architects.

THE ARCHITECT'S GUIDE: Being a Text-book of Useful Information for Architects, Engineers, Surveyors, Contractors, Clerks of Works, &c., &c. By FREDERICK ROGERS, Architect, Author of 'Specifications for Practical Architecture,' &c. With numerous Illustrations. Crown 8vo, 6s. cloth.

CARPENTRY, TIMBER, MECHANICS.

Tredgold's Carpentry, new and cheaper Edition.

THE ELEMENTARY PRINCIPLES OF CARPENTRY : a Treatise on the Pressure and Equilibrium of Timber Framing, the Resistance of Timber, and the Construction of Floors, Arches, Bridges, Roofs, Uniting Iron and Stone with Timber, &c. To which is added an Essay on the Nature and Properties of Timber, &c., with Descriptions of the Kinds of Wood used in Building ; also numerous Tables of the Scantlings of Timber for different purposes, the Specific Gravities of Materials, &c. By THOMAS TREDGOLD, C.E. Edited by PETER BARLOW, F.R.S. Fifth Edition, corrected and enlarged. With 64 Plates (11 of which now first appear in this edition), Portrait of the Author, and several Woodcuts. In 1 vol., 4to, published at 2*l.* 2*s.*, reduced to 1*l.* 5*s.*, cloth.

" ' Tredgold's Carpentry' ought to be in every architect's and every builder's library, and those who do not already possess it ought to avail themselves of the new issue."—*Builder.*

"A work whose monumental excellence must commend it wherever skilful carpentry is concerned. The Author's principles are rather confirmed than impaired by time, and, as now presented, combine the surest base with the most interesting display of progressive science. The additional plates are of great intrinsic value."—*Building News.*

Grandy's Timber Tables.

THE TIMBER IMPORTER'S, TIMBER MERCHANT'S, and BUILDER'S STANDARD GUIDE. By RICHARD E. GRANDY. Comprising :—An Analysis of Deal Standards, Home and Foreign, with comparative Values and Tabular Arrangements for Fixing Nett Landed Cost on Baltic and North American Deals, including all intermediate Expenses, Freight, Insurance, &c., &c. ; together with Copious Information for the Retailer and Builder. Second Edition. Carefully revised and corrected. 12mo, 3*s.* 6*d.* cloth.

"Everything it pretends to be : built up gradually, it leads one from a forest to a treenail, and throws in, as a makeweight, a host of material concerning bricks, columns, cisterns, &c.—all that the class to whom it appeals requires."—*English Mechanic.*

"The only difficulty we have is as to what is NOT in its pages. What we have tested of the contents, taken at random, is invariably correct."—*Illustrated Builder's Journal.*

Tables for Packing-Case Makers.

PACKING-CASE TABLES ; showing the number of Superficial Feet in Boxes or Packing-Cases, from six inches square and upwards. Compiled by WILLIAM RICHARDSON, Accountant. Second Edition. Oblong 4to, 3*s.* 6*d.*, cloth. [*Just Published.*

"Will save much labour and calculation to packing-case makers and those who use packing-cases."—*Grocer.* "Invaluable labour-saving tables."—*Ironmonger.*

Nicholson's Carpenter's Guide.

THE CARPENTER'S NEW GUIDE; or, BOOK of LINES for CARPENTERS : comprising all the Elementary Principles essential for acquiring a knowledge of Carpentry. Founded on the late PETER NICHOLSON's standard work. A new Edition, revised by ARTHUR ASHPITEL, F.S.A., together with Practical Rules on Drawing, by GEORGE PYNE. With 74 Plates, 4to, 1*l.* 1*s.* cloth.

Dowsing's Timber Merchant's Companion.

THE TIMBER MERCHANT'S AND BUILDER'S COM-PANION ; containing New and Copious Tables of the Reduced Weight and Measurement of Deals and Battens, of all sizes, from One to a Thousand Pieces, and the relative Price that each size bears per Lineal Foot to any given Price per Petersburgh Standard Hundred ; the Price per Cube Foot of Square Timber to any given Price per Load of 50 Feet; the proportionate Value of Deals and Battens by the Standard, to Square Timber by the Load of 50 Feet ; the readiest mode of ascertaining the Price of Scantling per Lineal Foot of any size, to any given Figure per Cube Foot.   Also a variety of other valuable information.   By WILLIAM DOWSING, Timber Merchant.   Third Edition, Revised and Corrected.   Crown 8vo, 3s. cloth.

"Everything is as concise and clear as it can possibly be made.   There can be no doubt that every timber merchant and builder ought to possess it."—*Hull Advertiser.*

Timber Freight Book.

THE TIMBER IMPORTERS' AND SHIPOWNERS' FREIGHT BOOK : Being a Comprehensive Series of Tables for the Use of Timber Importers, Captains of Ships, Shipbrokers, Builders, and all Dealers in Wood whatsoever.   By WILLIAM RICHARDSON, Timber Broker.   Crown 8vo, 6s., cloth.

Horton's Measurer.

THE COMPLETE MEASURER ; setting forth the Measure-ment of Boards, Glass, &c., &c. ; Unequal-sided, Square-sided, Octagonal-sided, Round Timber and Stone, and Standing Timber. With just allowances for the bark in the respective species of trees, and proper deductions for the waste in hewing the trees, &c.; also a Table showing the solidity of hewn or eight-sided timber, or of any octagonal-sided column.   Compiled for the accommodation of Timber-growers, Merchants, and Surveyors, Stonemasons, Architects, and others.   By RICHARD HORTON. Third edition, with considerable and valuable additions, 12mo, strongly bound in leather, 5s.

" Not only are the best methods of measurement shown, and in some instances illustrated by means of woodcuts, but the erroneous systems pursued by dishonest dealers are fully exposed. . . . . The work must be considered to be a valuable addi-tion to every gardener's library.—*Garden.*

Superficial Measurement.

THE TRADESMAN'S GUIDE TO SUPERFICIAL MEA-SUREMENT.   Tables calculated from 1 to 200 inches in length, by 1 to 108 inches in breadth.   For the use of Architects, Surveyors, Engineers, Timber Merchants, Builders, &c.   By JAMES HAW-KINGS.   Fcp. 3s. 6d. cloth.

Practical Timber Merchant.

THE PRACTICAL TIMBER MERCHANT, being a Guide for the use of Building Contractors, Surveyors, Builders, &c., comprising useful Tables for all purposes connected with the Timber Trade, Marks of Wood, Essay on the Strength of Timber, Remarks on the Growth of Timber, &c.   By W. RICHARDSON. Fcap. 8vo, 3s. 6d., cloth.

The Mechanic's Workshop Companion.

THE OPERATIVE MECHANIC'S WORKSHOP COM-
PANION, and THE SCIENTIFIC GENTLEMAN'S PRAC-
TICAL ASSISTANT. By WILLIAM TEMPLETON. Twelfth
Edition, with Mechanical Tables for Operative Smiths, Millwrights,
Engineers, &c.; and an Extensive Table of Powers and Roots,
&c., &c. 11 Plates. 12mo, 5s. bound.

"As a text-book of reference, in which mechanical and commercial demands are
judiciously met, TEMPLETON'S COMPANION stands unrivalled."—*Mechanics' Magazine.*

"Admirably adapted to the wants of a very large class. It has met with great
success in the engineering workshop, as we can testify; and there are a great many
men who, in a great measure, owe their rise in life to this little work."—*Building News.*

Engineer's Assistant.

THE ENGINEER'S, MILLWRIGHT'S, and MACHINIST'S
PRACTICAL ASSISTANT; comprising a Collection of Useful
Tables, Rules, and Data. Compiled and Arranged, with Original
Matter, by WILLIAM TEMPLETON. 6th Edition. 18mo, 2s. 6d.
cloth.

"So much varied information compressed into so small a space, and published at a
price which places it within the reach of the humblest mechanic, cannot fail to command the sale which it deserves. With the utmost confidence we commend this book
to the attention of our readers."—*Mechanics' Magazine.*

"A more suitable present to an apprentice to any of the mechanical trades could not
possibly be made."—*Building News.*

Designing, Measuring, and Valuing.

THE STUDENT'S GUIDE to the PRACTICE of MEA-
SURING, and VALUING ARTIFICERS' WORKS; containing
Directions for taking Dimensions, Abstracting the same, and bringing
the Quantities into Bill, with Tables of Constants, and copious
Memoranda for the Valuation of Labour and Materials in the respective Trades of Bricklayer and Slater, Carpenter and Joiner,
Painter and Glazier, Paperhanger, &c. With 43 Plates and Woodcuts. Originally edited by EDWARD DOBSON, Architect. New
Edition, re-written, with Additions on Mensuration and Construction, and useful Tables for facilitating Calculations and Measurements. By E. WYNDHAM TARN, M.A., 8vo, 10s. 6d. cloth.

"We have failed to discover anything connected with the building trade, from ex-
cavating foundations to bell-hanging, that is not fully treated upon."—*The Artizan.*

"Altogether the book is one which well fulfils the promise of its title-page, and we
can thoroughly recommend it to the class for whose use it has been compiled. Mr.
Tarn's additions and revisions have much increased the usefulness of the work, and
have especially augmented its value to students."—*Engineering.*

Plumbing.

PLUMBING; a text-book to the practice of the art or craft of the
plumber. With supplementary chapters upon house-drainage, embodying the latest improvements. By WILLIAM PATON BUCHAN,
Sanitary Engineer. 12mo., with about 300 illustrations. 3s. 6d.,
cloth.

"There is no other manual in existence of the plumber's art; and the volume will
be welcomed as the work of a practical master of his trade."—*Public Health.*

"The chapters on house-drainage may be usefully consulted, not only by plumbers,
but also by engineers and all engaged or interested in house-building."—*Iron.*

"A book containing a large amount of practical information, put together in a very
intelligent manner, by one who is well qualified for the task."—*City Press.*

MATHEMATICS, &c.

Gregory's Practical Mathematics.

MATHEMATICS for PRACTICAL MEN ; being a Common-place Book of Pure and Mixed Mathematics. Designed chiefly for the Use of Civil Engineers, Architects, and Surveyors. Part I. Pure Mathematics—comprising Arithmetic, Algebra, Geometry, Mensuration, Trigonometry, Conic Sections, Properties of Curves. Part II. Mixed Mathematics—comprising Mechanics in general, Statics, Dynamics, Hydrostatics, Hydrodynamics, Pneumatics, Mechanical Agents, Strength of Materials. With an Appendix of copious Logarithmic and other Tables. By Olinthus Gregory, LL.D., F.R.A.S. Enlarged by Henry Law, C.E. 4th Edition, carefully revised and corrected by J. R. Young, formerly Professor of Mathematics, Belfast College ; Author of "A Course of Mathematics," &c. With 13 Plates. Medium 8vo, 1*l*. 1*s*. cloth.

" As a standard work on mathematics it has not been excelled."—*Artizan.*
" The engineer or architect will here find ready to his hand, rules for solving nearly every mathematical difficulty that may arise in his practice. The rules are in all cases explained by means of examples, in which every step of the process is clearly worked out."—*Builder.*
"One of the most serviceable books to the practical mechanics of the country. . . . . . . In the edition just brought out, the work has again been revised by Professor Young. He has modernised the notation throughout, introduced a few paragraphs here and there, and corrected the numerous typographical errors which have escaped the eyes of the former Editor. The book is now as complete as it is possible to make it. It is an instructive book for the student, and a Text-book for him who having once mastered the subjects it treats of, needs occasionally to refresh his memory upon them."—*Building News.*

The Metric System.

A SERIES OF METRIC TABLES, in which the British Standard Measures and Weights are compared with those of the Metric System at present in use on the Continent. By C. H. Dowling, C. E. Second Edition, revised and enlarged. 8vo, 10*s*. 6*d*. strongly bound.

" Mr. Dowling's Tables, which are well put together, come just in time as a ready reckoner for the conversion of one system into the other."—*Athenæum.*
" Their accuracy has been certified by Prof. Airy, Astronomer-Royal."—*Builder.*
" Resolution 8.—That advantage will be derived from the recent publication of Metric Tables, by C. H. Dowling, C.E."—*Report of Section F, Brit. Assoc., Bath.*

Comprehensive Weight Calculator.

THE WEIGHT CALCULATOR; being a Series of Tables upon a New and Comprehensive Plan, exhibiting at one Reference the exact Value of any Weight from 1 lb. to 15 tons, at 300 Progressive Rates, from 1 Penny to 168 Shillings per cwt., and containing 186,000 Direct Answers, which with their Combinations, consisting of a single addition (mostly to be performed at sight), will afford an aggregate of 10,266,000 Answers ; the whole being calculated and designed to ensure Correctness and promote Despatch. By Henry Harben, Accountant, Sheffield, Author of 'The Discount Guide.' An entirely New Edition, carefully revised. Royal 8vo, strongly half-bound, 30*s*.

Comprehensive Discount Guide.

THE DISCOUNT GUIDE: comprising several Series of Tables
for the use of Merchants, Manufacturers, Ironmongers, and others,
by which may be ascertained the exact profit arising from any mode
of using Discounts, either in the Purchase or Sale of Goods, and
the method of either Altering a Rate of Discount, or Advancing a
Price, so as to produce, by one operation, a sum that will realise
any required profit after allowing one or more Discounts: to which
are added Tables of Profit or Advance from 1¼ to 90 per cent.,
Tables of Discount from 1¼ to 98¾ per cent., and Tables of Commission, &c., from ⅛ to 10 per cent. By HENRY HARBEN, Accountant,
Author of "The Weight Calculator." New Edition, carefully Revised and Corrected. In a handsome demy 8vo. volume (544 pp.),
strongly and elegantly half-bound, £1 5s.

Inwood's Tables, greatly enlarged and improved.

TABLES FOR THE PURCHASING of ESTATES, Freehold,
Copyhold, or Leasehold; Annuities, Advowsons, &c., and for the
Renewing of Leases held under Cathedral Churches, Colleges, or
other corporate bodies; for Terms of Years certain, and for Lives;
also for Valuing Reversionary Estates, Deferred Annuities, Next
Presentations, &c., together with Smart's Five Tables of Compound
Interest, and an Extension of the same to Lower and Intermediate
Rates. By WILLIAM INWOOD, Architect. The 20th edition, with
considerable additions, and new and valuable Tables of Logarithms
for the more Difficult Computations of the Interest of Money, Discount, Annuities, &c., by M. FÉDOR THOMAN, of the Société
Crédit Mobilier of Paris. 12mo, 8s. cloth.

"Those interested in the purchase and sale of estates, and in the adjustment of
compensation cases, as well as in transactions in annuities, life insurances, &c., will
find the present edition of eminent service."—*Engineering.*

"'Inwood's Tables' still maintain a most enviable reputation. The new issue has been
enriched by large additional contributions by M. Fédor Thoman, whose carefully
arranged Tables cannot fail to be of the utmost utility."—*Mining Journal.*

Geometry for the Architect, Engineer, &c.

PRACTICAL GEOMETRY, for the Architect, Engineer, and
Mechanic; giving Rules for the Delineation and Application of
various Geometrical Lines, Figures and Curves. By E. W. TARN,
M.A., Architect, Author of "The Science of Building," &c.
With 164 Illustrations. Demy 8vo. 12s. 6d., cloth.

"No book with the same objects in view has ever been published in which the
clearness of the rules laid down and the illustrative diagrams have been so satisfactory."—*Scotsman.*

Compound Interest and Annuities.

THEORY of COMPOUND INTEREST and ANNUITIES;
with Tables of Logarithms for the more Difficult Computations of
Interest, Discount, Annuities, &c., in all their Applications and
Uses for Mercantile and State Purposes. With an elaborate Intro-
duction. By FÉDOR THOMAN, of the Société Crédit Mobilier,
Paris. 3rd Edition, carefully revised and corrected. 12mo, 4s. 6d., cl.

"A very powerful work, and the Author has a very remarkable command of his
subject."—*Professor A. de Morgan.*

"We recommend it to the notice of actuaries and accountants."—*Athenæum.*

⁎ The above bound with LINTERN'S MINERAL SURVEYOR.
(See page 10.) Price 7s. 6d. cloth.

PUBLISHED BY CROSBY LOCKWOOD & CO. 23

SCIENCE AND ART.

Brewing.

A HANDBOOK FOR YOUNG BREWERS. By HERBERT EDWARDS WRIGHT, B.A. (Trin. Coll. Camb.). Crown 8vo, 3s. 6d., cloth. [*Just Published.*

"A thoroughly scientific treatise in popular language. It is evident that the author has mastered his subject in its scientific aspects."—*Morning Advertiser.*

"We would particularly recommend all teachers of the art to place a copy of it in every pupil's hands, and we feel sure its perusal will be attended with advantage."—*Brewer.*

The Military Sciences.

AIDE-MÉMOIRE to the MILITARY SCIENCES. Framed from Contributions of Officers and others connected with the different Services. Originally edited by a Committee of the Corps of Royal Engineers. Second Edition, most carefully revised by an Officer of the Corps, with many additions; containing nearly 350 Engravings and many hundred Woodcuts. 3 vols. royal 8vo, extra cloth boards, and lettered, 4l. 10s.

"A compendious encyclopædia of military knowledge."—*Edinburgh Review.*

"The most comprehensive work of reference to the military and collateral sciences."—*Volunteer Service Gazette.*

Field Fortification.

A TREATISE on FIELD FORTIFICATION, the ATTACK of FORTRESSES, MILITARY MINING, and RECONNOITRING. By Colonel I. S. MACAULAY, late Professor of Fortification in the R. M. A., Woolwich. Sixth Edition, crown 8vo, cloth, with separate Atlas of 12 Plates, 12s. complete.

Field Fortification.

HANDBOOK OF FIELD FORTIFICATION, intended for the Guidance of Officers preparing for Promotion, and especially adapted to the requirements of Beginners. By Major W. W. KNOLLYS, F.R.G.S., 93rd Sutherland Highlanders, &c. With 163 Woodcuts. Crown 8vo, 3s. 6d. cloth.

Dye-Wares and Colours.

THE MANUAL of COLOURS and DYE-WARES: their Properties, Applications, Valuation, Impurities, and Sophistications. For the Use of Dyers, Printers, Dry Salters, Brokers, &c. By J. W. SLATER. Post 8vo, 7s. 6d., cloth.

"A complete encyclopædia of the *materia tinctoria*. The information given respecting each article is full and precise, and the methods of determining the value of articles such as these, so liable to sophistication, are given with clearness, and are practical as well as valuable."—*Chemist and Druggist.*

Storms.

STORMS: their Nature, Classification, and Laws, with the Means of Predicting them by their Embodiments, the Clouds. By WILLIAM BLASIUS. With Coloured Plates and numerous Wood Engravings. Crown 8vo, 10s. 6d. cloth boards.

Light-Houses.
EUROPEAN LIGHT-HOUSE SYSTEMS ; being a Report of
a Tour of Inspection made in 1873. By Major GEORGE H.
ELLIOT, Corps of Engineers, U.S.A. Illustrated by 51 Engravings and 31 Woodcuts in the Text. 8vo, 21s. cloth.

Electricity.
A MANUAL of ELECTRICITY ; including Galvanism, Magnetism, Diamagnetism, Electro-Dynamics, Magno-Electricity, and
the Electric Telegraph. By HENRY M. NOAD, Ph.D., F.C.S.,
Lecturer on Chemistry at St. George's Hospital. Fourth Edition,
entirely rewritten. Illustrated by 500 Woodcuts. 8vo, 1l. 4s. cloth.
" The commendations already bestowed in the pages of the *Lancet* on the former
editions of this work are more than ever merited by the present. The accounts given
of electricity and galvanism are not only complete in a scientific sense, but, which is a
rarer thing, are popular and interesting."—*Lancet.*

Text-Book of Electricity.
THE STUDENT'S TEXT-BOOK OF ELECTRICITY. By
HENRY M. NOAD, Ph.D., Lecturer on Chemistry at St. George's
Hospital. New Edition, revised and enlarged, with additions on
Telegraphy, the Telephone, Phonograph, &c., by G. E. PREECE,
Esq., of the Telegraph Department, General Post Office, London.
Upwards of 400 Illustrations.
[*Nearly ready.*

Rudimentary Magnetism.
RUDIMENTARY MAGNETISM : being a concise exposition
of the general principles of Magnetical Science, and the purposes
to which it has been applied. By Sir W. SNOW HARRIS, F.R.S.
New and enlarged Edition, with considerable additions by Dr.
NOAD, Ph.D. With 165 Woodcuts. 12mo, cloth, 4s. 6d.
" As concise and lucid an exposition of the phenomena of magnetism as we believe
it is possible to write."—*English Mechanic.*
" The best possible manual on the subject of magnetism."—*Mechanics' Magazine.*

Chemical Analysis.
THE COMMERCIAL HANDBOOK of CHEMICAL ANA-
LYSIS ; or Practical Instructions for the determination of the Intrinsic or Commercial Value of Substances used in Manufactures,
in Trades, and in the Arts. By A. NORMANDY, Author of " Practical Introduction to Rose's Chemistry," and Editor of Rose's
" Treatise on Chemical Analysis." *New Edition.* Enlarged, and
to a great extent re-written, by HENRY M. NOAD, Ph. D., F.R.S.
With numerous Illustrations. Cr. 8vo, 12s. 6d. cloth.
" We recommend this book to the careful perusal of every one ; it may be truly
affirmed to be of universal interest, and we strongly recommend it to our readers as a
guide, alike indispensable to the housewife as to the pharmaceutical practitioner."—
Medical Times.
" Essential to the analysts appointed under the new Act. The most recent results
are given, and the work is well edited and carefully written."—*Nature.*

Mollusca.
A MANUAL OF THE MOLLUSCA ; being a Treatise on
Recent and Fossil Shells. By Dr. S. P. WOODWARD, A.L.S.
With Appendix by RALPH TATE, A.L.S. F.G.S. With numer-
ous Plates and 300 Woodcuts. 3rd Edition. Cr. 8vo, 7s. 6d. cloth.

Clocks, Watches, and Bells.

RUDIMENTARY TREATISE on CLOCKS, and WATCHES, and BELLS.   By Sir EDMUND BECKETT, Bart. (late E. B. Denison), LL.D., Q.C., F.R.A.S., Author of "Astronomy without Mathematics," &c.   Sixth edition, thoroughly revised and enlarged, with numerous Illustrations.   Limp cloth (No. 67, Weale's Series), 4s. 6d.; cloth boards, 5s. 6d.

"As a popular and practical treatise it is unapproached."—*English Mechanic.*
"The best work on the subject probably extant.   The treatise on bells is undoubtedly the best in the language.   To call it a rudimentary treatise is a misnomer, at least as respects clocks and bells.   It is the most important work of its kind in English."—*Engineering.*
"The only modern treatise on clock-making."—*Horological Journal.*
"This admirable treatise on clocks, by the most able authority on such a subject, is completely perfect of its kind."—*Standard.*

Gold and Gold-Working.

THE PRACTICAL GOLD-WORKER; or, The Goldsmith's and Jeweller's Instructor.   The Art of Alloying, Melting, Reducing, Colouring, Collecting and Refining.   The processes of Manipulation, Recovery of Waste, Chemical and Physical Properties of Gold, with a new System of Mixing its Alloys; Solders, Enamels, and other useful Rules and Recipes, &c.   By GEORGE E. GEE.   Crown 8vo, 7s. 6d., cloth.   [*Just Published.*

"A good, sound, technical educator, and will be generally accepted as an authority. It gives full particulars for mixing alloys and enamels, is essentially a book for the workshop, and exactly fulfils the purpose intended."—*Horological Journal.*
"The best work yet printed on its subject for a reasonable price.   We have no doubt that it will speedily become a standard book which few will care to be without."—*Jeweller and Metalworker.*

Silver and Silver Working.

THE SILVERSMITH'S HANDBOOK, containing full Instructions for the Alloying and Working of Silver, including the different modes of refining and melting the metal, its solders, the preparation of imitation alloys, methods of manipulation, prevention of waste, instructions for improving and finishing the surface of the work, together with other useful information and memoranda.   By GEORGE E. GEE, Jeweller, &c.   Crown 8vo, with numerous illustrations, 7s. 6d., cloth.   [*Just Published.*

"This work is destined to take up as good a position in technical literature as the *Practical Goldworker,* a book which has passed through the ordeal of critical examination and business tests with great success."—*Jeweller and Metalworker.*

Science and Scripture.

SCIENCE ELUCIDATIVE OF SCRIPTURE, AND NOT ANTAGONISTIC TO IT; being a Series of Essays on—1. Alleged Discrepancies; 2. The Theory of the Geologists and Figure of the Earth; 3. The Mosaic Cosmogony; 4. Miracles in general—Views of Hume and Powell; 5. The Miracle of Joshua—Views of Dr. Colenso: The Supernaturally Impossible; 6. The Age of the Fixed Stars—their Distances and Masses.   By Professor J. R. YOUNG.   Fcap. 8vo, 5s. cloth.

"Distinguished by the true spirit of scientific inquiry, by great knowledge, by keen logical ability, and by a style peculiarly clear, easy, and energetic."—*Nonconformist.*

Practical Philosophy.

A SYNOPSIS of PRACTICAL PHILOSOPHY.   By Rev. J. CARR, M.A., late Fellow of Trin. Coll., Camb. 2nd Ed. 18mo, 5s. cl.

DR. LARDNER'S POPULAR WORKS.
Dr. Lardner's Museum of Science and Art.
THE MUSEUM OF SCIENCE AND ART. Edited by
DIONYSIUS LARDNER, D.C.L., formerly Professor of Natural Philosophy and Astronomy in University College, London. With upwards of 1200 Engravings on Wood. In 6 Double Volumes.
Price £1 1s., in a new and elegant cloth binding, or handsomely
bound in half morocco, 31s. 6d.

"The 'Museum of Science and Art' is the most valuable contribution that has
ever been made to the Scientific Instruction of every class of society."—*Sir David
Brewster in the North British Review.*
"Whether we consider the liberality and beauty of the illustrations, the charm of
the writing, or the durable interest of the matter, we must express our belief that
there is hardly to be found among the new books, one that would be welcomed by
people of so many ages and classes as a valuable present."—*Examiner.*

*** Separate books formed from the above, suitable for Workmen's
Libraries, Science Classes, &c.*

COMMON THINGS EXPLAINED. Containing Air, Earth, Fire,
Water, Time, Man, the Eye, Locomotion, Colour, Clocks and
Watches, &c. 233 Illustrations, cloth gilt, 5s.
THE MICROSCOPE. Containing Optical Images, Magnifying
Glasses, Origin and Description of the Microscope, Microscopic
Objects, the Solar Microscope, Microscopic Drawing and Engrav-
ing, &c. 147 Illustrations, cloth gilt, 2s.
POPULAR GEOLOGY. Containing Earthquakes and Volcanoes,
the Crust of the Earth, etc. 201 Illustrations, cloth gilt, 2s. 6d.
POPULAR PHYSICS. Containing Magnitude and Minuteness, the
Atmosphere, Meteoric Stones, Popular Fallacies, Weather Prog-
nostics, the Thermometer, the Barometer, Sound, &c. 85 Illus-
trations, cloth gilt, 2s. 6d.
STEAM AND ITS USES. Including the Steam Engine, the Lo-
comotive, and Steam Navigation. 89 Illustrations, cloth gilt, 2s.
POPULAR ASTRONOMY. Containing How to Observe the
Heavens. The Earth, Sun, Moon, Planets. Light, Comets,
Eclipses, Astronomical Influences, &c. 182 Illustrations, 4s. 6d.
THE BEE AND WHITE ANTS: Their Manners and Habits.
With Illustrations of Animal Instinct and Intelligence. 135 Illus-
trations, cloth gilt, 2s.
THE ELECTRIC TELEGRAPH POPULARISED. To render
intelligible to all who can Read, irrespective of any previous Scien-
tific Acquirements, the various forms of Telegraphy in Actual
Operation. 100 Illustrations, cloth gilt, 1s. 6d.

Scientific Class-Books, by Dr. Lardner.
NATURAL PHILOSOPHY FOR SCHOOLS. By DR. LARDNER.
328 Illustrations. Sixth Edition. 1 vol. 3s. 6d. cloth.
"Conveys, in clear and precise terms, general notions of all the principal divisions
of Physical Science."—*British Quarterly Review.*
ANIMAL PHYSIOLOGY FOR SCHOOLS. By DR. LARDNER.
With 190 Illustrations. Second Edition. 1 vol. 3s. 6d. cloth.
"Clearly written, well arranged, and excellently illustrated."—*Gardeners' Chronicle.*

DR. LARDNER'S SCIENTIFIC WORKS.

Astronomy.

THE HANDBOOK OF ASTRONOMY. 4th Edition. Edited by EDWIN DUNKIN, F.R.S., Rl. Observatory, Greenwich. With 38 plates and upwards of 100 Woodcuts. Cr. 8vo, 9s. 6d. cloth.
" Probably no other book contains the same amount of information in so compendious and well-arranged a form."—Athenæum.

Animal Physics.

THE HANDBOOK OF ANIMAL PHYSICS. With 520 Illustrations. New edition, small 8vo, cloth, 7s. 6d. 732 pages.
" We have no hesitation in cordially recommending it."—Educational Times.

Electric Telegraph.

THE ELECTRIC TELEGRAPH. New Edition. By E. B. BRIGHT, F.R.A.S. 140 Illustrations. Small 8vo, 2s. 6d. cloth.
One of the most readable books extant on the Electric Telegraph."—Eng. Mechanic.

LARDNER'S COURSE OF NATURAL PHILOSOPHY.

Mechanics.

THE HANDBOOK OF MECHANICS. Enlarged and almost rewritten by BENJAMIN LOEWY, F.R.A.S. With 378 Illustrations. Post 8vo, 6s. cloth. [Just Published.
' The perspicuity of the original has been retained, and chapters which had become obsolete, have been replaced by others of more modern character. The explanations throughout are studiously popular, and care has been taken to show the application of the various branches of physics to the industrial arts, and to the practical business of life."—Mining Journal.

Heat.

THE HANDBOOK OF HEAT. Edited and almost entirely Re-written by BENJAMIN LOEWY, F.R.A.S. etc. 117 Illustrations. Post 8vo, 6s. cloth. [Just Published.
" The style is always clear and precise, and conveys instruction without leaving any cloudiness or lurking doubts behind."—Engineering.

Hydrostatics and Pneumatics.

THE HANDBOOK of HYDROSTATICS and PNEUMATICS. New Edition, Revised and Enlarged by BENJAMIN LOEWY, F.R.A.S. With 236 Illustrations. Post 8vo, 5s. cl.
" For those ' who desire to attain an accurate knowledge of physical science without the profound methods of mathematical investigation,' this work is not merely intended, but we adapted."—Chemical News.

Electricity, Magnetism, and Acoustics.

THE HANDBOOK of ELECTRICITY, MAGNETISM, and ACOUSTICS. New Edition. Edited by GEO. CAREY FOSTER, B.A., F.C.S. With 400 Illustrations. Post 8vo, 5s. cloth.
" The book could not have been entrusted to any one better calculated to preserve the terse and lucid style of Lardner, while correcting his errors and bringing up his work to the present state of scientific knowledge."—Popular Science Review.

Optics.

THE HANDBOOK OF OPTICS. New Edition. Edited by T. OLVER HARDING, B.A. 298 Illustrations. Post 8vo, 5s. cloth.
" Written by one of the ablest English scientific writers, beautifully and elaborately illustrated."—Mechanic's Magazine.

✱ The above 5 Vols. form A COMPLETE COURSE OF NATURAL PHILOSOPHY.

Pictures and Painters.

THE PICTURE AMATEUR'S HANDBOOK AND DIC-
TIONARY OF PAINTERS: being a Guide for Visitors to
Public and Private Picture Galleries, and for Art-Students, including an explanation of the various methods of Painting; Instructions for Cleaning, Re-Lining, and Restoring Oil Paintings;
A Glossary of Terms; an Historical Sketch of the Principal Schools
of Painting; and a Dictionary of Painters, giving the Copyists
and Imitators of each Master. By PHILIPPE DARYL, B.A.   Crown
8vo, 3s. 6d., cloth.   [Just Published.

" Useful as bringing together in a compendious form an almost complete bio-
graphical stock of information respecting the painters of the world."—Mayfair.
" The bulk of the book is occupied by a dictionary of painters which, considering
its small compass, is really admirable ; where only a few words are devoted to an artist,
his speciality is well indicated ; and the utility of a table of dates of painters in so
portable a form is unquestionable. We cordially recommend the book."—Builder.

Popular Work on Painting.

PAINTING POPULARLY EXPLAINED; with Historical
Sketches of the Progress of the Art. By THOMAS JOHN GULLICK,
Painter, and JOHN TIMBS, F.S.A.   Fourth Edition, revised and
enlarged. With Frontispiece and Vignette. In small 8vo, 6s. cloth.

, This Work has been adopted as a Prize-book in the Schools of
Art at South Kensington.

" Much may be learned, even by those who fancy they do not require to be taught,
from the careful perusal of this unpretending but comprehensive treatise."—Art Journal.
" Contains a large amount of original matter, agreeably conveyed, and will be found of
value, as well by the young artist seeking information as by the general reader."—Builder.

Grammar of Colouring.

A GRAMMAR OF COLOURING, applied to Decorative
Painting and the Arts. By GEORGE FIELD.   New edition, enlarged and adapted to the use of the Ornamental Painter and
Designer, by ELLIS A. DAVIDSON. With new Coloured Diagrams
and numerous Engravings on Wood.   12mo, 3s. 6d. cloth boards.

" One of the most useful of student's books, and probably the best known of the
few we have on the subject."—Architect.
" The book is a most useful résumé of the properties of pigments."—Builder,

Geology and Genesis Harmonised.

THE TWIN RECORDS of CREATION; or, Geology and
Genesis, their Perfect Harmony and Wonderful Concord. By
G. W. VICTOR LE VAUX. With numerous Illus. Fcap. 8vo, 5s. cl.

" We can recommend Mr. Le Vaux as an able and interesting guide to a popular
appreciation of geological science."—Spectator.

Geology, Physical and Historical.

A CLASS-BOOK of GEOLOGY, PHYSICAL and HIS-
TORICAL. With more than 250 Woodcuts. By RALPH TATE,
A.L.S., F.G.S.   12mo, 5s., cloth boards.

" The fulness of the matter has elevated the book into a manual."—School Board
Chronicle.

Wood-Carving.

INSTRUCTIONS in WOOD-CARVING, for Amateurs; with
Hints on Design. By A LADY. In emblematic wrapper, handsomely printed, with Ten large Plates, 2s. 6d.

" The handicraft of the wood-carver, so well as a book can impart it, may be learnt
from ' A Lady's ' publication."—Athenæum.

Delamotte's Works on Illumination & Alphabets.

A PRIMER OF THE ART OF ILLUMINATION; for the use of Beginners: with a Rudimentary Treatise on the Art, Practical Directions for its Exercise, and numerous Examples taken from Illuminated MSS., printed in Gold and Colours. By F. DELA-MOTTE. Small 4to, 9*s.* Elegantly bound, cloth antique.

"A handy book, beautifully illustrated; the text of which is well written, and calculated to be useful. . . . The examples of ancient MSS. recommended to the student, which, with much good sense, the author chooses from collections accessible to all, are selected with judgment and knowledge, as well as taste."—*Athenæum.*

ORNAMENTAL ALPHABETS, ANCIENT and MEDIÆVAL; from the Eighth Century, with Numerals; including Gothic, Church-Text, large and small, German, Italian, Arabesque, Initials for Illumination, Monograms, Crosses, &c. &c., for the use of Architectural and Engineering Draughtsmen, Missal Painters, Masons, Decorative Painters, Lithographers, Engravers, Carvers, &c. &c. &c. Collected and engraved by F. DELAMOTTE, and printed in Colours. Royal 8vo, oblong, 4*s.* cloth.

"A well-known engraver and draughtsman has enrolled in this useful book the result of many years' study and research. For those who insert enamelled sentences round gilded chalices, who blazon shop legends over shop-doors, who letter church walls with pithy sentences from the Decalogue, this book will be useful."—*Athenæum.*

EXAMPLES OF MODERN ALPHABETS, PLAIN and ORNA-MENTAL; including German, Old English, Saxon, Italic, Perspective, Greek, Hebrew, Court Hand, Engrossing, Tuscan, Riband, Gothic, Rustic, and Arabesque; with several Original Designs, and an Analysis of the Roman and Old English Alphabets, large and small, and Numerals, for the use of Draughtsmen, Surveyors, Masons, Decorative Painters, Lithographers, Engravers, Carvers, &c. Collected and engraved by F. DELAMOTTE, and printed in Colours. Royal 8vo, oblong, 4*s.* cloth.

"To artists of all classes, but more especially to architects and engravers, this very handsome book will be invaluable. There is comprised in it every possible shape into which the letters of the alphabet and numerals can be formed, and the talent which has been expended in the conception of the various plain and ornamental letters is wonderful."—*Standard.*

MEDIÆVAL ALPHABETS AND INITIALS FOR ILLUMI-NATORS. By F. DELAMOTTE, Illuminator, Designer, and Engraver on Wood. Containing 21 Plates, and Illuminated Title, printed in Gold and Colours. With an Introduction by J. WILLIS BROOKS. Small 4to, 6*s.* cloth gilt.

"A volume in which the letters of the alphabet come forth glorified in gilding and all the colours of the prism interwoven and intertwined and intermingled, sometimes with a sort of rainbow arabesque. A poem emblazoned in these characters would be only comparable to one of those delicious love letters symbolized in a bunch of flowers well selected and cleverly arranged."—*Sun.*

THE EMBROIDERER'S BOOK OF DESIGN; containing Initials, Emblems, Cyphers, Monograms, Ornamental Borders, Ecclesiastical Devices, Mediæval and Modern Alphabets, and National Emblems. Collected and engraved by F. DELAMOTTE, and printed in Colours. Oblong royal 8vo, 2*s.* 6*d.* in ornamental boards.

AGRICULTURE, &c.

Youatt and Burn's Complete Grazier.

THE COMPLETE GRAZIER, and FARMER'S and CATTLE-BREEDER'S ASSISTANT. A Compendium of Husbandry. By WILLIAM YOUATT, ESQ., V.S. 12th Edition, enlarged by ROBERT SCOTT BURN, Author of "The Lessons of My Farm," &c. One large 8vo volume, 860 pp. with 244 Illustrations. 1*l.* 1*s.* half-bd.

"The standard and text-book, with the farmer and grazier."—*Farmer's Magazine.*
"A treatise which will remain a standard work on the subject as long as British agriculture endures."—*Mark Lane Express.*

Spooner on Sheep.

SHEEP; THE HISTORY, STRUCTURE, ECONOMY, AND DISEASES OF. By W. C. SPOONER, M.R.V.C., &c. Third Edition, considerably enlarged ; with numerous fine engravings, including some specimens of New and Improved Breeds. Fcp. 8vo, 366 pp., 6*s.* cloth.

"The book is decidedly the best of the kind in our language."—*Scotsman.*
"Mr. Spooner has conferred upon the agricultural class a lasting benefit by embodying in this work the improvements made in sheep stock by such men as Humphreys, Rawlence, Howard, and others."—*Hampshire Advertiser.*
'The work should be in possession of every flock-master."—*Banbury Guardian.*

Scott Burn's System of Modern Farming.

OUTLINES OF MODERN FARMING. By R. SCOTT BURN. Soils, Manures, and Crops—Farming and Farming Economy—Cattle, Sheep, and Horses—Management of the Dairy, Pigs, and Poultry—Utilisation of Town-Sewage, Irrigation, &c. New Edition. In 1 vol. 1250 pp., half-bound, profusely illustrated, 12*s.*

"There is sufficient stated within the limits of this treatise to prevent a farmer from going far wrong in any of his operations."—*Observer.*

Good Gardening.

A PLAIN GUIDE TO GOOD GARDENING ; or, How to Grow Vegetables, Fruits, and Flowers. With Practical Notes on Soils, Manures, Seeds, Planting, Laying-out of Gardens and Grounds, &c. By S. WOOD. Second Edition, with considerable Additions, &c., and numerous Illustrations. Cr. 8vo, 5*s.*, cloth.

"A very good book, and one to be highly recommended as a practical guide. The practical directions are excellent."—*Athenæum.*
"A thoroughly useful guidebook for the amateur gardener who may want to make his plot of land not merely pretty, but useful and profitable."—*Daily Telegraph.*

Profitable Gardening.

MULTUM-IN-PARVO GARDENING; or, How to make One Acre of Land produce £620 a year, by the Cultivation of Fruits and Vegetables ; also, How to Grow Flowers in Three Glass Houses, so as to realise £176 per annum clear Profit. By SAMUEL WOOD, Author of "Good Gardening," &c. 2nd Edition, revised. With Wood Engravings. Cr. 8vo, 2*s.*, cloth. [*Just Published.*

"We are bound to recommend it as not only suited to the case of the amateur and gentleman's gardener, but to the market grower."—*Gardener's Magazine.*

Horton's Underwood and Woodland Tables.

TABLES FOR PLANTING and Valuing Underwood and Woodland. By R. HORTON. 12mo., 2*s.* bound¶

Donaldson and Burn's Suburban Farming.

SUBURBAN FARMING. A Treatise on the Laying Out and Cultivation of Farms, adapted to the produce of Milk, Butter and Cheese, Eggs, Poultry, and Pigs. By the late Professor JOHN DONALDSON. With considerable Additions, Illustrating the more Modern Practice by ROBERT SCOTT BURN. With numerous Illustrations. Crown 8vo, 6s., cloth. [*Just Published.*
"An admirable treatise on all matters connected with the laying-out and cultivation of dairy farms."—*Live Stock Journal.*

Ewart's Land Improver's Pocket-Book.

THE LAND IMPROVER'S POCKET-BOOK OF FORMULÆ, TABLES, and MEMORANDA, required in any Computation relating to the Permanent Improvement of Landed Property. By JOHN EWART, Land Surveyor and Agricultural Engineer. Royal 32mo, oblong, leather, gilt edges, with elastic band, 4s.
"Admirably calculated to serve its purpose."—*Scotsman.*
"A compendious and handy little volume."—*Spectator.*

Hudson's Tables for Land Valuers.

THE LAND VALUER'S BEST ASSISTANT: being Tables, on a very much improved Plan, for Calculating the Value of Estates. With Tables for reducing Scotch, Irish, and Provincial Customary Acres to Statute Measure ; also, Tables of Square Measure, and of the Dimensions of an Acre by which the Contents of any Plot of Ground may be ascertained without the expense of a regular Survey ; &c. By R. HUDSON, C.E. New Edition, royal 32mo, oblong, leather, gilt edges, with elastic band, 4s.
"Of incalculable value to country gentlemen and professional men."—*Farmer's Journal.*

Complete Agricultural Surveyor's Pocket-Book.

THE LAND VALUER'S AND LAND IMPROVER'S COMPLETE POCKET-BOOK ; consisting of the above two works bound together, leather, gilt edges, with strap, 7s. 6d.
☞ *The above forms an unequalled and most compendious Pocket Vade-mecum for the Land Agent and Agricultural Engineer.*
" We consider Hudson's book to be the best ready-reckoner on matters relating to the valuation of land and crops we have ever seen, and its combination with Mr. Ewart's work greatly enhances the value and usefulness of the latter-mentioned . . It is most useful as a manual for reference."—*North of England Farmer.*

The Management of Estates.

LANDED ESTATES MANAGEMENT : Treating of the Varieties of Lands, Peculiarities of its Farms, Methods of Farming, the Setting-out of Farms and their Fields, Construction of Roads, Fences, Gates, and Farm Buildings, of Waste or Unproductive Lands, Irrigation, Drainage, Plantation, &c. By R. SCOTT BURN, Fcp. 8vo. numerous Illustrations, 3s. 6d.

Scott Burn's Introduction to Farming.

THE LESSONS of MY FARM : a Book for Amateur Agriculturists, being an Introduction to Farm Practice, in the Culture of Crops, the Feeding of Cattle, Management of the Dairy, Poultry, Pigs, &c. By R. SCOTT BURN. With numerous Illus. Fcp. 6s. cl.
" A complete introduction to the whole round of farming practice."—*John Bull.*

"*A Complete Epitome of the Laws of this Country.*"

EVERY MAN'S OWN LAWYER; a Handy-Book of the Principles of Law and Equity. By A BARRISTER. 15th Edition, Revised to the end of last Session. Including a Summary of the Judicature Acts, and the principal Acts of the past Session, viz. —The Canal Boats' Act, The Destructive Insects' (or Colorado Beetle) Act, The Fisheries' (Oyster, Crab, and Lobster) Act, and the Fisheries' (Dynamite) Act, &c., &c.   With Notes and References to the Authorities.   Crown 8vo, price 6s. 8d. (saved at every consultation), strongly bound.

COMPRISING THE LAWS OF

BANKRUPTCY—BILLS OF EXCHANGE—CONTRACTS AND AGREEMENTS—COPYRIGHT —DOWER AND DIVORCE—ELECTIONS AND REGISTRATION—INSURANCE—LIBEL AND SLANDER—MORTGAGES—SETTLEMENTS—STOCK EXCHANGE PRACTICE— TRADE MARKS AND PATENTS—TRESPASS, NUISANCES, ETC.—TRANSFER OF LAND, ETC.—WARRANTY—WILLS AND AGREEMENTS, ETC. Also Law for Landlord and Tenant—Master and Servant—Workmen and Apprentices—Heirs, Devisees, and Legatees—Husband and Wife—Executors and Trustees—Guardian and Ward—Married Women and Infants—Partners and Agents—Lender and Borrower—Debtor and Creditor—Purchaser and Vendor—Companies and Associations—Friendly Societies—Clergymen, Churchwardens—Medical Practitioners, &c.—Bankers—Farmers—Contractors—Stock and Share Brokers—Sportsmen and Gamekeepers—Farriers and Horse-Dealers—Auctioneers, House-Agents—Innkeepers, &c.—Pawnbrokers—Surveyors—Railways and Carriers, &c. &c.

" No Englishman ought to be without this book."—*Engineer.*
"What it professes to be—a complete epitome of the laws of this country, thoroughly intelligible to non-professional readers."—*Bell's Life.*

Auctioneer's Assistant.

THE APPRAISER, AUCTIONEER, BROKER, HOUSE AND ESTATE AGENT, AND VALUER'S POCKET ASSISTANT, for the Valuation for Purchase, Sale, or Renewal of Leases, Annuities, and Reversions, and of property generally; with Prices for Inventories, &c. By JOHN WHEELER, Valuer, &c. Fourth Edition, enlarged, by C. NORRIS. Royal 32mo, cloth, 5s.

" A neat and concise book of reference, containing an admirable and clearly-arranged list of prices for inventories, and a very practical guide to determine the value of furniture, &c."—*Standard.*

Pawnbroker's Legal Guide.

THE PAWNBROKER'S, FACTOR'S, and MERCHANT'S GUIDE to the LAW of LOANS and PLEDGES. By H. C. FOLKARD, Esq., Barrister-at-Law, Author of the "Law of Slander and Libel," &c. With Additions and Corrections to 1876. 12mo, cloth boards, 3s. 6d.

House Property.

HANDBOOK OF HOUSE PROPERTY : a Popular and Practical Guide to the Purchase, Mortgage, Tenancy, and Compulsory Sale of Houses and Land ; including the Law of Dilapidations and Fixtures; with Explanations and Examples of all kinds of Valuations, and useful Information and Advice on Building. By EDWARD LANCE TARBUCK, Architect and Surveyor. 12mo, 5s. cloth boards.

"We are glad to be able to recommend it."—*Builder.*
" The advice is thoroughly practical."—*Law Journal.*

www.ingramcontent.com/pod-product-compliance
Lightning Source LLC
Chambersburg PA
CBHW021342210326
41599CB00011B/715